SPECIAL EFFECTS SOURCEBOOK

SPECIAL EFFECTS SOURCEBOOK

ROBERT E. McCARTHY

Focal Press
Boston London

Focal Press is an imprint of Butterworth-Heinemann.

Library of Congress Cataloging-in-Publication Data
McCarthy, Robert E. 1931–
    Special effects sourcebook/by Robert E. McCarthy.
      p.   cm.
    ISBN 0-240-80147-4
    1. Cinematography—Special effects—Equipment and supplies—
        Directories.   I. Title.
TR885.M33 1992
791.43'024'025—dc20   92–115
                                          CIP

British Library Catologing in Publication Data
McCarthy, Robert E.
    Special effects sourcebook
    I. Title
    778.5

    ISBN 0-240-80147-4

Butterworth–Heinemann
80 Montvale Avenue
Stoneham, MA 02180

10 9 8 7 6 5 4 3 2 1

Printed in the United States of America

To my dear friend and brother-in-law
Fritz Wenslovas

Always loved...
Always there...

Who never left us enough time
To tell how much we cared.

CONTENTS

INTRODUCTION

During the writing of *Secrets of Hollywood Special Effects,* my editors asked me a question which I have so often been confronted with, especially when finding myself working in a state or country for the first time: "Where do you find the basic equipment and materials needed to construct the elaborate effects you've created?"

The answer to that question is this sourcebook. It is intended to be a resource for anyone in the special effects business whether in the United States or abroad. I have attempted to make it accurate and versatile, usable both in the office during preproduction and on location. Doubtless there are omissions. No compilation of this nature, however assiduously compiled, can be completely accurate. However, within the constraints of time and range of subject, I believe this to be the most extensive and thorough special effects directory available. In this effort, I would like to thank the countless film commissions, companies and incividuals who have graciously given their time and effort.

It is my hope that updated and corrected editions of this sourcebook will be made available in future years by Focal Press. Toward that end, I invite corrections and additions to be mailed to me, Robert E. McCarthy, at 18333 Lahey St., Northridge, CA 91326. Attention: Special Effects Sourcebook.

SPECIAL EFFECTS SOURCEBOOK

THE UNITED STATES

THE UNITED STATES FILM COMMISSIONS

ALABAMA

ALABAMA FILM OFFICE
340 NORTH HULL ST.
MONTGOMERY, 36130
PHONE: (800) 633-5898/(205)
242-4195
FAX: (205) 265-5078

ALASKA

ALASKA MOTION PICTURE AND TV PRODUCTION SERVICES
3601 "C" ST., STE 722
ANCHORAGE, 99503
PHONE: (907) 563-2167

ARIZONA

ARIZONA MOTION PICTURE OFFICE
3800 N. CENTRAL AVE.,
BUILDING D
PHOENIX, 85013
PHONE: (602) 280-1380
CITY OF SCOTTSDALE FILM OFFICE
3939 CIVIC CENTER PLAZA
SCOTTSDALE, 85251
PHONE: (602) 994-7809
NAVAJO NATION, MP OFFICE
P.O. BOX 2310
WINDOW ROCK, 86515
PHONE: (602) 871-6655
NOGALES FILM COMMISSION
1018 GRAND AVE
NOGALES, 85621
PHONE: (602) 287-4655
PARKER FILM COMMISSION
1217 CALIFORNIA AVE
PARKER, 85344
PHONE: (602) 669-2174
PAYSON FILM COMMISSION
303 N BEELINE HWY
PAYSON, 85541
PHONE: (602) 474-5676
PHOENIX MOTION PICTURE COMMERCIAL COORDINATION
251 WEST WASHINGTON
PHOENIX, 85003
PHONE: (602) 262-4850
PHOENIX MOTION PICTURE OFFICE
251 W WASHINGTON
PHOENIX, 85003
PHONE: (602) 262-4850
PRESCOTT FILM COMMITTEE
P.O. BOX 1147
PRESCOTT, 86302
PHONE: (602) 445-2000

SIERRA VISTA FILM INDUSTRY COMM.
P.O. BOX 1886
SIERRA VISTA, 85635
PHONE: (602) 378-6404
TUSCON FILM OFFICE
255 W. ALAMEDA
TUSCON, 85726
PHONE: (602) 791-4000
FAX: (602) 791-4017
VERDE VALLEY FILM COMMISSION
1010 S MAIN ST
COTTONWOOD, 86326
PHONE: (602) 634-7593
YUMA FILM COMMISSION
P.O. BOX 230
YUMA, 85364
PHONE: (602) 782-2567

ARKANSAS

ARKANSAS MOTION PICTURE DEVELOPMENT OFFICE
ONE STATE CAPITOL MALL,
RM 2C-200
LITTLE ROCK,72201
PHONE: (501) 682-7676
FAX: (501) 682-7691
NORTHWEST ARKANSAS MOTION PICTURE
P.O. BOX 4216
FAYETTEVILLE, 72702
PHONE: (501) 442-6554

CALIFORNIA

CALIFORNIA FILM COMMISSION
6922 HOLLYWOOD BL #600
HOLLYWOOD, 90028
PHONE: (213) 736-2465
ANTELOPE VALLEY FILM COMMISSION
P.O. BOX 1641
LANCASTER, 93539
PHONE: (805) 945-1292
BAKERSFIELD/KERN COUNTY BOARD OF TRADE
P.O. BOX 1312
BAKERSFIELD, 93306
PHONE: (805) 861-2367
BASS LAKE & MADERA COUNTY
P.O. BOX 126
BASS LAKE, 93604
PHONE: (209) 642-3676
BIG BEAR LAKE FILM COMMISSION
P.O. BOX 2860
BIG BEAR LAKE, 92315
PHONE: (714) 886-6190
CHAMBER OF COMMERCE
P.O. BOX 2860
BIG BEAR LAKE, 92315
PHONE: (714) 866-4607

BRAWLEY CHAMBER OF COMMERCE
P.O. BOX 218
BRAWLEY, 92227
PHONE: (619) 344-3160
BUTTE COUNTY ECONOMIC DEVELOPMENT
275 FAIRSHILD #104
CHICO, 95926
PHONE: (916) 893-8732
CALAVERAS COUNTY CHAMBER OF COMMERCE
P.O. BOX 177
SAN ANDREAS, 95249
PHONE: (209) 754-1821
CAMARILLO CHAMBER OF COMMERCE
350 N LANTANA #224
CAMARILLO, 93010
PHONE: (805) 484-4383
CATALINA/CITY OF AVALON
P.O. BOX 707
AVALON, 90704
PHONE: (213) 510-2626
CHICO CHAMBER OF COMMERCE
P.O. BOX 3038
CHICO, 95927
PHONE: (916) 891-5559
COLUSA COUNTY CHAMBER OF COMMERCE
P.O. BOX 1027
COLUSA, 95932
PHONE: (916) 458-2541
CONEJO VALLEY CHAMBER OF COMMERCE
401 WEST HILLCREST DR
THOUSAND OAKS, 91360
PHONE: (805) 497-1621
CONTRA COSTA COUNTY CONVENTION
2151 SALVIO ST. #N
CONCORD, 94520
PHONE: (415) 685-1184
DEL NORTE COUNTY CHAMBER OF COMMERCE
1001 FRONT ST.
CRESCENT CITY, 95531
PHONE: (707) 464-3174
DESERT HOT SPRINGS CHAMBER OF COMMERCE
P.O. BOX 848
DESERT HOT SPRINGS, 92240
PHONE: (619) 329-6403
EL DORADO COUNTY CHAMBER OF COMMERCE
542 MAIN ST
PLACERVILLE, 95667
PHONE: (916) 621-5885
FILLMORE CHAMBER OF COMMERCE
501 SANTA CLARA ST
FILLMORE, 93015
PHONE: (805) 524-0351

FOREST SERVICE NATIONAL MEDIA OFFICE - USDA
444 E BONITA AVE
SAN DIMAS, 91773
PHONE: (213) 681-1400
FRESNO COUNTY CONVENTION & VISITORS BUREAU
808 E BONITA AVE
FRESNO, 93717
PHONE: (800) 543-8488
GRASS VALLEY/NEVADA COUNTY CHAMBER OF COMMERCE
248 MILL ST
GRASS VALLEY, 95945
PHONE: (916) 273-4667
HANFORD CHAMBER OF COMMERCE
213 W 7TH ST
HANFORD, 93230
PHONE: (209) 582-0483
HUMBOLDT COUNTY VISITOR & CONVENTION BUREAU
1034 2ND ST
EUREKA,95501
PHONE: (800) 338-7352
IMPERIAL COUNTY REGIONAL ECONOMIC DEVELOPMENT
2031 WENSKY
EL CENTRO, 92243
PHONE: (619) 344-3560
KERN COUNTY/EAST KERN FILM OFFICE
P.O. BOX 878
MOJAVE, 93501
PHONE: (805) 824-2595
KERN COUNTY BOARD OF TRADE
P.O. BOX 1312
BAKERSFIELD, 93306
PHONE: (805) 861-2367
KINGS COUNTY CROWN DEVELOPMENT CORP
1222 W LACEY BL #101
HANFORD, 93230
PHONE: (209) 582-4326
<T3>LAKE ARROWHEAD
P.O. BOX 640
LAKE ARROWHEAD, 92352
PHONE: (714) 337-2533
LAKE COUNTY VISITORS INFORMATION CENTER
875 LAKEPORT BL
LAKEPORT, 95453
PHONE: (800) 525-3743
LAKE TAHOE CONVENTION & VISITORS BUREAU
P.O. BOX 5578
TAHOE CITY, 95730
PHONE: (916) 583-3494
VISITORS AUTHORITY
P.O. BOX 15299
SOUTH LAKE TAHOE, 95706
PHONE: (916) 544-5050

LANCASTER CHAMBER OF COMMERCE
44943 10TH ST WEST
LANCASTER, 93534
PHONE: (805) 948-4518

LASSEN COUNTY
P.O. BOX 1206
WESTWOOD, 96137
PHONE: (916) 256-3619

LONE PINE
P.O. DRAWER H
LONE PINE, 93534
PHONE: (619) 876-5433

LONG BEACH/CITY OF LONG BEACH
300 EAST OCEAN BL
LONG BEACH, 90802
PHONE: (213) 436-3636

LOS ANGELES CITY BOARD OF PUBLIC WORKS
200 N SPRING ST #373
LOS ANGELES, 90012
PHONE: (213) 485-3377

L.A. COUNTY FILM OFFICE
6922 HOLLYWOOD BL #606
HOLLYWOOD, 90028
PHONE: (213) 957-1000

MAMMOTH LAKES/MAMMOTH LOCATION SERVICES
P.O. BOX 24
MAMMOTH LAKES, 93546
PHONE: (619) 934-2571

MARYSVILLE/YUBA CITY CHAMBER OF COMMERCE
P.O. BOX 1429
MARYSVILLE, 95901
PHONE: (916) 743-6501

MENDOCINO COUNTY COUNTY DEVELOPMENT CORP
320 S. STATE ST
UKIAH, 95482
PHONE: (707) 463-0860

MERCED VISITORS & CONVEN-TION BUREAU
P.O. BOX 3107
MERCED, 95344
PHONE: (209) 384-3333

MODESTO FILM COMMISSION
P.O. BOX 844
MODESTO, 95353
PHONE: (408) 577-5757

MODOC FILM COMMISSION
P.O. BOX 470
ALTURAS, 96101
PHONE: (916) 233-3133

MOJAVE CHAMBER OF COMMERCE
P.O. BOX 878
MOJAVE, 93501
PHONE: (805) 824-2595

MONO COUNTY CHAMBER OF COMMERCE
RTE 2 P.O. BOX 315
BISHOP, 93514
PHONE: (619) 387-2723

MONTEREY COUNTY FILM COMMISSION
P.O. BOX 111
MONTEREY, 93942
PHONE: (408) 646-0910

MORRO BAY CHAMBER OF COMMERCE
P.O. BOX 876
MORRO BAY 93442
PHONE: (805) 772-4467

MOTION PICTURE COORDINA-TION OFFICE
6922 HOLLYWOOD BL #602
HOLLYWOOD, 90028
PHONE: (213) 485-5324

NEVADA CITY CHAMBER OF COMMERCE
132 MAIN ST
NEVADA CITY, 95959
PHONE: (916) 273-4667

OAKLAND MAYOR'S FILM LIAISON
ONE CITY HALL PLAZA
OAKLAND, 94612
PHONE: (415) 273-3109

CITY OF OJAI PUBLIC WORKS
401 SOUTH VENTURA ST
OJAI, 93023
PHONE: (805) 646-5581

ORANGE COUNTY PUBLIC INFORMATION OFFICE
10 CIVIC CENTER PLAZA
SANTA ANA, 92701
PHONE: (714) 834-2069

OROVILLE CHAMBER OF COMMERCE
1789 MONTGOMERY ST
OROVILLE, 95965
PHONE: (916) 533-2542

OXNARD ECONOMIC DEVELOPMENT DEPT
305 W 3RD ST
OXNARD, 93030
PHONE: (805) 984-4611

PALM SPRINGS CONVENTION & VISITORS BUREAU
255 N EL CIELLO #315
PALM SPRINGS, 92262
PHONE: (619) 327-8411

PALMDALE CHAMBER OF COMMERCE
712 EAST PALMDALE BL
PALMDALE, 93550
PHONE: (805) 273-3232

PARADISE CHAMBER OF COMMERCE
5800 CLARK RD #3
PARADISE, 95969
PHONE: (916) 877-9356

PISMO BEACH CHAMBER OF COMMERCE
581 DOLLIVER ST
PISMO BEACH, 93449
PHONE: (805) 773-4382

PITTSBURG CHAMBER OF COMMERCE
2010 RAILROAD AVE
PITTSBURG, 94565
PHONE: (415) 432-7301

PLACER COUNTY OFFICE OF ECONOMIC DEVELOPMENT
P.O. BOX 749
NEWCASTLE, 95658
PHONE: (916) 663-2062

PLUMAS COUNTY CHAMBER OF COMMERCE
P.O. BOX 1018
QUINCY, 95971
PHONE: (916) 283-6345

REDDING CONVENTION & VISITORS BUREAU
777 AUDITORIUM DR
REDDING, 96001
PHONE: (916) 225-4100

REEDLEY FILM COMMISSION
1613 12TH ST
REEDLEY, 93654
PHONE: (209) 638-2500

RIVERSIDE ECONOMIC COMMUNITY DEVELOPMENT
3499 10TH ST
RIVERSIDE, 92501
PHONE: (714) 788-9770

ROSEVILLE CHAMBER OF COMMERCE
700 VERNON ST
ROSEVILLE, 95678
PHONE: (916) 783-8136

SACRAMENTO FILM & VIDEO COMMISSION
1126 2ND ST
SACRAMENTO, 95814
PHONE: (916) 446-1584

SACRAMENTO METROPOLITAN CHAMBER OF COMMERCE
P.O. BOX 1017
SACRAMENTO, 05805
PHONE: (916) 443-3771

SAN BENITO COUNTY FILM COUNCIL
1629-B CLEARVIEW DR
HOLLISTER, 95023
PHONE: (408) 637-7096

SAN BERNADINO COUNTY BOARD OF SUPERVISORS
385 N ARROWHEAD
SAN BERNADINO, 92415
PHONE: (714) 387-4828

SAN DEIGO MOTION PICTURE & TELEVISION BUREAU
110 WEST "C" ST., STE. 1600
SAN DIEGO, 92101
PHONE: (619) 234-3456

SAN FRANCISCO FILM COMMISSION
CITY HALL ROOM 200
SAN FRANCISCO, 94102
PHONE: (415) 554-6144

SAN JOSE FILM & VIDEO COMMISSSION
333 W SAN CARLOS ST #1000
SAN JOSE, 95110
PHONE: (408) 295-9600

SAN JUAN BAUTISTA CHAMBER OF COMMERCE
201 3RD ST
SAN JUAN BAUTISTA, 95045
PHONE: (408) 623-2454

SAN LUIS OBISPO VISITORS & CONVENTION BUREAU
1041 CHORRO ST #E
SAN LUIS OBISPO, 93401
PHONE: (805) 543-8000

SAN MATEO COUNTY FILM COMMISSION
601 GATEWAY BL #970
SAN FRANCISCO, 94080
PHONE: (415) 952-7600

SANTA BARBARA FILM COUNCIL
201 N SALSIPUEDES #202
SANTA BARBARA, 93103
PHONE: (805) 968-7356

SANTA CLARITA CHAMBER OF COMMERCE
23920 VALENCIA BL #125
SANTA CLARITA, 91355
PHONE: (805) 259-4787

SANTA CRUZ CONVENTION & VISITORS COUNCIL
701 FRONT ST
SANTA CRUZ, 95060
PHONE: (408) 425-4787

SANTA CRUZ FILM BOARD
P.O. BOX 176
SANTA CRUZ, 95061
PHONE: (408) 423-6927

SANTA MONICA MOUNTAINS NATIONAL REC AREA, NATIONAL PARK SERVICE
30401 AGOURA RD #101
AGOURA HILLS, 91301
PHONE: (805) 597-1036

SANTA PAULA CHAMBER OF COMMERCE
P.O. BOX 1
SANTA PAULA, 93060
PHONE: (805) 525-5561

SHASTA CASCADE WONDERLAND
1250 PARKVIEW AVE
REDDING, 96001
PHONE: (916) 243-2643

SIERRA COUNTY ECONOMIC DEVELOPMENT DIS
1230 HIGH ST #224
AUBURN, 95603
PHONE: (916) 823-4703

SISKIYOU COUNTY CHAMBER OF COMMERCE
1000 S MAIN ST
YREKA, 96097
PHONE: (916) 842-1649

SOLANO COUNTY
P.O. BOX 357
VACAVILLE, 95688
PHONE: (707) 446-2092

SONOMA COUNTY CONVENTION & VISITORS BUREAU
10 4TH ST #100
SANTA ROSA, 95401
PHONE: (707) 575-1191

SONORA MOTION PICTURE ASSOC
P.O. BOX 382
SONORA, 95370
PHONE: (209) 532-3344

STATE OF CALIFORNIA FILM OFFICE
6922 HOLLYWOOD BLVD., STE. 600
HOLLYWOOD, 90028
PHONE: (213) 736-2465

STOCKTON CHAMBER OF COMMERCE
445 WEST WEBER
STOCKTON, 95203
PHONE: (209) 466-7066

TEHAMA COUNTY CHAMBER OF COMMERCE
1108 SOLANO ST
CORNING, 96021
PHONE: (916) 824-3802

TOULUMNE COUNTY VISITORS BUREAU
P.O. BOX 4020
SONORA, 95370
PHONE: (209) 553-4420

TULARE COUNTY PLANNING & DEVELOPMENT OFFICE
COURTHOUSE RM 103
VISALIA, 93291
PHONE: (209) 733-6303

VENTURA CHAMBER OF COMMERCE
785 SOUTH SEWARD AVE
VENTURA, 93001
PHONE: (805) 648-2875

WOODLAND HILLS/SANTA MONICA MOUNTAINS NATIONAL RECREATION
22900 VENTURA BL #140
WOODLAND HILLS, 91364
PHONE: (818) 888-3440

CONNECTICUT

CONNECTICUT FILM COMMISION
865 BROOK ST.
ROCKY HILL, 06067-3405
PHONE: (203) 258-4301
FAX: (203) 563-4877

DELAWARE

DELAWARE DEVELOPMENT OFFICE - FILM DIVISION
DELAWARE TOURISM OFFICE
99 KINGS HIGHWAY, BOX 1401
DOVER, 19903
PHONE: (302) 736-4271/(800) 441-8846

DISTRICT OF COLUMBIA

DISTRICT OF COLUMBIA MAYOR'S OFFICE OF MOTION PICTURE & TELEVISION DEVELOPMENT
1111 E. STREET NW, STE. 6600,
WASHINGTON DC, 20004
PHONE: (202) 727-6600

FLORIDA

AMELIA ISLAND-FERNANDINA BEACH CHAMBER OF COMMERCE
P.O. BOX 472
FERNANDINA BEACH, 32034
PHONE: (904) 261-3248

APALACHICOLA BAY CHAMBER OF COMMERCE
128 MARKET ST.
APALACHICOLA, 32320
PHONE: (904) 653-9419

BAHAMAS FILM BUREAU/MIAMI
255 ALHAMBRA CIR., STE. 414
CORAL GABLES, 33134
PHONE: (305) 444-8440
FAX: (305) 444-1080

BAY COUNTY CHAMBER OF COMMERCE
P.O. BOX 1850
PANAMA CITY, 32402
PHONE: (904) 785-5206

BAY COUNTY FILM COMMISSION
310 W. 6TH ST.
PANAMA CITY, 32401
PHONE: (904) 784-4000

BREVARD COUNTY TOURIST DEVELOPMENT COUNCIL
2235 N. COURTNEY PARKWAY
MERITT ISLAND, 32953
PHONE: (407) 453-2211
FAX: (407) 459-1969

BROWARD COUNTY MOTION PICTURE & TELEVISION OFFICE
BROWARD ECONOMIC DEVELOPMENT BOARD
1 EAST BROWARD BLVD.,
STE. 1604
FORT LAUDERDALE, 33301
PHONE: (305) 524-3113

CITY OF MIAMI DEPARTMENT OF DEVELOPMENT
DUPONT PLAZA CENTER,
300 BISCAYNE BLVD. WAY,
STE 400
MIAMI, 33131
PHONE: (305) 579-3366
FAX: (305) 371-9710

CITY OF TAMPA MOTION PICTURE AND TV DEVELOPMENT
306 JACKSON ST.- 7E
TAMPA, 33602
PHONE: (813) 223-8419
FAX: (813) 223-8127

CLAY COUNTY CHAMBER OF COMMERCE
P.O. BOX 1441,
1734 KINGSLEY AVE.
ORANGE PARK, 32760-1441
PHONE: (904) 264-2651

ECONOMIC DEVELOPMENT COUNCIL COLLIER COUNTY
4501 TAMIAMI TRAIL NORTH,
STE. 106
NAPLES, 33940
PHONE: (813) 263-8989

ECONOMIC DEVLEOPMENT COUNCIL OCALA/MARION COUNTY
P.O. DRAWER 459
OCALA, 32678
PHONE: (904) 629-2757
FAX: (904) 629-1581

FLORIDA FILM BUREAU/ FLORIDA DEPARTMENT OF COMMERCE/DIV. OF ECONOMIC DEVELOPMENT
430 COLLINS BLDG., STE. K-1,
107 W. GAINES ST.
TALLAHASSEE, 32399-2000
PHONE: (904) 487-1100
EXT. 106
FAX: (904) 922-5943

FLORIDA MOTION PICTURE AND TELEVISION BUREAU
COLLINS BUILDING
TALLAHASSEE, 32301
PHONE: (904) 487-1100

FORT MYERS PUBLIC INFORMATION OFFICER
P.O. BOX 398
FT. MYERS, 33902-0398
PHONE: (813) 335-2481
FAX: (813) 335-2262

FORT WALTON BAEACH PUBLIC RELATIONS/SPEC. PROJECTS
P.O. BOX DRAWER 640
FT. WALTON BEACH, 32549
PHONE: (904) 244-8191

GADSDEN COUNTY CHAMBER OF COMMERCE
P.O. BOX 389, QUINCY, 32351
PHONE: (904) 627-9231

GAINSVILLE AREA CHAMBER OF COMMERCE
P.O. BOX 1187,
GAINSVILLE, 32602
PHONE: (904) 336-7100
FAX: (904) 336-7141

GREAT LAKE PLACID CHAM- BER OF COMMERCE
P.O. BOX 187
LAKE PLACID, 33852
PHONE: (813) 465-6809

GREATER TARPON SPRINGS CHAMBER OF COMMERCE
210 SOUTH PINELLA AVE., STE.
120
TARPON SPRINGS, 34689
PHONE: (813) 937-6109

INFORMATION RESOURCE SERVICES
P.O. BOX 398
FT. MYERS, 33902
PHONE: (813) 335-2481

JACKSONVILLE CITY HALL
ROOM 413, 220 EAST BAY ST.
JACKSONVILLE, 32202
PHONE: (904) 630-1073
FAX: (904) 630-2903

JACKSONVILLE MOTION PICTURE & TELEVISION LIAISON OFFICE
ROOM 103, CITY HALL, 220
EAST BAY ST.
JACKSONVILLE, 32202
PHONE: (904) 630-1073

KEY WEST CHAMBER OF COMMERCE
410 WALL STREET
KEY WEST, 33040
PHONE: (305) 294-2587

MADISON COUNTY CHAMBER OF COMMERCE
105 NORTH RANGE
MADISON, 32340
PHONE: (305) 973-2788

MANATEE COUNTY FILM COMMISSION
1111 THIRD AVENUE WEST,
#180
BRADENTON, 34206
PHONE (813) 746-5989
FAX: (813) 747-7459

MIAMI-DADE OFFICE OF FILM & TV CO-ORDINATOR
111 NORTH WEST FIRST ST.,
25 FLOOR,
MIAMI, FL 33128
PHONE: (305) 375-3456
FAX (305) 375-5667

MOTION PICTURE & TV ECONOMIC DEVELOPMENT COMMISSION OF MID-FLORIDA, INC.
315 EAST ROBINSON ST., #510
ORLANDO, 32801
PHONE: (407) 422-7159
FAX: (407) 843-9514

MOTION PICTURE & TV OFFICE BROWARD ECONOMIC DEVELOPMENT BOARD
1 EAST BROWARD BLVD., #1604
FT. LAUDERDALE, 33301
PHONE: (305) 524-3113
FAX: (305) 524-3167

OCALA/MARION COUNTY FILM COMMISSION
110 EAST SILVER SPRINGS BLVD.
OCALA, 32670
PHONE: (904) 629-2757
FAX: (904) 629-1581

ORLANDO FILM OFFICE
315 E. ROBINSON ST., #510
ORLANDO, 32801
PHONE: (407) 422-7159 & (407) 896-2933
FAX: (407) 843-9514

PALM BEACH COUNTY FILM LIAISON OFFICE
1555 PALM BEACH LAKES BLVD.,
#204
WEST PALM BEACH, 33401
PHONE: (407) 471-3995
FAX: (407) 471-3990

PALM BEACH COUNTY MOTION PICTURE & TV OFFICE
1555 PALM BEACH LAKES
BLVD., #204
WEST PALM BEACH, 33401
PHONE: (407) 471-3995
FAX: (407) 471-3990

PANAMA CITY FILM COMMISSION
P.O.BOX 1880
PANAMA CITY, 32402
PHONE: (904) 872-3010

POLK COUNTY FILM COORDINATOR
MPTV ECONOMIC DEVELOPMENT COUNCIL
P.O. BOX 1839
BARTOW, 33830
PHONE: (813) 534-6089
FAX: (813) 533-1247

PRIVATE INDUSTRY COUNCIL OF PASCO COUNTY, THE
10730 US HIGHWAY 19 NORTH,
EVERGREEN MALL
PORT RICHEY, 33568
PHONE: (813) 862-7621

ST. AUGUSTINE/ST. JOHNS COUNTY CHAMBER OF COMMERCE
1 RIBERIA ST.
SAINT AUGUSTINE, 32084-3508
PHONE: (904) 829-5681
FAX: (904) 829-6477

SARASOTO COMMITTEE OF 100
P.O. BOX 308
SARASOTA, 34230-0308
PHONE: (813) 955-8187

TAMPA COORDINATOR, FILM & TELEVISION DEVELOPMENT
CITY HALL PLAZA, 8N
TAMPA, 33602
PHONE: (813) 223-8419
FAX: (813) 223-8127

TAMPA MOTION PICTURE & TELEVISION DEVELOPMENT
306 EAST JACKSON ST.
TAMPA, 33602
PHONE: (813) 223-8419
FAX: (813) 223-8127

TOURISM & CONVENTION DEVELOPMENT
PENSACOLA AREA CHAMBER OF COMMERCE
1401 EAST GREGORY ST.
PENSACOLA, 32501
PHONE: (904) 434-1234

VOLUSIA COUNTY BUSINESS DEVELOPMENT CORPORATION
101 CORSAIR DR., #103
DAYTONA BEACH
SHORES, 32014
PHONE: (904) 255-8883
FAX: (904) 255-1604

GEORGIA

GEORGIA FILM & VIDEOTAPE OFFICE
230 PEACHTREE STREET NORTH,
WEST, #650
ATLANTA, 30301
PHONE: (404) 656-3591
TELEX 54-2586 GA INTL ATL
FAX: (404) 656-9063

HAWAII

HAWAII FILM OFFICE
P O. BOX 2359
HONOLULU, 96804
TEL: (808) 548-4535

IDAHO

IDAHO FILM BUREAU
CAPITOL BUILDING, ROOM 108
BOISE, 83720
PHONE: (208) 334-4357

ILLINOIS

CHICAGO CONVENTION & TOURISM BUREAU
MCCORMICK PL
CHICAGO, 60616
PHONE: (312) 567-8500

CHICAGO FILM OFFICE
174 W. RANDOLPH, THIRD FLOOR
CHICAGO, 60601
PHONE;(312) 744-6415
FAX: (312)744-1378

CHICAGO PARK DISTRICT
425 E. MCFETRIDGE
CHICAGO, 60605
PHONE: (312) 294-2200

CHICAGO REGIONAL PORT DISTRICT
95TH ST & WATERFRONT
CHICAGO, 60617
PHONE: (312) 646-4400

CHICAGO TRANSIT AUTHORITY
P.O. BOX 3555
CHICAGO, 60654
PHONE: (312) 664-7200

MAYOR'S OFFICE OF FILM & ENTERTAINMENT
1221 N LASALLE #703
CHICAGO, 60602
PHONE. (312) 814-3500

STATE OF ILLINOIS FILM OFFICE
100 W RANDOLPH #3-400
CHICAGO, 60601
PHONE: (312) 814-3600

INDIANA

INDIANA FILM COMMISSION
1 NORTH CAPITOL, #700
INDIANAPOLIS, 46204
PHONE: (317) 232-8829
FAX (317) 232-4146

IOWA

IOWA DEVELOPMENT COMMISSION
600 EAST COURT AVE., #A
DES MOINES, 50309
PHONE: (515) 281-8319

KENTUCKY

GREATER LOUISVILLE OFFICE
2905 PERIMETER DR.
JEFFERSON, 47130
KENTUCKY FILM OFFICE
BERRY HILL, US60 LOUISVILLE RD.
FRANKFORT, 40601
PHONE: (502) 564-3456

LOUISIANA

CITY OF KENNER FILM LIAISON
624 WILLIAMS BLVD.
KENNER,70062
PHONE: (504) 468-7221
LOUISIANA FILM COMMISSION
P.O. BOX 44320
BATON ROUGE, 70804-4320
PHONE: (504) 342-8150
MONROE-WEST MONROE CONVENTION AND VISTOR
1333 STATE FARM DR.
MONROE, 71202
PHONE: (318) 387-5691
NEW ORLEANS FILM COMMISSION
822 PERDIDO, STE. 402
NEW ORLEANS, 70112
PHONE: (504) 565-7497

MAINE

MAINE FILM COMMISSION
189 STATE STREET STATION 59
AUGUSTA 04333
PHONE: (207) 289-5710

MARYLAND

BALTIMORE FILM COMMISSION
303 EAST FAYETTE, #300
BALTIMORE, 21202
PHONE: (301) 396-4550
FAX: (301) 396-8422
MARYLAND FILM COMMISSION
217 E. REDWOOD ST., NINTH FLOOR
BALTIMORE, 21202
PHONE: (301) 333-6633
PRINCE GEORGE'S CITY MEDIA /FILM OFFICE
9475 LATTSFORD RD.,#130
LANDOVER, 20785
PHONE: (301) 386-3456/ 577-5785
FAX: (301) 322-6132

MASSACHUSETTS

MASSACHUSETTS FILM OFFICE
TRANSPORTATION BUILDING,
10 PARK PLAZA, #2310
BOSTON, 02116
PHONE: (617) 973-8800
FAX: (617) 973-8810

MICHIGAN

MAYORS OFFICE OF FILM/TV
1126 CITY/COUNTY BUILDING
DETROIT, 48226
PHONE: (313) 224-4733
FAX: (313) 224-7157

MICHIGAN FILM OFFICE
525 W. OTTAWA, P.O. BOX 30107
LANSING, 48909
PHONE: (517) 373-FILM & (800) 477-FILM
FAX: (517) 373-3872

MINNESOTA

MINNESOTA MOTION PICTURE & TELEVISION BOARD INC.
401 NORTH THIRD ST., STE 460
MINNEAPOLIS, 55401
PHONE: (612) 332-6493

MISSISSIPPI

COLUMBUS FILM COMMISSION
P.O. BOX 789
COLUMBUS, 39703
PHONE: (800) 327-2686/ (601) 329-1191
FAX: (601) 327-3417
MISSISSIPPI FILM COMMISSION
P.O. BOX 849
JACKSON, 39205
PHONE: (601) 359-3297
NATCHEZ FILM COMMISSION
P.O. BOX 794
NATCHEZ, 39120
PHONE: (601) 446-6345
VICKSBURG FILM COMMISSION
P.O. BOX 110
VICKSBURG, 39180
PHONE: (601) 636-9421/ (800) 221-3536

MISSOURI

CONVENTION/TOURISM BUREAU
601 NORTH KINGSHIGHWAY
CAPE GIRARDEAU, 63701
PHONE: (314) 335-3312
FAX: (314) 335-4686
MISSOURI FILM COMMISSION
P.O. BOX 118
JEFFERSON CITY, 65102
PHONE: (314) 751-9050
ST. LOUIS FILM PARTNERSHIP
100 SOUTH FOURTH STREET
ST. LOUIS, 63102
PHONE: (314) 231-5555
FAX: (314) 444-1174

MONTANA

MONTANA FILM & TV OFFICE
1424 NINTH AVE.
HELENA, 59620
PHONE: (406) 444-2654

NEBRASKA

NEBRASKA FILM OFFICE
BOX 94666
LINCOLN, 68509
PHONE: (402) 471-3791

NEVADA

CARSON CITY
5151 S CARSON ST 2ND FL
CARSON CITY, 89710
PHONE: (702)885-4325
NEVADA MOTION PICTURE & TELEVISION DIVISION
2501 EAST SAHARA AVE.,#101
LAS VEGAS, 89104
PHONE: (702) 386-5287/ 361-1164

NEVADA MOTION PICTURE & TV DIVISION
CARSON CITY-CAPITOL COMPLEX
CARSON CITY, 89710
PHONE: (702) 885-4321/ 791-0839
NEVADA MOTION PICTURE & TV DIVISION
3770 HOWARD HUGHES PKWY #295
LAS VEGAS, 89158
PHONE: (702) 486-7150
FAX: (702) 486-7155
NEVADA MOTION PICTURE DIVISION
MCCARRAN INTERNATIONAL AIRPORT,
ESPLANADE LEVEL, 2ND FLOOR
LAS VEGAS, 89158
PHONE: (702) 486-7150/ 791-0839
FAX: (702) 486-4059

NEW HAMPSHIRE

NEW HAMPSHIRE FILM & TELEVISION BUREAU
P.O. BOX 856
CONCORD, 03301
PHONE: (603) 271-2598

NEW MEXICO

ALBUQUERQUE FILM COMMISSION
OFFICE OF THE MAYOR, P.O. BOX 1293
ALBUQUERQUE, 87103
PHONE: (505) 768-3000
NEW MEXICO FILM COMMISSION
1050 OLD PECOS TRAIL
SANTA FE, 87501
PHONE: (505) 827-8680

NEW YORK - NEW JERSEY

NASSAU COUNTY FILM OFFICE
C/O COMMERCE & INDUSTRY,
1550 FRANKLIN AVE.
MINEOLA, 11501
PHONE: (516) 535-4160
FAX: (516) 535-4229
NEW JERSEY MOTION PICTURE & TELEVISION COMMISSION
ONE GATEWAY CENTER, #510
NEWARK, 07102
PHONE: (201) 648-6279
FAX: (201) 648-7350
NEW YORK CITY MAYOR'S OFFICE OF MOTION PICTURES AND TELEVISION
254 WEST 54TH ST.
NEW YORK, 10019
PHONE: (212) 489-6710
NEW YORK STATE MOTION PICTURE & TELEVISION DEVELOPMENT
1515 BROADWAY, 32ND FLOOR
NEW YORK, 10169
PHONE: (212) 575-6570
FAX: (212) 840-7149
SUFFOLK COUNTY FILM COMMISSION
DENNISON BUILDING 11TH FLOOR, VETS HIGHWAY
HAUPPAUGE, 11788
PHONE: (516) 360-4800
FAX: (516) 360-4888

NORTH CAROLINA

CITY MANAGER'S OFFICE
P.O. BOX 1810
WILMINGTON, 28402
PHONE: (919) 341-7810
FAX: (919) 341-7887
NORTH CAROLINA FILM OFFICE
430 NORTH SALISBURY ST.
RALEIGH, 27611
PHONE: (919) 733-9900
FAX: (919) 733-0110

OHIO

GREATER CINCINNATI FILM COMMISSION
264 MCCORMICK PL.
CINCINNATI, 45219
PHONE: (513) 784-1744
OHIO FILM BUREAU
77 SOUTH HIGH ST., 28TH FLOOR
COLUMBUS, 43266-0101
PHONE: (614) 466-2284/(800) 848-1300
FAX: (614) 644-1789

OKLAHOMA

FILM INDUSTRY TASK FORCE
500 WILL ROGERS BUILDING
OKLAHOMA CITY, 73105
PHONE: (405) 521-3525
OKLAHOMA FILM OFFICE
6601 BROADWAY EXTENSION
OKLAHOMA CITY, 73116-8214
PHONE: (405) 841-5135
FAX: (405) 841-5199

OREGON

OREGON FILM & VIDEO BUREAU
595 COTTAGE ST., NORTH EAST
SALEM, 97310
PHONE: (503) 373-1232

PENNSYLVANIA

PENNSYLVANIA FILM BUREAU
455 FORUM BUILDING
HARRISBURG, 17120
PHONE: (717) 783-3456
PHILADELPHIA FILM OFFICE
120 MUNICIPAL SERVICES BUILDING
PHILADELPHIA, 19102
PHONE: (215) 686-2668

PUERTO RICO

PUERTO RICO FILM COMMISSION, ECONOMIC DEVELOPMENT ADMINISTRATION
P.O. BOX 2350
SAN JUAN, 00936
PHONE: (809) 758-4747

RHODE ISLAND

RHODE ISLAND FILM COMMISSION
OLD STATE HOUSE, 150 BENEFIT ST.
PROVIDENCE, 02903
PHONE: (401) 277-3456

SOUTH CAROLINA

MYRTLE BEACH AREA FILM OFFICE
P.O. BOX 2115
MYRTLE BEACH, 29578
PHONE: (803) 626-7444
FAX: (803) 448-3010

SOUTH CAROLINA FILM OFFICE
P.O. BOX 927
COLUMBIA, 29201
PHONE: (803) 737-0400
FAX: (803) 737-0418

SOUTH DAKOTA

SOUTH DAKOTA DEPARTMENT OF DEVELOPMENT
320 SIXTH AVENUE NORTH, SEVENTH FLOOR
PIERRE, 57501
PHONE: (605) 773-3301

TENNESSEE

MEMPHIS/SHELBY COUNTY FILM, TAPE-MUSIC
BEALE STREET LANDING, 245 WAGNER PL., #4
MEMPHIS, 38103
PHONE: (901) 527-8300
FAX: (901) 527-8326

TENNESSEE FILM, ENTERTAINMENT & MUSIC COMMISSION
RACHEL JACKSON BUILDING, SEVENTH FLOOR, 320 6TH AVE. NORTH
NASHVILLE, 37219-5308
PHONE: (615) 741-FILM (IN NASHVILLE)
(800) 251-8594 (OUT OF STATE)
(800) 342-8470 (IN STATE)
FAX: (615) 741-5829

TEXAS

HOUSTON CONVENTION & VISITORS COUNCIL
3300 MAIN ST.
HOUSTON, 77002
PHONE: (713) 523-5050

SAN ANTONIO CONVENTION & VISITOR'S BUREAU
P.O. BOX 2277
SAN ANTONIO, 78298
PHONE: (512) 270-8700

TEXAS FILM COMMISSION
410 EAST FIFTH ST., P.O. BOX 12728
AUSTIN, 78711
PHONE: (512) 469-9111

UTAH

BUREAU OF LAND MANAGEMENT
2370 SOUTH 2300 WEST
SALT LAKE CITY, 84119
PHONE: (801) 977-4300

CACHE COUNTY CORPORATION
120 NORTH 100 WEST
LOGAN, 84321
PHONE: (801) 752-5935

CENTRAL UTAH FILM COMMISSION
51 S UNIVERSITY AVE
PROVO, 84606
PHONE: (801) 370-8390

KANAB FILM COMMISSSION
60 E CENTER
KANAB, 84741
PHONE: (801) 644-2452

MOAB FILM DEVELOPMENT COMMISSION
59 S MAIN 2ND FL
MOAB, 84532
PHONE: (801) 259-6388

OGDEN FILM LIAISON
2540 WASHINGTON BL 6TH FL
OGDEN, 84401
PHONE: (801) 629-8915

PARK CITY FILM COMMISSION
P.O. BOX 1630
PARK CITY, 84060
PHONE: (801) 453-1350

SALT LAKE CITY POLICE DEPARTMENT
315 EAST 200 SOUTH
SALT LAKE CITY, 84111
PHONE: (801) 799-3806

UTAH FILM COMMISSION
324 S STATE #230
SALT LAKE CITY, 84111
PHONE: (801) 538-8740

UTAH FILM DEVELOPMENT
6150 STATE OFFICE BUILDING, CAPITOL MALL
SALT LAKE CITY, 84114
PHONE: (801) 533-5041/(800) 453-8824

UTAH HIGHWAY PATROL
4501 SOUTH 2700 WEST
SALT LAKE CITY, 84119
PHONE: (801) 965-4558

WASHINGTON COUNTY FILM OFFICE
425 S 700 E
ST GEORGE, 84770
PHONE: (800) 869-6635

VERMONT

VERMONT FILM BUREAU
134 STATE ST.
MONTPELIER, 05602
PHONE: (802) 828-3236
FAX: (802) 828-3233

VIRGINIA

VIRGINIA FILM OFFICE
DEPARTMENT OF ECONOMICAL, DEVELOPMENT
1000 WASHINGTON BUILDING, RICHMOND, 23219
PHONE: (804) 786-8204
FAX: (804) 786-1121

VIRGIN ISLANDS

U.S. VIRGIN ISLANDS FILM PROMOTION OFFICE
DEPT. OF ECON. DEV. & AGRICULTURE
P.O. BOX 6400,
ST. THOMAS, 00804
PHONE: (809) 774-8784

WEST VIRGINIA

WEST VIRGINIA DEPARTMENT OF COMMERCE, TOURISM DIV.
CAPITOL COMPLEX 2101 E. WASHINGTON ST.
CHARLESTON, 25305
PHONE: (304) 348-2286

WISCONSIN

MILWAUKEE FILM OFFICE
DEPARTMENT OF CITY DEVELOPMENT, CITY OF MILWAUKEE
P.O. BOX 324,
MILWAUKEE, 53201
PHONE: (414) 223-5818

WISCONSIN FILM OFFICE DIVISION OF TOURISM DEVELOPMENT
P.O. BOX 7970,
MADISON, 53707
PHONE: (608) 267-3456
FAX: (608) 266-3403

WYOMING

WYOMING FILM OFFICE
I-25 AND COLLEGE DR.,
CHEYENNE, 82002
PHONE: (307) 777-7777

ALABAMA

STUDIOS/SOUND STAGES

AIR-MOBILE PRODUCTIONS
95 ROBERT JEMISON ROAD
BIRMINGHAM, 35209
PHONE: (205) 942-7023
STRAIGHT FURROW PRODUCTION COMPANY
2829 SEVENTH AVENUE SOUTH
BIRMINGHAM, 35233
PHONE: (205) 252-5625

ARIZONA

SPECIAL EFFECTS

ACCURACY SYSTEMS INC.
15203 NORTH CAVE CREEK ROAD
PHOENIX, AZ.85032
PHONE: (602) 971-1991
ARIZONA CINE EQUIPMENT INC
2125 EAST 20TH ST
TUCSON, AZ.85719
PHONE: (602) 623-8268
BAKERS BOBALU RANCH
OUTFITTERS & PACKERS
590 WEST HAWK WAY #1
AMADO, AZ.85645
PHONE: (602) 398-9611
BILLEY, PHILLIP G.
5642 NORTH CAMINO DE LA NOCHE
TUCSON, AZ.85717
PHONE: (602) 299-6423
BREWER, JERRY
3624 WEST BUTTERFLY
TUCSON, AZ.85741
PHONE: (602) 744-4288
ROBERT DIEPENBROCK
1133 W. LOS LAGOS VISTA CIRCLE
MESA, AZ 85210
PHONE: (602) 730-0380
DOUBLE F PRODUCTIONS
8225 NORTH RANCHO CATALINA
TUCSON, AZ.85704
PHONE: (602) 742-3161
DUDLEY, MICHAEL C.
2299 NORTH SILVERBELL #8125
TUCSON, AZ.85745
PHONE: (602) 624-6072
GLAS-TEC INC
3417 EAST MICHIGAN
TUCSON, AZ.85714
PHONE: (602) 889-0181
GRAPEVINE PRODUCTIONS LTD
5055 EAST BROADWAY #214
TUCSON, AZ.85711
PHONE: (602) 747-3115
L.O.G. PRODUCTION
3201 EAST GREENLEE
TUCSON, AZ.85716
PHONE: (602) 323-1963
MOTION PICTURE VEHICLES
6055 SOUTH PALOMINO RD
TUCSON, AZ.85746
PHONE: (602) 883-4144
PADILLA, RAY
1004 EAST ALTA VISTA
TUCSON, AZ 85719
PHONE: (602) 792-1781
PATIO POOLS OF TUCSON, INC
7960 EAST 22ND ST
TUCSON, AZ.86710
PHONE: (602) 886-1211

PHOTO ICE CUBE COMPANY
BOX 5216
PHOENIX, AZ.85010
PHONE: (602) 273-60
PIONEER PAINT OF ARIZONA
3755 EAST 43RD PL
TUCSON, AZ.85713
PHONE: (602) 571-1800
FAX: (602) 745-8678
PYRO-SPECTACULARS
P.O. BOX 16306
PHOENIX, AZ.85011
PHONE: (602) 241-1034 &
581-3348
RARE FILM PRODUCTIONS INC
5755 EAST RIVER RD #2618
TUCSON, AZ. 85715
PHONE: (602) 529-2937
FAX: (602) 577-3619
RAVENLOCK PRODUCTIONS
P.O. BOX 31632
PHOENIX, AZ 85046
PHONE: (602) 867-8387 &
941-1240
RHONDA GRAPHICS, INC
2235 W. ALICE
PHOENIX,AZ. 85021
PHONE: (602) 263-3939/3577
SAFE SHOT CUSTOM FIRING BOXES
2701 SOUTH BRANDYWINE LANE
TUCSON, AZ.85730
PHONE: (602) 721-2310/886-5090
SKEELES ENGINEERING INC
770 EAST EVANS BLVD
TUCSON, AZ.85713
PHONE: (602) 792-4122
SNOWSKINS
640 E PURDUE
PHOENIX, AZ 85021
PHONE: (602) 944-0083
SOUTHWEST SCENIC GROUP, INC
2615 S. INDUSTRIAL PARK AVE
TEMPE, AZ. 85282
PHONE: (602) 968-5595
SPECIAL EFFECT STUDIO
204 N. FRASER DR. EAST
MESA, AZ.85023
PHONE: (602) 964-9606
STUNTS & EFFECTS
6301 WEST OREGON
GLENDALE, AZ 85301
PHONE: (602) 846-2544
SUNBELT SCENIC STUDIOS
8980 S. MCKEMY ST
PHOENIX, AZ 85284
PHONE: (602) 598-0181
CARLOS TERRAZAS PRODUCTIONS
2800 NORTH PARK AVE
TUCSON, AZ.85719
PHONE: (602) 882-7826
TRA-LA FIBERGLASS INC
4315 NORTH SULLINGER
TUCSON, AZ.85705-2013
PHONE: (602) 887-1308

EXPENDABLES

A & A LEASING INC
P.O. BOX 7591
PHOENIX AZ.85011
PHONE: (602) 274-5992
ABC THEATRICAL RENTALS & SALES
825 N SEVENTH ST
PHOENIX AZ.85004
PHONE: (602) 258-5204
ACME LEASING CORP.
P.O. BOX 50204
TUCSONAZ.85703
PHONE: (602) 883-6883

ADVANCED PRODUCTION SERVICES INC
1035 S TYNDALL
TUCSON,AZ.85719
PHONE: (602) 884-8550
AFFILIATED CINE SERVICES
2949 W INDIAN SCHOOL RD
PHOENIX, AZ.85017
PHONE: (602) 266-4198
AMERICAN TELEPRODUCTIONS
17602 N BLACK CYN HWY #111
PHOENIX, AZ.85023
PHONE: (602) 886-0162
ARIZONA AERIAL EQUIPMENT COMPANY
3708 E MIAMI ST
PHOENIX, AZ.85040
PHONE: (602) 437-3955
ARIZONA CINE EQUIPMENT
2125 E 20TH ST
TUCSON, AZ.85719
PHONE: (602) 623-8268
ARIZONA HYDRO CRANE SERVICE
2915 E ILLINI ST
PHOENIX, AZ.85040
PHONE: (602) 242-468
ARIZONA MOTION PICTURE SERVICES
P.O. BOX 33622
PHOENIX, AZ.85067
PHONE: (602) 230-0107
ARIZONA TENT TABLE & CHAIR RENTAL
1345 N 22ND AVE
PHOENIX, AZ.85009
PHONE: (602) 252-8368
ARIZONA THEATRE EQUIP & SUPPLY
1410 E WASHINGTON
PHOENIX, AZ.85036
PHONE: (602) 254-0215
ARIZONA TWO-WAY COMMUNICATIONS
1251 S TYDALL AVE #111
TUSCON, AZ.85713
PHONE: (602) 973-1610
ATCO STRUCTURES INC
2229 W ROOSEVELT
PHOENIX, AZ.85009
PHONE: (602) 254-9000
AUDIO VIDEO RECORDERS
3830 N SEVENTH ST.
PHOENIX, AZ.85009
PHONE: (602) 277-4723
AZTEC RENTAL CORPORATION
1310 W 23RD ST
TEMPE, AZ., 85282
PHONE: (602) 968-8735
B & B VIDEO INC
1806 W GRANT RD #104
TUCSON, AZ.85745
PHONE: (602) 623-8201
BLUE WATER ADVENTURES
697 N NAVAJO DR
PAGE, AZ.86040
PHONE: (602) 645-3087
BT SCAFFOLD & EQUIPMENT CO. INC.
3003 N CENTRAL AVE #600
PHOENIX, AZ.85012
PHONE: (602) 265-5597
BURL'S INC
1636 S. 7TH ST
PHOENIX, AZ.85004
PHONE: (602) 258-1829
CENTERLINE LIGHTING INC
1933 THIRD ST
TEMPE, AZ.85281
PHONE: (602) 967-5321
CINE UTILITY
P.O. BOX 834
SCOTTSDALE, AZ.85252
PHONE: (602) 948-3638

DATAPHAZ COMPUTER RENTALS
15002 N 25TH DR #4
PHOENIX, AZ.85023
PHONE: (602) 351-2850
E.A.R. PROFESSIONAL AUDIO
2641 E MCDOWELL
PHOENIX, AZ.85008
PHONE: (602) 267-0600
GELCO SPACE
2940 N 29TH DR
PHOENIX, AZ.85017
PHONE: (602) 269-9377
HENLEY PRODUCTION SVC & EQUIP, LARRY
4271 N MILLER RD
SCOTTSDALE, AZ.85251
PHONE: (602) 994-9081
HILL PRODUCTIONS & SERVICES
14425 N 36TH PLACE
PHOENIX, AZ.85032
PHONE: (602) 971-1128
ICS WAREHOUSE INC
1802 E 18TH ST
TUCSON, AZ.85719
PHONE: (602) 882-3853
INTERFACE COMMUNICATIONS
1828 E UNIVERSITY #9
TEMPE, AZ 85281
PHONE: (602) 829-7447
J-C EQUIPMENT COMPANY
5061 E ROADRUNNER
MESA, AZ 85205
PHONE: (602) 985-4343
J.S. JAMES ENTERPRISES, INC
P.O. BOX 9201
SCOTTSDALE, AZ 85252
PHONE: (602) 949-5661
KAGE INC
532 E YALE
PHOENIX, AZ 85008
PHONE: (602) 840-9290
KEYLITE/PSI PHOENIX
3823 N 34TH AVE
PHOENIX, AZ 85017
PHONE: (602) 272-0881
KIRST EQUIPMENT CO.
14425 N 79TH ST #H
SCOTTSDALE, AZ.85260
PHONE: (602) 951-9116
K.M. & ASSOCIATES
3649 W LAWRENCE RD
PHOENIX, AZ.85019
PHONE: (602) 274-5834
KPHO-TV 5 PRODUCTIONS
4016 N BLACK CYN HWY
PHOENIX, AZ.85016
PHONE: (602) 996-9847
MARCO EQUIPMENT COMPANY
221 S 35TH AVE
PHOENIX,AZ.85009
PHONE: (602) 272-2671
MEDIA WORKS OF ARIZONA
843 W ELNA RAE ST
TEMPE, AZ.85281
PHONE: (602) 968-4392
MILE HIGH ENTERPRISES
P.O. BOX 5785
BISBEE, AZ.85603
PHONE: (602) 432-4212
MJMK PRODUCTION SERV/ EQUIP RENTAL
6050 N 79TH ST
SCOTTSDALE, AZ. 85253
PHONE: (602) 948-6330
MODULAIRE/WMI SERVICES
2902 S 44TH ST
PHOENIX, AZ. 85040
PHONE: (602) 437-1177
NITE FLYER
P.O. BOX 23318
PHOENIX, AZ.85063
PHONE: (602) 248-0929

PHOENIX VIDEOFILMS
2949 W INDIAN SCHOOL RD
PHOENIX, AZ.85017
PHONE: (602) 266-4198

PHOTO & SOUND COMPANY
4246 E WOOD ST #560
PHOENIX, AZ.85040

PRODUCTION TRANSPORT SERVICES
P.O. BOX 61234
PHOENIX, AZ.85082
PHONE: (602) 220-1460

RAVEN LOCKE PRODUCTIONS
P.O.BOX 31632
PHOENIX, AZ.85046
PHONE: (602) 867-8387

MAKE-UP SUPPLIERS

LOWNS COSTUMES, INC
2524 N. CAMPBELL AVE
TUCSON, AZ.85719
PHONE: (602) 795-5467

SPECIAL EFFECT STUDIO
204 N FRASER DR. EAST
MESA, AZ. 85023
PHONE: (602) 964-9606B

PROPS

AAA-TRIPLE A PROPS/ COSTUMES
209 N PERRY LANE
TEMPE, AZ 85281
PHONE: (602) 461-1773/498-4990

ACE CASINO EQUIPMENT
2112 NORTH DRAGON #20
TUCSON, AZ85745
PHONE: (602) 624-4162

AERO MERIDIAN PRODUCTIONS
14806 NORTH 74TH STREET
SCOTTSDALE ,AZ.85260
PHONE: (602) 948-7626

AFRICAN ARTS LTD
3025 NORTH CAMPBELL #151
TUCSON, AZ.85719
PHONE: (602) 795-1997

ANTIQUE AUTO CLUB OF AMERICA
PHOENIX, AZ
PHONE: (602) 569-1288/991-3666

ARIZONA REENACTORS ASSOC.
9029 N. 43RD AVE., SUITE 229
PHOENIX, AZ 85051-3267
PHONE: (602) 864-6766/8249

ARIZONA HATTERS
3600 NORTH FIRST AVE
TUCSON, AZ.85719
PHONE: (602) 292-1320

ARTISTIC GLASS & MIRROR CO
2029 EAST 14TH ST
TUCSON, AZ.85719
PHONE: (602) 624-4984

AUNT MINNIES
5120 SOUTH JULIAN DR #160
TUCSON, AZ.85706
PHONE: (602) 889-7805

AVRA VALLEY AIR
11700 WEST AVRA VALLEY RD
MARANA, AZ.85653
PHONE: (602) 792-2536

BAKERS BOBALU RANCH
OUTFITTERS & PACKERS
590 WEST HAWK WAY #1
AMADO, AZ 85645
PHONE: (602) 398-9611

BALLOON-A-TIC
3256 EAST GRANT RD
TUCSON, AZ.85716
PHONE: (602) 795-7006

BARNETT & DEYOE CONTRACTORS
701 WEST SILVERLAKE RD
TUCSON, AZ.85713
PHONE: (602) 623-2662

BILLEY, PHILLIP G.
5642 NORTH CAMINO DE LA NOCHE
TUCSON, AZ 85717
PHONE: (602) 299-6423

BISHOP TAXIDERMY
5409 EAST 30TH ST
TUCSON, AZ 85711
PHONE: (602) 790-9694

BONEWITZ CANVAS & AWNING CO.
2606 NORTH STONE AVE
TUCSON, AZ.85705
PHONE: (602) 623-7041

BRICK TREE FARM
P.O. BOX 511
POMERENE, AZ.85627
PHONE: (602) 586-7531/596-5247

BROKEN CIRCLE BAR G
P.O. BOX 406
POMERENE, AZ 85627
PHONE: (602) 298-2974/586-5247

BUDGET FURNITURE
1120 SOUTH SWAN
TUCSON, AZ.85711
PHONE: (602) 747-7368

BROKEN CIRCLE/ BAR G MOVIE RANCH
P.O. BOX 406
POMERENE, AZ 85627
PHONE: (602) 586-2530

BUFFALO RICKS WILD WEST, INC
P.O. BOX 255
CAVE CREEK, AZ 85331
PHONE: (602) 992-4578

CARAVAN WEST PRODUCTIONS
7125 E. 2ND ST.. #101A
SCOTTSDALE, AZ 85251
PHONE: (602) 994-5442

CHAMPLIN AIR MUSEUM
MESA, AZ
PHONE: (602) 830-4540

CHAMELEON-PROPS PROD. SRVS.
8647 EAST GOLF LINKS RD
TUCSON, AZ.85730
PHONE: (602) 885-1829/298-5646/326-4392

CHARNIE
2702 NORTH 28 STREET
PHOENIX, AZ 85008
PHONE: (602) 955-1539

CLASSIC CARRIAGE HOUSE
5552 EAST WASHINGTON
PHOENIX, AZ 85034
PHONE: (602) 275-6825

CONTENTS: CONTEMPORARY & SOUTHWESTERN FURNITURE
4380 EAST GRANT RD
TUCSON, AZ.85712
PHONE: (602) 881-6900

CORONADO HOME HEALTH CARE
3939 SOUTH PARK AVE
TUCSON, AZ.85714
PHONE: (602) 748-2273

COUNTRY HOME FURNITURE
2330 EAST BROADWAY
TUCSON, AZ.85719
PHONE: (602) 629-9979

COVINGTON FINE ARTS GALLERY
4951 EAST GRANT RD #107
TUCSON, AZ.85712
PHONE: (602) 326-6111

CUSTOM BRANDS INC
1966 WEST PRINCE
TUCSON, AZ.85705
PHONE: (602) 887-7111

DAVILA, DEBBIE
P.O. BOX 40542
TUCSON, AZ.85717
PHONE: (602) 742-1712

DESERT EXOTICS
RT. 1, BOX 720
WILCOX, AZ.85643
PHONE: (602) 384-3053

DO-WAH DIDDY PRODUCTIONS
3642 EAST THOMAS ROAD
PHOENIX , AZ.85018
PHONE: (602) 957-3874

DOLLAR PRINCESS MINING SALOON
P.O. BOX 886
OATMAN, AZ 86433
PHONE: (602) 768-2760

DOUBLE F PRODUCTIONS
8225 NORTH RANCHO CATALINA
TUCSON, AZ.85704
PHONE: (602) 742-3161

EARHART FORD TRACTOR INC
4350 SOUTH PALO VERDE
TUCSON, AZ.85714
PHONE: (602) 889-6396

EISELLE RAILWAY CORP
1125 N. LEHMBERG
CASA GRANDE, AZ 85222
PHONE: (602) 836-3062/7483

ELBO ANTIQUES
5605 W. GLENDALE AVE
GLENDALE, AZ 85301
PHONE: (602) 842-0220/3161

ELEPHANTS TRUNK ANTIQUE EMPORIUM
1111 N. MAIN
TUCSON, AZ 85705
PHONE: (602) 623-2926

EMCEE DISPLAY CO.
3612 NORTH 7TH STREET
PHOENIX, AZ 85014
PHONE: (602) 266-1740

ERICKSON DESIGN & PRODUCTION
620 EAST 19TH ST #150
TUCSON,AZ.85719
PHONE: (602) 628-8867

E TROOP - SIXTH CALVARY
6220 EAST SECOND ST
TUCSON, AZ.85711
PHONE: (602) 745-3747/296-4551

FINE LAMPS & SHADES
2919 EAST GRANT RD
TUCSON, AZ.85716
PHONE: (602) 327-0012

FIREHOUSE ANTIQUES
6522 EAST 22ND ST
TUCSON, AZ.85710
PHONE: (602) 571-177

FLUORESCO LIGHTING - SIGN MAINTENANCE CORPORATION
3131 EAST 46TH ST
TUCSON, AZ.85713
PHONE: (602) 623-79

FLYING SAUCERS
P.O. BOX 1432
BISBEE, AZ.85603
PHONE: (602) 434-4858

FURNITURE CORAL UPHOLSTERY
2604 NORHT STONE AVE
TUCSON, AZ.85705
PHONE: (602) 884-752

GARIGAN ENTERPRISES
P.O. BOX 17359
TUCSON, AZ.85731
PHONE: (602) 296-6739

GATZWILLER, WILLIAM A.
700 SOUTH INDIAN AGENCY RD
TUCSON, AZ.85746
PHONE: (602) 294-1839

GEBHART, DICK
5326 EAST SIXTH ST
TUCSON, AZ.85711
PHONE: (602) 745-1882/326-3973

GIZINSKI, JULIE
3649 WEST BUTTERFLY LANE
TUCSON, AZ.85741
PHONE: (602) 744-3747

GLASS TEC INC
3417 EAST MICHIGAN
TUCSON, AZ.85714
PHONE: (602) 889-0181

GOLDEN EAGLE ANTIQUES & GIFTS
2320 NORTH SWAN RD
TUCSON, AZ.85712
PHONE: (602) 790-3333

GRAND OLD CARS
201 WEST APACHE TRAIL
APACHE JUNCTION, AZ 85220
PHONE: (602) 982-3500

GROSSMAN, DAVE
P.O. BOX 44062
TUCSON, AZ 85733
PHONE: (602) 881-2615

GUNS FOR HIRE INC.
2526 EAST INDIAN SCHOOL ROAD
PHOENIX, AZ 85016
PHONE: (602)955-0930 & 948-9741

HOUSE N GARDEN FURNITURE INC
4310 EAST BROADWAY
TUCSON, AZ.85711
PHONE: (602) 327-5675

HUNLEY CONTRUCTION CO
1831 WEST LUCERO RD
TUCSON, AZ.85737
PHONE: (602) 297-1376

JLC CUSTOM LEATHER
5220 EAST PIMA ST
TUCSON, AZ.
PHONE: (602) 795-0405

LAMIS AUTO UPHOLSTERY
4734 EAST SPEEDWAY
TUCSON, AZ.85712
PHONE: (602) 323-362

KONING, MARI LOU
10131 EAST TIERRA ALTA PL
TUCSON, AZ.85712
PHONE: (602) 749-5240

LAUGHON, CHARLOTTE
9740 EAST LORAIN PL
TUCSON, AZ.85748
PHONE: (602) 721-0790

LOCK, STOCK, AND BARRELL
365 W. GRANT RD
TUCSON, AZ.85705
PHONE: (602) 628-1321/822-5077

LOMAX AUTO TECH ETC
P.O. BOX 55203
PHOENIX, AZ 85073
PHONE: (602) 953-1909

LOST & FOUND APPLIANCES
351 NORTH FOURTH AVE
TUCSON, AZ.85705
PHONE: (602) 792-3474

LOST & FOUND FURNITURE
340 N FOURTH AVE
TUCSON, AZ.85705
PHONE: (602) 792-9319

MARTIN & MARTIN DESIGN
203 WEST CUSHING ST
TUCSON, AZ., 85701
PHONE: (602) 792-9903

MATOS FURNITURE RENTAL
4380 EAST BROADWAY
TUCSON, AZ 85711
PHONE: (602) 881-0070

MENDEZ, EDDIE
222 EAST ROGER
TUCSON, AZ.85705
PHONE: (502) 887-2587

MENDES, LOUIS
6750 NORTH WAY CROSS
TUCSON, AZ.85743
PHONE: (602) 744-2457/888-8880

MIRACLE WIGS BY DORIS
P.O. BOX 17371
TUCSON, AZ.85371
PHONE: (602) 298-6442
MOORE, SYDNEY
7789 WEST BOPP RD
TUCSON,AZ 85746
PHONE: (602) 883-9886
MOUNTAIN VIEW TOURS INC
4283 SOUTH SANTA RITA
TUCSON, AZ.85714
PHONE: (602) 741-1188
FAX: (602) 294-0555
MULLER MUSIC CENTER
1010 EAST BROADWAY
TUCSON, AZ.85719
PHONE: (602) 623-0525
**NELSON, JANET FILM & PHOTO
SERVICES**
P.O. BOX 143
TEMPE, AZ 85281
PHONE: (602) 968-3771
OFFICE CLUB
5251 EAST SPEEDWAY
TUCSON, AZ.85712
PHONE: (602) 881-2582
ORIENTAL RUG CENTER
3525 NORTH CAMPBELL
TUCSON, AZ.85716
PHONE: (602) 881-6667
PAPAGO RIDING STABLES
400 NORTH SCOTTSDALE ROAD
TEMPE, AZ 85281
PHONE: (602) 966-9292
PARTY CAROUSEL
903 NORTH SWAN
TUCSON, AZ.85711
PHONE: (602) 327-1945
PARTY CONCEPTS
4219 EAST SPEEDWAY
TUCSON, AZ.85712
PHONE: (602) 323-3313
PASCUA PUEBLO FIRE DEPT.
747 SOUTH CAMINO DE OESTE
TUCSON, AZ.85746
PHONE: (602) 883-4700
PATIO POOLS
7960 EAST 22ND ST
TUCSON, AZ.85710
PHONE: (602) 886-1211
MARY PEACHIN ART COMPANY
3955 EAST SPEEDWAY
TUCSON, AZ.85712
PHONE: (602) 881-1311
PEARLE VISION EXPRESS
4500 NORTH ORACLE
TUCSON, AZ.85705
PHONE: (602) 293-5671
PIMA AIR MUSEUM
TUCSON, AZ
PHONE: (602) 574-9658/0462
PIONEER PAINT OF ARIZONA
3755 EAST 43RD PLACE
TUCSON, AZ.85713
PHONE: (602) 571-1800
FAX: (602) 745-8678
THE PLANTSMAN
4024 EAST SPEEDWAY
TUCSON, AZ.85712
PHONE: (602) 323-0996
3302 NORTH COUNTRY CLUB
TUCSON, AZ.85716
PHONE: (602) 795-4130
PRECISION MINIATURES
2202 N. EDGEMERE
PHOENIX, AZ 85006
PHONE: (602) 222-8105
PRODUCTION SCENIC STUDIOS
1027 S. TYNDALL
TUCSON, AZ 85719
PHONE: (602) 622-2349/742-4440
QUIROZ, ARMANDO S.
8069 STREAMSIDE AVE
TUCSON, AZ.85741
PHONE: (602) 579-0942

R A S ENTERPRISES
4003 N. 45TH PLACE
PHOENIX, AZ 85018
PHONE: (602) 840-4501
RAINBOW JEWELERS
3635 NORTH CAMPBELL AVE
TUCSON, AZ.85719
PHONE: (602) 325-2150
RAWHIDE
23023 N. SCOTTSDALE RD
SCOTTSDALE, AZ 85255
PHONE: (602) 563-5600
REDDINGTON LAND & CATTLE
14772 EAST REDDINGTON
TUCSON, AZ.
PHONE: (602) 749-5555
**SCREEN DESIGNS & SIGNS BY
CARL**
4235 ROBERT E. LEE
PHOENIX, AZ 85032
PHONE: (602) 971-1912
THE SPORTS PAGE
1680 COUNTRY CLUB
TUCSON, AZ.85716
PHONE: (602) 326-5001
TERRITORIAL GUNLEATHER
P.O. BOX 57698
TUCSON, AZ.85732
PHONE: (881-0847)
THOSE WERE THE DAYS
516 SOUTH MILL AVENUE
TEMPE, AZ.85281
PHONE: (602) 967-4729
**THUNDERBOLT-GILA RIVER
LEATHER**
P.O. BOX 1263
PEORIA, AZ.85345
PHONE: (602) 878-6206
**TOONE & CO. INTERIOR
DESIGNS**
5930 EAST PIMA ST #144
TUCSON, AZ 85712
PHONE: (602) 298-7557
TUCSON OIL INC
1057 EAST MILL ST
TUCSON, AZ.85712
PHONE: (602) 298-7557
TUCSON QUILTER
2341 FRIEBUS #16
TUCSON, AZ.85713
PHONE: (602) 745-6224
**TUCSON RODEO PARADE
COMMITTEE & MUSEUM**
P.O. BOX 1788
TUCSON, AZ.85702
PHONE: (602) 747-9417
WEAPON SPECIALTIES
8101 WEST HEATHERBRAE DRIVE
PHOENIX, AZ 85033
PHONE: (602) 849-5199
WESTERN RENTALS
3010 CAMINO DEL YUCCA
SAHUARITA, AZ 85629
PHONE: (602) 625-8582
WESTERN UPHOLSTERY INC
125 EAST GRANT RD
TUCSON, AZ.85705
PHONE: (602) 882-8107
YPICO DESERT FURNISHING
6450 WEST HIDDEN CANYON DR
TUCSON, AZ.85745
PHONE: (602) 743-3681/293-8325
DON YUNKER DESIGN INC
1211 WEST LINDA VISTA
TUCSON, AZ.85737
PHONE: (602) 742-1788
ZIPS RECORDS & TAPES
7091 EAST SPEEDWAY
TUCSON, AZ.85710
PHONE: (602) 885-7799

PROPS - VEHICLES

ADLER, DONALD
841 EAST LINDA VISTA DR
TUCSON, AZ.85737
PHONE: (602) 742-0875
ARIZONA HISTORICAL SOCIETY
949 EAST SECOND ST
TUCSON, AZ 85719
PHONE: (602) 628-5774
**ARIZONA STAGECOACH
ARRANGER**
4520 EAST GRANT
TUCSON, AZ.85712
PHONE: (602) 881-4474
BAILEY, KEVIN D.
4310 EAST GRANT
TUCSON, AZ.85712
PHONE: (602) 326-4848
CLERLOCK, GORDON
947 NORTH NINTH AVE
TUCSON, AZ.85705
PHONE: (602) 622-3032
DESERT KAWASAKI & SUZUKI
3110 NORTH FIRST AVE
TUCSON, AZ.85719
PHONE: (602) 884-0801
DUNN, BRIAN D.
2603 EAST FLORENCE
TUCSON, AZ 85716
PHONE: (602) 881-6772/325-3376
FRONTIER TOWING SERVICES
3630 SOUTH DODGE
TUCSON, AZ.86713
PHONE: (602) 748-110
GLAS-TEC INC
3417 EAST MICHIGAN
TUCSON, AZ.85714
PHONE: (602) 889-0181
HOLMES TUTTLE FORD
TUCSON AUTO MALL
TUCSON, AZ. 85705
PHONE: (602) 292-3600
HONDA SUZUKI OF TUCSON
7075 EAST 22ND ST
TUCSON, AZ 85710
PHONE: (602) 747-9141
KENDALL IMPORTS
3145 NORTH FIRST AVE
TUCSON, AZ.85719
PHONE: (602) 628-1847
KLOSTER, ROCKY R.
69 FORREST FEEZOR
TUCSON, AZ.85747
PHONE: (602) 762-5506
KORDS AMBULANCE SERVICES
1881 EAST 18TH ST
TUCSON, AZ.85719
PHONE: (602) 795-5900
LORDS & LADIES LIMOUSINES
2402 NORTH INDIAN RUINS RD
TUCSON, AZ.85715
PHONE: (602) 721-0020/296-1115
MATIN, DAVID
351 NORTH FOURTH AVE
TUCSON, AZ.85705
PHONE: (602) 792-3473
MCCROREY, HARRY
7791 NORTH STEEL DR
TUCSON, AZ.85743
PHONE: (602) 682-7049
MOTION PICTURE VEHICLES
6055 SOUTH PALOMINO RD
TUCSON, AZ.85746
PHONE: (602) 883-4144
**MOUNTAIN VIEW
TRANSPORTATION**
4275 SOUTH SANTA RITA
TUCSON, AZ.85714
PHONE: (602) 622-4488/741-1188
OLD PUEBLO TROLLEY
P.O. BOX 1373
TUCSON, AZ.85702
PHONE: (602) 298-608

OREILLY CHEVROLET INC
6100 EAST BROADWAY
TUCSON, AZ.85711
PHONE: (602) 747-8000 EXT 246
SCOTT PARSONS
4909 EAST 28TH
TUCSON, AZ.85711
PHONE: (602) 790-1286
RADMACHER ENTERPRISES INC
604 EAST BLACKLIDGE DR
TUCSON, AZ.85705
PHONE: (602) 624-664
REDDINGTON LAND & CATTLE
14772 EAST REDDINGTON
TUCSON, AZ.
PHONE: (602) 749-5555
RICHARDSON, ELAINE
415 EAST SECOND ST
TUCSON, AZ.85705
PHONE: (602) 577-6339
RIESTER, TERRY
5608 EAST LEE
TUCSON, AZ.85712
PHONE: (602) 886-6661
RING, RAY
2855 WEST TIPPECANOE TRAIL
TUCSON, AZ.85745
PHONE: (602) 622-7534
SCHLOSSER, DAN
9260 EAST MADGELENA
TUCSON, AZ 85710
PHONE: (602) 885-9334
SCOTT, WES
5049 SOUTH JOSEPH
TUCSON, AZ.85746
PHONE: (602) 883-7447
TUCSON OIL PRODUCTS INC
1047 EAST MILL ST
TUCSON, AZ.85710
PHONE: (602) 882-9331
**TUCSON RODEO PARADE
COMMITTEE AND MUSEUM**
P.O. BOX 1788
TUCSON, AZ.86702
PHONE: (602) 747-9417
VAGASKY, DONALD & TETA
2220 SOUTH PLACITA
PERLOZZO
TUCSON, AZ.85748
PHONE: (602) 296-1130
VOGLER, RALPH
3238 EAST LINCOLN
TUCSON, AZ.85714
PHONE: (602) 294-8222
WELCH, RICHARD
6154 EAST ADOBE PL
TUCSON, AZ.85718
PHONE: (602) 792-9985/296-
5563
WEST, AUGUSTIN M.
8232 EAST 30TH ST
TUCSON, AZ.
PHONE: (602) 628-9790/624-
4020
**WENDT WEST AUTO
ENTERPRISES**
800 NORTH FOURTH AVE
TUCSON, AZ.85705
PHONE: (602) 624-3100
ZENITH TRANSFER
494 WEST SECOND ST
TUCSON, AZ 85705
PHONE: (602) 790-6622

PYROTECHNICS

PRO RISK
24 E PAPAGO DR
TEMPE, AZ 85281
PHONE: (602) 994-3492
**SAFE SHOT CUSTOM FIRING
BOXES**
2701 SOUTH BRANDYWINE LANE
TUCSON, AZ85730
PHONE: (602) 721-2310/886-5090

STUDIOS/SOUND STAGES

AMBERJACK LTD.
BOX 2336
CAREFREE, AZ. 85377
PHONE: (602) 488-9767

C.J.S. FILM STUDIOS
2005 N. 103RD AVE
TOLLESON, AZ. 85039
PHONE: (602) 936-5449/
264-2539

CAREFREE STUDIO
P.O. BOX 2336
CAREFREE, AZ. 85337
PHONE: (602) 488-9767

CAREFREE STUDIOS CITY OF SCOTTSDALE
3939 CIVIC CENTER PLAZA
SCOTTSDALE, AZ. 85251
PHONE: (602) 994-2422 &
944-2707

CENTERLINE STAGE/ STUDIO LIGHTING
1933 W. 3RD ST
PHOENIX, AZ. 85281
PHONE: (602) 967-5321

DESERT PRODUCTION CENTER
2235 W. ALICE
PHOENIX, AZ. 85021
PHONE: (602) 861-2666

KDTU-TV 18 CREATIVE SERVICES
1855 N. 6TH AVE
TUSCON, AZ. 85705
PHONE: (602) 624-0180

KPHO-TV 5 PRODUCTIONS
4016 NORTH BLACK CANYON
HIGHWAY
PHOENIX, AZ. 85017
PHONE: (602) 264-1000

MARKOW SOUTHWEST
2222 E. MCDOWELL
PHOENIX, AZ. 85006
PHONE: (602) 273-1651

MEDIA PEOPLE INC.
6736 EAST AVALON
SCOTTSDALE, AZ. 85251
PHONE: (602) 941-8701

OLD TUCSON
201 S. KINNEY RD
RUCSON, AZ. 86705
PHONE: (602) 883-0100

PRODUCTION MASTERS INC
834 N. 7TH AVE
PHOENIX, AZ. 85007
PHONE: (602) 254-1600

SIEGEL PHOTGRAPHIC
224 NORTH FIFTH AVENUE
PHOENIX, AZ. 85003
PHONE: (602) 257-9509

TELEMATION PRODUCTIONS
834 NORTH 7TH AVENUE
PHOENIX, AZ. 85007
PHONE: (602) 254-1600

TENTH STREET DANCE WORKS
738 NORTH 5TH AVENUE,
STUDIO 131
TUSCON, AZ. 85705
PHONE: (602) 628-8880

THI STUDIOS
7531 E. 2ND ST
SCOTTSDALE, AZ. 85251
PHONE: (602) 947-3395

VERMILLION PHOTGRAPHIC
124 W. MCDOWELL
PHOENIX, AZ. 85003
PHONE: (602) 253-6005

VIDEO IN PHOENIX
2235 WEST ALICE
PHOENIX, AZ. 85014
PHONE: (602) 995-4448

UNIONS AND ASSOCIATIONS

IATSE, LOCAL 336
3918 E GLENROSA
PHOENIX. AZ. 85018
PHONE: (602) 253-4145

INT. BROTHERHOOD OF ELECTRICAL WORKERS, INDOOR, LOCAL 640
5808 N SEVENTH ST
PHOENIX, AZ. 85012
PHONE: (602) 264-4506

INT. BROTHERHOOD OF ELECTRICAL WORKERS, OUTDOOR, LOCAL769
3232 N 20TH ST
PHOENIX, AZ. 85016
PHONE: (602) 264-1878

M.P. MACHINE OPERATORS U.S./CANADA
2213 N 37TH PL
PHOENIX, AZ. 85008
PHONE: (602) 275-9774

PROJECTIONISTS LOCAL 294
2213 N 37TH PL
PHOENIX, AZ. 85008
PHONE: (602) 275-9774

SCREEN ACTORS GUILD (SAG/AFTRA)
1616 E INDIAN SCHOOL RD
#330
PHOENIX, AZ. 265-2712

STAGE HANDS/ IATSE LOCAL 336
3918 E GLENROSA
PHOENIX, AZ. 85018
PHONE: (602) 253-4145

TEAMSTERS LOCAL 104
238 W ELM ST
TUCSON, AZ. 85705
PHONE: (602) 622-3616

ARKANSAS

SPECIAL EFFECTS

ARNOLD FIREWORKS INC.
P.O. BOX 873
NORTH LITTLE ROCK 72115
PHONE: (501) 758-2524

BYLITES
1601 WESTPARK, SUITE 8
LITTLE ROCK, 72204
PHONE: (501)666-2631

STUDIOS/SOUND STAGES

BYLITES
1601 WESTPARK, SUITE 8
LITTLE ROCK, 72204
PHONE: (501) 666-2631

JACK, BEN STUDIOS
5 SOUTH 13TH STREET
VAN BUREN, 72956
PHONE: (501) 474-1543

JERO PRODUCTIONS
900 SOUTH GREENWOOD,
SUITE C
FORT SMITH, 72901
PHONE: (501) 452-0262

JOHNSON TRIMBLE & CO., INC.
612 WALNUT
LITTLE ROCK, 72205
PHONE: (501) 666-8742

JONES PRODUCTIONS INC.
517 CHESTER
LITTLE ROCK, 72201
PHONE: (800) 880-1981/(501)
372-1981

KARK - TV 4
300 LOUISIANA
LITTLE ROCK, 72201
PHONE: (501) 376-2481

KATV SUPERSEVEN
401 MAIN STREET
LITTLE ROCK, 72201
PHONE: (501)372-7777

<KLRT CHANNEL 16
11711 WEST MARKHAM
LITTLE ROCK, 72211
PHONE: (501) 375-1616

KTHV-11
P.O. BOX 269, 8TH STREET AND
OZARD
LITTLE ROCK, 72203
PHONE: (501) 376-1111

KING OPERA HOUSE
427 MAIN STREET
VAN BUREN, 72956
PHONE: (501) 474-2426

UNIVERSITY OF ARKANSAS
DRAMA DEPT., FAYETTELVILLE,
AR. 72701
PHONE: (501) 575-2953

CALIFORNIA

SPECIAL EFFECTS

A & A SPECIAL EFFECTS
7021 HAYVENHURST AVE
VAN NUYS, 91406
PHONE: (818) 782-6558

ACTION JETS F/X
6546 HOLLYWOOD BLVD #201
HOLLYWOOD, 90028
PHONE: (213) 464-8381

ADAMS & ROBERT CALVERT
17402 CHASE ST
NORTHRIDGE, 91325
PHONE: (818) 345-7703

ALFONSO'S BREAKAWAY GLASS
8070 SAN FERNANDO RD
SUN VALLEY, 91352
PHONE: (818) 768-7402

ALL EFFECTS CO
10845 VANOWEN ST SUITE D
NORTH HOLLYWOOD, 91605
PHONE: (818) 769-7300
FAX: (818) 769-7442

ANIMAL MAKERS
85 DANDELION COURT
THOUSAND OAKS, 91320
PHONE: (805) 499-9779
FAX: (805) 499-3454

APOGEE PRODUCTIONS INC
6842 VALJEAN AVE
VAN NUYS, 91406
PHONE: (818) 989-5757
FAX: (818) 781-6671

APOLLO EFFECTS, LTD
13105 SATICOY ST
NORTH HOLLYWOOD, 91605
PHONE: (818) 982-9398

ARTEFFEX/ DANN O'QUINN
5419 CLEON ST
NORTH HOLLYWOOD 91601
PHONE: (818) 506-5353
FAX: (818) 506-3171

ASSOCIATED MODEL SERVICES
1044 PIONEER WAY SUITE G
EL CAJON, 92020
PHONE: (619) 447-1445

ASTRO PYROTECHNICS
13449 EXCELSIOR DR
NORWALK, 90650
PHONE: (213) 921-6418/(714)
521-1424

B.A.D. COMPANY, INC., THE
11174 FLEETWOOD ST
SUN VALLEY, 91352
PHONE: (818) 504-2404

GARY F. BENTLEY SPECIAL EFFECTS SYSTEMS
26846 OAK AVE., SUITE J
CANYON COUNTRY, 91350
PHONE: (805) 251-1333/
523-4002
FAX: (805) 251-6619

BLUE LINE SCENERY & DESIGN, INC
5218 VINELAND AVE
NORTH HOLLYWOOD, 91601
PHONE: (818) 508-1913
FAX: (818) 508-1907

BOSS FILM CORPORATION
13335 MAXELLA AVE
MARINA DEL REY, 90292
PHONE: (213) 305-8576

BRANAM ENTERPRISES INC
216 CHATSWORTH DR
SAN FERNANDO, 91340
PHONE: (818) 361-5030

BUENA VISTA STUDIOS
500 SOUTH BUENA VISTA ST
BURBANK, 91521
PHONE: (818) 560-00

BURNETT PRODUCTIONS, MICHAEL
3210 LANKERSHIM BL #11
NORTH HOLLYWOOD, 91605
PHONE: (818) 768-5103

CALVERT PYROTECHNICS
17402 CHASE ST
NORTHRIDGE, 91325
PHONE: (818) 345-7703

CARLUCCI, LOU
11022 CARDAMINE DR
TUJUNGA, 91402
PHONE: (818) 353-1723

CHANDLER GROUP, THE
4121 REDWOOD AVE
LOS ANGELES, 90066
PHONE: (213) 305-7431
FAX: (213) 306-2532

CINEMA RESEARCH CORP
6860 LEXINGTON AVE
HOLLYWOOD, 90038
PHONE: (213) 460-4111

CINNABAR
1040 NORTH LAS PALMAS
LOS ANGELES, 90038
PHONE: (213) 469-8910
FAX (213) 462-0515CMI
612 HAMPTON DR
VENICE, 90291
PHONE: (213) 392-8771

CONTINENTAL SCENERY, INC
1022 N LA BREA AVE
HOLLYWOOD, 90038
PHONE: (213) 461-4139
FAX: (213) 461-1424

CONTROLLED AIRSTREAMS-LA
32180 MULHOLLAND HWY
MALIBU, 90265
PHONE: (818) 597-1977

COONEY, M. KAM
P.O. BOX 5235
MISSION HILLS, 91395
PHONE: (818) 886-2233

COPLEN, TED
2383 PANORAMA TERRACE
LOS ANGELES, 90039
PHONE: (213) 663-8748

CREATIVE EFFECTS,INC
760 ARROYO AVE
SAN FERNANDO, 91340
PHONE: (818) 365-0555
FAX: (818) 365-0651

THE CREATURE SHOP
PHONE: (818) 989-0220
FAX: (818) 989-0419

CUSTOM CHARACTERS, INC
820 THOMPSON AVE., SUITE 6
GLENDALE, 91201
PHONE: (818) 507-5940
FAX: (818) 507-1619

CUSTOM SERVICES, CO.
PHONE: (818) 785-1248/993-3756

D.C. SPECIAL EFFECTS CO
8601 CASABA
CANOGA PARK, 91306
PHONE: (818) 998-4619

DAVE'S MARINE SERVICES, INC
1400 WEST 260TH ST
HARBOR CITY, 90701
PHONE: (213) 539-1072
FAX: (213) 539-2763

DILIGENT DWARVES EFFECTS LAB
7119 LAUREL CYN BLVD., SUITE 4
NORTH HOLLYWOOD, 91605
PHONE: (818) 503-9273/458
FAX: (818) 503-9459

DREAMATION
10021 1/2 AVE
CHATSWORTH, 91311
PHONE: (818) 341-2131
FAX: (818) 718-2862

DRY ICE COMPANY, THE
4510 SPERRY ST
LOS ANGELES, 90039
PHONE: (818) 246-9233

EAGLERY-RC MOVIE MODELS
13000 HARTSOOK ST
SHERMAN OAKS, 91423
PHONE: (818) 907-1442

EFFECTS/GENE YOUNG
517 W. WINDSOR RD
GLENDALE, 91204
PHONE: (818) 848-7471/243-8593

ELECTRO EFFECTS SUPPLY
65 W. EASY ST., SUITE 202
SIMI VALLEY, 93065
PHONE: (805) 527-3317
FAX: (805) 527-8911

ENERGY PRODUCTIONS
12700 VENTURA BLVD., 4TH FLOOR
STUDIO CITY, 91604
PHONE: (818) 508-1444
FAX: (818) 508-1293

EQUIPMENT EXPRESS - STUDIO CRANE RENTALS
11826 BALBOA BLVD., SUITE 398
GRANADA HILLS, 91344
PHONE: (818) 367-1006/7
FAX: (818) 367-1008

FILMTRIX, INC
11054 CHANDLER BLVD
NORTH HOLLYWOOD, 91601
PHONE: (818) 980-3700
FAX: (818) 980-3703

FIORITTO, LARRY
1067 EAST ORANGE GROVE
BURBANK, 91501
PHONE: (818) 954-9828

FIRST TAKE
8070 SAN FERNANDO RD
SUN VALLEY, 91352
PHONE: (818) 767-8261
FAX: (818) 767-6964

ROGER GEORGE RENTALS, INC
14525 1/2 BESSEMER ST
VAN NUYS, 91411
PHONE: (818) 994-3049/762-6478
FAX: (818) 994-9432

GILDERFLUKE & CO
820 THOMPSON AVE., SUITE 35
GLENDALE, 91201
PHONE: (818) 546-1618
FAX: (818) 546-1619

NORMAN GRIFFIN STUDIO
919 ISABEL ST
BURBANK,91506
PHONE: (818) 840-8913

HAMILTON FX
9570 WEST WASHINGTON BLVD
LOS ANGELES, 90016
PHONE: (213) 840-4793
FAX: (213) 857-9321

RUFUS HERRICK
PHONE: (213) 462-3543/
(206) 232-7100

HOLLYWOOD BREAKAWAY
15125 CALIFA ST., SUITE B
VAN NUYS, 91411
PHONE: (818) 781-0621

HOLLYWOOD WELDING/ SHEPHERDS SHOP
1045 N HUDSON AVE
HOLLYWOOD, 90038
PHONE: (213) 465-3137/467-7320

IMAGE CREATORS
2712 SIXTH ST
SANTA MONICA, 90405
PHONE: (213) 392-3583/392-5680

IMAGINE THAT
28220 AVENUE CROCKER, SUITE 400
VALENCIA, 91355
PHONE: (805) 294-0061

IMAGE ENGINEERING
632 N VICTORY BL
BURBANK, 91502
PHONE: (818) 846-5865

IMAGINATION EFFECTS & RENTALS
13726 VALLEYHEARST DR
SHERMAN OAKS, 91423

INDUSTRIAL ARTISTS
815 WESTERN AVE., SUITE 4
GLENDALE, 91201
PHONE: (818) 244-4100
FAX: (818) 244-9339

INDUSTRIAL LIGHT & MAGIC
P.O. BOX 2459
SAN RAFAEL, 94912
PHONE: (415) 258-2000
FAX: (415) 456-0883

INNOVISION OPTICS
1318 SECOND ST., SUITE 31
SANTA MONICA, 90401
PHONE: (213) 394-5510
FAX: (213) 395-2941

DAVID D. JOHNSON PRODUCTIONS
210 PASADENA AVE.
SOUTH PASADENA, 91030
PHONE: (818) 441-4869

JOHNSON, RAY, STUDIO
5434 DENNY AVE
NORTH HOLLYWOOD, 91601
PHONE: (818) 508-7348
FAX: (818) 508-8156

KAMDON & ASSOCIATES
PHONE: (818) 886-2233
FAX: (818) 886-6652

KNOTT LIMITED SPECIAL EFFECTS
6919 TREASURE TRAIL
LOS ANGELES, 90068
PHONE: (213) 876-9724
FAX: (213) 876-2356

LANDMARK ENTERTAINMENT
6834 HOLLYWOOD BLVD
LOS ANGELES, 90028
PHONE: (213) 960-1420

LASER IMAGES, INC
6911 HAYVENHURST AVE
VAN NUYS,91406
PHONE: (818) 997-6611

LASER MEDIA, INC
6383 ARIZONA CIRCLE
LOS ANGELES, 90045
PHONE: (213) 338-9200
FAX: (213) 338-9221

LASERFX
7633 SALE AVE
WEST HILLS, 91307
PHONE: (818) 704-0423

LAZARUS LIGHTING DESIGN
4718 SAN FERNANDO RD., SUITE 1
GLENDALE, 91204
PHONE: (800) 553-5554/
(818)956-3211
FAX: (818) 956-3233

LIGHTING EFFECTS, INC
905 TERRACE 49
LOS ANGELES, 90022
PHONE: (213) 256-3604

LUCASFILM COMMERCIAL PRODUCTIONS
1040 NORTH LAS PALMAS
LOS ANGELES, 90038
PHONE: (213) 960-8545
FAX: (213) 960-8540

M.E.L. INC.
7110 LAUREL CYN BLVD., SUITE E
NORTH HOLLYWOOD, 91605
PHONE: (818) 982-1483
FAX: (818) 982-5712

MAGICRAFT DESIGN & FABRICATION GROUP
1218 S. GERHART AVE
LOS ANGELES, 90022
PHONE: (213) 724-2279
FAX: (213) 724-2693

MAKEUP & MONSTERS
BRIAN PENIKAS & ASSOCIATES
10555 OKLAHOMA AVE
CHATSWORTH, 91311
PHONE: (818) 407-0197
FAX: (818) 709-MKUP

MARIAS, MARTY
6345 FOUNTAIN AVE
HOLLYWOOD, 90028
PHONE: (213) 962-2032/ (818) 784-6522
FAX: (213) 856-4971

STEPHEN J. MCHALE
PHONE: (818) 785-5683/593-1229
FAX: (818) 782-6978

MEININGER/DESIGN: BRAZIL STUDIO
4522 BRAZIL ST
LOS ANGELES, 90039
PHONE: (818) 500-8098
FAX: (818) 500-1467

MINI-FOG/JIM FOX
PHONE: (213) 462-2272

RILEY MORGAN
PHONE: (988-4914)

MOTION PICTURE MODELS
15357 LEADWELL ST
VAN NUYS, 91406
PHONE: (818) 780-5827

NEW HORIZON EFFECTS
600 SOUTH MAIN STREET
VENICE, 90291
PHONE: (213) 399-9152

NEOTEK
211 WEST PALM AVE
BURBANK, 91502
PHONE: (818) 840-8225/849-1502
FAX: (818) 840-8301

ODYSSEY IN ILLUSION/FRANZ HARARY
1000 N. SPAULDING AVE
LOS ANGELES, 90046
PHONE: (213) 871-1796

PACIFIC ART & LIGHT
15831 ROMAR ST
SEPULVEDA, 91343
PHONE: (818) 895-7662
FAX: (818) 891-6411

PATINO, STEVE S.P.F.X. STUDIOS
P.O. BOX 546
WHITTIER, 90601
PHONE: (213) 696-2441

DONALD PENNINGTON, INC
PHONE: (818) 705-29

PLAYERS SPECIAL EFFECTS
1128 PASO ROBLES AVE.
GRANADA HILLS, 91344
PHONE: (818) 360-4558

POST GROUP, THE
6335 HOMEWOOD AVE
LOS ANGELES, 90028
PHONE: (213) 462-2300
FAX: (213) 462-0836

PRAXIS FILM WORKS
6918 TUJUNGA AVE
NORTH HOLLYWOOD, 91605
PHONE: (818) 508-0402
FAX: (818) 508-0988

PROP-ART
PHONE: (213) 871-2320

PROPFX
6850 VINELAND AVE., SUITE E
NORTH HOLLYWOOD, 91605
PHONE: (818) 762-1225

QUANTUM LEAP UNLIMITED
6111 LEXINGTON
LOS ANGELES, 90038
PHONE: (213) 461-6451

QUANTUM ROBOTICS
5629 MESMER AVE
CULVER CITY, 90230
PHONE: (213) 391-4484

RAMBALDI, CARLO ENTERPRISES
18409 BRYANT ST
NORTHRIDGE, 91325
PHONE: (818) 701-5658

REEL EFX, INC
5300 MELROSE AVE., SUITE 201D
HOLLYWOOD, 90038
PHONE: (213) 960-4500
FAX: (213) 960-4577

ROBOSHOP
25571 RYE CYN
VALENCIA, 91355
PHONE: (805) 295-8263
FAX: (805) 295-8453

ROBOTIC ASSISTANCE COMPANY
2500 SOUTH FAIRVIEW
SANTA ANA, 92704
PHONE: (714) 979-9462
FAX: (714) 979-1526

SAFARI ANIMATION & EFFECTS
28210 AVENUE CROCKER, SUITE 300
VALENCIA, 91355
PHONE: (805) 295-8119
FAX: (805) 295-1686

SCENIC EXPRESS
3025 FLETCHER DR
LOS ANGELES, 90065
PHONE: (213) 245-4351

SERRURIER & ASSOCIATES
61 W. MOUNTAIN ST
PASADENA, 91103
PHONE: (818) 798-0951/ (213) 681-3711
FAX: (818) 798-3040

SHARP, DAVID B. PRODUCTIONS
12743 DARYL AVE
GRANADA HILLS, 91344
PHONE: (818) 366-2169

MONTGOMERY SHOOK
PHONE: (818) 887-9750

SHORT, ROBERT, PRODUCTIONS, INC
4228 GLENCOE AVE
MARINA DEL REY, 90291
PHONE: (213) 306-6842

SHOW-MOTION DESIGN
1242 LOS ANGELES ST
GLENDALE, 91204
PHONE: (818) 545-8604

SOLEX ENVIRONMENTAL SYSTEMS, INC
10805 PARAMOUNT BLVD., SUITE B
DOWNEY, 90241
PHONE: (213) 861-1401/
(800) 848-0484

SPECIAL EFFECTS SERVICES
1067 EAST ORANGE GROVE
BURBANK, 91501
PHONE: (818) 954-9828

SPECIAL EFFECTS SYSTEMS
26846 OAK AVE UNIT J
CANYON COUNTRY, 91351-2473
PHONE: (805) 251-1333
FAX: (805) 251-6619

SPECIAL EFFECTS UNLIMITED INC
752 N CAHUENGA BLVD
HOLLYWOOD, 90038
PHONE: (213) 466-3361

SPECIALTIES, INC
9074 DEGARMO AVE., UNIT A
SUN VALLEY, 91352
PHONE: (818) 768-1571
FAX: (818) 768-1806

STAGE MECHANIX
1634 E 23RD ST
LOS ANGELES, 90011
PHONE: (213) 232-3663
FAX: (23) 232-8177

STARLIGHT EFFECTS
923 N. LOUISE ST., SUITE C
GLENDALE, 91207
PHONE: (818) 765-7739/760-3392
FAX: (818) 760-6701

STETSON VISUAL SERVICES, INC
5200 W. 83RD ST
LOS ANGELES, 90045
PHONE: (213) 645-8822
FAX: (213) 645-5363

STIPES PRODUCTIONS INC, DAVID
10665 VANOWEN ST
BURBANK, 91505
PHONE: (818) 753-9093

ED SUSSMAN/ THE PURE IMAGINATION CO
7940 YOLANDA AVE
RESEDA, 91335
PHONE: (818) 609-9629

TELSA TECHNOLOGY RESEARCH
2527 TREELANE AVE
MONROVIA, 91016
PHONE: (818) 359-1373

TRI-ESS SCIENCES, INC
1020 WEST CHESTNUT ST
BURBANK, 91506
PHONE: (818) 274-6910/
(213) 245-7685
FAX: (818) 848-3521

TRISTANO, MIKE, SPECIAL EFFECTS MAKE-UP
4520 VAN NUYS BLVD.,
SUITE 702
SHERMAN OAKS, 91403
PHONE: (818) 981-7860

ULTRASPECIAL MATERIALS CO.
P.O. BOX 310
COTATI, 94931
PHONE: (707) 795-3272

UNIVERSAL STUDIOS, SPECIAL EFFECTS
100 UNIVERSAL CITY PLAZA
UNIVERSAL CITY, 91608
PHONE: (818) 777-1656

JOE VISKOCIL
PHONE: (818) 246-0577

WARD FILM SERVICES
4747 W. SIERRA HWY
ACTON, 93510
PHONE: (805) 269-4321/
944-0292
FAX: (805) 269-4322

WARD, SHELLY, ENTERPRISES
32151 TRAIL RD
AGUA DULCE, 91350
PHONE: (805) 252-0069
FAX: (805) 252-0055

WAYNE, DON MAGIC EFFECTS
10929 HARTSOOK ST
NORTH HOLLYWOOD, 91601
PHONE: (818) 763-3192
FAX: (818) 985-4953

WHAT A SET
11345 CHANDLER BLVD
NORTH HOLLYWOOD, 91601
PHONE: (818) 760-1478
FAX: (818) 760-0258

WILDFIRE ULTRA-VIOLET LIGHT, INC
152 LASKY DR., SUITE 105
BEVERLY HILLS, 90212
PHONE: (213) 858-0229
PHONE: (213) 821-5149

WIZARDS INC.SPECIAL EFFECTS & FLYING EFFECTS
18333 LAHEY ST
NORTHRIDGE, 91326
PHONE: (818-368-5084)
FAX: (818)-368-5084

WONDER WORKS, INC
7231 REMMET AVE
CANOGA PARK, 91303

EXPENDABLES

ASTROCAL PRODUCTION BOARDS
5909 MELROSE, SUITE 20
LOS ANGELES, 90038
PHONE: (213) 651-4220

BIRNS & SAWYER
1026 NORTH HIGHLAND AVE
LOS ANGELES, 90038
PHONE: (213) 466-8211
TELEX: 673280
FAX: (213) 466-7049

CHRISTIE ELECTRICAL CORP.
18120 SOUTH BROADWAY
GARDENA, 90248
PHONE: (213) 715-1402

CINE 60 INC.
1050 CAHUENGA BLVD
HOLLYWOOD, 90038
PHONE: (213) 461-304

CINE RENT WEST
991 TENNESSEE ST
SAN FRANCISCO, 91407
PHONE: (415) 864-4644
FAX: (415) 826-4522

COOL LIGHT CO. INC.
5723 AUCKLAND AVE
NORTH HOLLYWOOD, 91601
PHONE: (818) 761-6116

EXPENDABLE SUPPLY STORE INC., THE
1316 NORTH WESTERN BLVD
HOLLYWOOD, 90027
PHONE: (213) 465-3191
FAX: (818) 768-2422
7830 NORTH SAN FERNANDO RD.
SUN VALLEY, 91352
PHONE: (818) 767-5065/(213) 875-2409

G & M POWER PRODUCTS
943 NORTH ORANGE DRIVE
LOS ANGELES, 90028
PHONE: (213) 850-6800

GASSER, ADOLPH INC
750 BRYANT ST
SAN FRANCISCO, 91407
PHONE: (415) 543-3388
FAX: (415) 543-3438

GENERAL CAMERA WEST
6779 HAWTHORNE AVE
HOLLYWOOD, 90028
PHONE: (213) 464-3800
TELEX: 69641-01

GORDON, ALAN ENTERPRISES
P.O. BOX 315, 1430 CAHUENGA BLVD
HOLLYWOOD, 90028
PHONE: (213) 466-3561
FAX: (213) 871-2193

L.T.M. CORPORATION OF AMERICA
11646 PENDELTON ST
SUN VALLEY, 91352
PHONE: (213) 460-6166/
(818) 767-1313

LEONETTI
5609 SUNSET BLVD
HOLLYWOOD 90028
PHONE: (213) 469-2987

LIGHTING & SHADOWS INC.
290 DIVISADERO ST
SAN FRANCISCO, 94117
PHONE: (415) 861-2808

MASON STUDIO SERVICES INC.
430 COLOMA ST
SAUSALITO, 94965
PHONE: (415) 332-4230

PRODUCTION BOARDS BY JACK CASH
650 NORTH BRONSON AVE,
SUITE 126
HOLLYWOOD, 90004
PHONE: (213) 462-5885

RALEIGH FILM & TV STUDIOS
650 NORTH BRONSON AVE
HOLLYWOOD, 90004
PHONE: (213) 466-3111

ROGER GEORGE RENTALS, INC
14525 1/2 BESSEMER ST
VAN NUYS, 91411
PHONE: (818) 994-3049/762-6478
FAX: (818) 994-9432

SEQUOIA EQUIPMENT RENTAL CO.
11601 PENDLETON ST
SUN VALLEY, 91352
PHONE: (818) 78=68-6200

SPECIAL EFFECTS UNLIMITED INC
752 N CAHUENGA BLVD
HOLLYWOOD, 90038
PHONE: (213) 466-3361

STUDIO SPECTRUM INC
1056 NORTH LAKE ST
BURBANK, 91502
PHONE: (818) 843-1510

TRI-ESS SCIENCES, INC
1020 WEST CHESNUT ST
BURBANK, 91411
PHONE: (818) 274-6910/ (213) 245-7685
FAX: (818) 848-3521

FLYING EFFECTS/FLYING HARNESSES

PLAYERS SPECIAL EFFECTS
11028 PASO ROBLES AVE.
GRANADA HILLS CA.91344
PHONE: 818- 360-4558

WIZARDS INC.
18333 LAHEY ST.
NORTH RIDGE CA. 91326
PHONE (818) 368-5084
FAX (818) 368-5084

INTERNATIONAL FILM LIAISONS

BAHAMAS FILM PROMOTION BUREAU
10 UNIVERSAL CITY PLAZA,
SUITE 2600
UNIVERSAL CITY, 91608
PHONE: (818) 509-1840
FAX: (818) 509-1973

CINE AUSTRIA
11601 WILSHIRE BLVD.,
SUITE 2480
LOS ANGELES, 90025
PHONE: (213) 477-3332/
(800) 252-0468
FAX: (213) 477-5141

CONNOISSEUR VIDEO - ARGENTINA
8436 W. THIRD ST., SUITE 600
LOS ANGELES, 90048
PHONE: (213) 653-8873

COSTA RICAN FILM COMMISSION
9000 W. SUNSET BLVD.,
SUITE 1000
LOS ANGELES, 90069
PHONE: (213) 271-5858
FAX: (213) 273-5566

EUROCOM - HOLLAND
223 STRAND ST., SUITE K
SANTA MONICA, 90405
PHONE: (213) 399-1101

FRENCH CULTURAL SERVICES OFFICE
10990 WILSHIRE BLVD.,
SUITE 300
LOS ANGELES, 90024
PHONE: (213) 658-7924/5
FAX: (213) 651-0752

HONG KONG TOURIST ASSOCIATION
10940 WILSHIRE BLVD.,
SUITE 1220
LOS ANGELES, 90024
PHONE: (213) 208-4582
FAX: (213) 208-1869

INTERNATIONAL PRODUCTION CONSULTING
132 S. MANSFIELD AVE
LOS ANGELES, 90036
PHONE: (213) 935-2342/827-5641

MEXICAN GOVERNMENT TOURISM OFFICE
10100 SANTA MONICA BLVD.,
SUITE 224
LOS ANGELES, 90067
PHONE: (213) 203-8191

NEW ZEALAND TOURISM/ PUBLICITY OFFICE
10950 WILSHIRE BLVD.,
SUITE 1530
LOS ANGELES, 90024
PHONE: (213) 477-8241
FAX: (213) 473-5621

PERU NATIONAL INSTITUTE OF CULTURE
P.O. BOX 1787
HOLLYWOOD, 90028
PHONE: (213) 465-8900

TAHITI FILM OFFICE
9841 AIRPORT BLVD.,
SUITE 1108
LOS ANGELES, 90045
PHONE: (213) 649-2884
FAX: (213) 649-3825

TELEFILM CANADA
144 S. BEVERLY DR
BEVERLY HILLS, 90212
PHONE: (213) 859-0268

MAKE-UP

ABBOTT, DAVID
1425 N CATALINA
BURBANK, 91505
PHONE: (818) 567-2327

BALL BEAUTY SUPPLIES
415 N. FAIRFAX AVE
LOS ANGELES, 90036
PHONE: (213) 655-2330

BLASCO, JOE MAKE-UP CENTER & SCHOOL
1708 HILLHURST AVE
LOS ANGELES, 90027
PHONE: (213) 467-4949

CAL- EAST IMPORTS
232 S. BEVERLY DR., SUITE 211
BEVERLY HILLS, 90212
PHONE: (213) 278-2520

CASTEX RENTALS
1044 N. COLE AVE.
HOLLYWOOD, 90038
PHONE: (213) 462-1468

CINEMA SECRETS INC
4400 RIVERSIDE DR
BURBANK, 91505ZZZ
PHONE: (818) 846-0579

COLUMBIA STAGE & SCREEN COSMETICS
1440 N. GOWER ST.
HOLLYWOOD, 90028
PHONE: (213) 464-7555

FRIENDS BEAUTY SUPPLY
5270 LAUREL CANYON BLVD
NORTH HOLLYWOOD, 91607
PHONE: (818) 769-3834/ (213) 877-4828

GORDON, ALAN ENTERPRISES
P.O. BOX 315, 1430 CAHUENGA BLVD
HOLLYWOOD, 90028
PHONE: (213) 466-3561
FAX: (213) 871-2193

GREENSPOON, DR. MORTON K. INC
14607 VENTURA BLVD
SHERMAN OAKS, 91403
PHONE: (818) 789-3311

HOFFMAN'S INTERNATIONAL BARBER BEAUTY SUPPLY
8126 SANTA MONICA BLVD
WEST HOLLYWOOD, 90046
PHONE: (213) 654-6840

INSTITUTE OF STUDIO MAKEUP LTD, THE
3497 CAHUENGA BLVD WEST
HOLLYWOOD, 90068
PHONE: (213) 850-6661

INVINCIBLE
2303 W. NINTH ST
LOS ANGELES, 90006
PHONE: (213) 383-1681

KRYOLAN COPR. U.S.A.
132 NINTH ST
SAN FRANCISCO, 94103
PHONE: (415) 928-5825
TELEX: 277029 KRYOLAN SFO

MAKE-UP EFFECTS LABS
7110 LAUREL CANYON BLVD, SUITE E
NORTH HOLLYWOOD, 91605

NYE, BEN MAKE-UP
5935 BOWCROFT ST
LOS ANGELES, 90016
PHONE: (213) 839-1984
FAX: (213) 839-2640

OLESEN
1535 IVAR AVE
HOLLYWOOD, 90028
PHONE: (213) 461-4631

PENNINGTON-HOLZ, WENDY
3640 MONON ST #210
LOS ANGELES, 90027
PHONE: (213) 666-3983

SPERLING BEAUTY SUPPLIES
13639 VANOWEN ST
VAN NUYS, 91405
PHONE: (818) 781-6300

WARNER BROS. STUDIOS FACILITIES - MAKEUP DEPT.
4000 WARNER BLVD
BURBANK, 91522
PHONE: (818) 954-2151

MAKE-UP EFFECTS

ART & MAGIC
7338 VARNA #5
NORTH HOLLYWOOD, 91605
PHONE: (818) 765-0662
FAX: (818) 765-3965

BALSCO, JOE, SPECIAL MAKE-UP EFFECTS & PROSTHETICS LAB
1708 HILLHURST AVE
HOLLYWOOD, 90027
PHONE: (213) 467-4949

CREATURE SHOP, THE
15424 CABRITO RD #7
VAN NUYS, 91406
PHONE: (818) 989-0220

DILIGENT DWARVES EFFECTS LAB
7119 LAUREL CYN BLVD., SUITE 4
NORTH HOLLYWOOD, 91605
PHONE: (818) 503-9273

MAKEUP & MONSTERS
10555 OKLAHOMA AVE
CHATSWORTH, 91311
PHONE: (818) 407-0197

M.E.L., INC
7110 LAUREL CANYON BLVD., SUITE E
NORTH HOLLYWOOD, 91605
PHONE: (818) 982-1483

MILLER, DAVID, CREATIONS
12547 EAST SHERMAN WAY
NORTH HOLLYWOOD, 91605
PHONE: (818) 982-0293

RCMA
P.O.BOX 850
SOMIS,.93066
PHONE:

ROBOSHOP
25571 RYE CANYON
VALENCIA, 91355
PHONE: (805) 295-8263
FAX: (805) 295-8453

SCREAMING MAD GEORGE, INC.
11750 ROSCOE BLVD #11
SUN VALLEY, 91352
PHONE: (818) 767-1631
FAX: (818) 768-3698

SHORT, ROBERT, PRODUCTIONS INC
4228 GLENCOE AVE
MARINA DEL REY, 90291
PHONE: (213) 306-6842

TRISTANO SPECIAL MAKE-UP FX, MIKE
4520 VAN NUYS BL #702
SHERMAN OAKS, 91403
PHONE: (818) 981-7860

WILDFIRE, INC., VIOLET LIGHT TECHNOLOGY
152 SOUTH LASKY DR
BEVERLY HILLS, 90212
PHONE: (213) 858-0229

ZENOBIA AGENCY, INC
130 SOUTH HIGHLAND AVE
LOS ANGELES, 90036
PHONE: (213) 937-1010
FAX: (213) 937-1131

MINIATURES AND MODELS

ALLEN, DAVID, PRODUCTIONS
918 WEST OAK ST
BURBANK, 91506
PHONE: (818) 845-9270/848-0303
FAX: (818) 886-1586

ALPHA OMEGA PRODUCTIONS
13045 APRIL DR
RIVERSIDE, 92503
PHONE: (714) 279-3481

AMERICAN VAC-U-FORM
18555 EDDY ST
NORTHRIDGE, 91324
PHONE: (818) 886-1585
FAX: (818) 886-1586

ANIMAL MAKERS
85 DANDELION COURT
THOUSAND OAKS, 91320
PHONE: (805) 499-9779

APOGEE PRODUCTIONS, INC
6842 VALJEAN AVE
VAN NUYS, 91406
PHONE: (818) 989-5757
FAX: (818) 781-6671

ARTEFFEX/ DANN O'QUINN
5419 CLEON ST
NORTH HOLLYWOOD, 91601
PHONE: (818) 506-5358
FAX: (818) 506-3171

ASSOCIATED MODEL SERVICES
1044 PIONEER WAY, SUITE G
EL CAJON, 92020
PHONE: (619) 447-1445

CINNEBAR
1040 NORTH LAS PALMAS
LOS ANGELES, 90038
PHONE: (213) 462-8910
FAX: (213) 462-0515

CIRCLE K PRODUCTS
20814 S. NORMANDIE AVE
TORRANCE, 90502
PHONE: (213) 320-4218

CONTINENTAL SCENERY, INC
1022 N. LA BREA AVE
HOLLYWOOD, 90038
PHONE: (213) 461-4139

DESIGN MODELS, INC
10300 GLASGOW PL
LOS ANGELES, 90045
PHONE: (213) 645-0796

DOWDESIGN
872 W. 18TH ST
COSTA MESA, 92627
PHONE: (714) 650-3000/ (213) 248-0172

EAGLERY
13000 HARTSOOK ST
SHERMAN OAKS, 91423
PHONE: (818) 907-1442
PHONE: (818) 907-1442

ENVIRONMENTAL ARTS/KARL WEST
29446 TRAILWAY LN
AGOURA HILLS, 91301
PHONE: (818) 991-8861
FAX: (818) 991-2162

FILMTRIX
11054 CHANDLER BLVD
NORTH HOLLYWOOD, 91601
PHONE: (818) 980-3700
FAX: (818) 980-3703

FIRST TAKE
8070 SAN FERNANDO RD
SUN VALLEY, 91352
PHONE: (818) 767-8261

FOAM-TEC, INC
1100 EILINITA AVE
GLENDALE, 91208
PHONE: (818) 248-3692

MIMI GREENBERG
348 SOUTH ORANGE DR
LOS ANGELES, 90036
PHONE: (213) 936-1131

GRIFFIN, NORMAN, STUDIO
919 ISABEL ST
BURBANK, 91506
PHONE: (818) 840-8913

HAMILTON EFFECTS
5570 W. WASHINGTON BLVD
PHONE: (213) 936-1131

GRIFFIN, NORMAN, STUDIO
919 ISABEL ST
BURBANK, 91506
PHONE: (818) 840-8913

HAMILTON EFFECTS
5570 W. WASHINGTON BLVD
LOS ANGELES, 90016
PHONE: (213) 933-7528

HAMILTON PROTOTYPES & MINIATURES
9570 WEST WASHINGTON BLVD
LOS ANGELES, 90016
PHONE: (213) 840-4793
FAX: (213) 857-9321LOS ANGELES, 90016

HOT PROPS
7504 FOUNTAIN AVE
WEST HOLLYWOOD, 90046
PHONE: (213) 281-8334
FAX: (213) 874-4026

CLARK HUNTER & ASSOCIATES
1945 N HOOVER ST
LOS ANGELES, 90027
PHONE: (213) 666-9816

IMAGE DESIGN & MARKETING
23 14H ST
HERMOSA BEACH, 90254
PHONE: (213) 379-0374

INNOVASION OPTICS
1318 SECOND ST., #31
SANTA MONICA, 90401
PHONE: (213) 394-5510
FAX: (213) 395-2941

JOHNSON, DAVID, PRODUCTIONS
210 PASADENA AVE
SOUTH PASADENA, 91030
PHONE: (818) 441-4869

JOHNSON, RAY, STUDIO
5434 DENNY AVE
NORTH HOLLYWOOD, 91601
PHONE: (818) 508-7348
FAX: (818) 508-8156

JOYCE, ROBERT, STUDIO
7017 CANOGA AVE., SUITE E
CANOGA PARK, 91303
PHONE: (818) 716-9464

RAMON JUNCAL, INC
3446 W. FIRST ST
LOS ANGELES, 90004
PHONE: (213) 385-0474
FAX: (213) 384-4557

KNOTT LIMITED SPECIAL EFFECTS
14755 VENTURA BLVD
SUITE 1-401
SHERMAN OAKS, 91403
PHONE: (213) 969-2322
FAX: (213) 876-2356

KOPP, TOM
14349 GERMAIN ST
MISSION HILLS, 91345
PHONE: (818) 899-3743

LANDMARK ENTERTAINMENT GROUP
11044 WEDDINGTON ST
NORTH HOLLYWOOD, 91601
PHONE: (818) 753-6777

LUCASARTS ATTRACTIONS
P.O. BOX 2459
SAN RAFAEL, 94912
PHONE: (415) 258-2000
FAX: (415) 453-7891

MAKEUP & EFFECTS LABORATORIES, INC (MEL)
7110 LAUREL CANYON BLVD., SUITE E
NORTH HOLLYWOOD, 91605
PHONE: (818) 982-1483
FAX: (818) 982-5712

MAQUITTE
5104 NORTH CAHUENGA BLVD
NORTH HOLLYWOOD, 91601
PHONE: (818) 980-7253

HARVEY MAYO DESIGNS
6353 TEESDALE AVE
NORTH HOLLYWOOD, 91606
PHONE: (818) 985-8636

MEININGER/DESIGN: BRAZIL STUDIO
4522 BRAZIL ST
LOS ANGELES, 90039
PHONE: (818) 500-8098
FAX: (818) 500-1467

MINI-MOVER MOTION CONTROL TABLES
1318 SECOND ST #31
SANTA MONICA, 90401
PHONE: (213) 394-5510
FAX: (213) 395-2941

MINIATURE ESTATES
1451 SOUTH ROBERTSON BLVD
LOS ANGELES, 90035
PHONE: (213) 552-2200

MINIATURE TOWNE USA
17624 SHERMAN WAY
VAN NUYS, 91406
PHONE: (818) 996-3330

MODELWERKES
28206 TAMBORA DR
CANYON COUNTRY, 91351
PHONE: (805) 298-0627

MOSTLY MINIATURES
13759 VENTURA BLVD
SHERMAN OAKS, 91423
PHONE: (818) 990-6713

NEOTEK
211 WEST PALM AVE
BURBANK, 91502
PHONE: (818) 840-8225/849-1502
FAX: (818) 840-8301

ORIOL PRODUCTIONS
175 N. SYCAMORE ST
LOS ANGELES, 90036
PHONE: (213) 933-1812

PACIFIC ART & LIGHT
15831 ROMAR ST
SEPULVEDA, 91343
PHONE: (818) 780-6289
FAX: (818) 891-6411

PENNINGTON, DONALD, INC
PHONE: (818) 705-2956

PROP ART
PHONE: (213) 871-2320

PROP MASTERS, INC
420 SOUTH FIRST ST
BURBANK, 91502
PHONE: (818) 846-3915
FAX: (818) 846-1278

PURE IMAGINATION COMPANY
7940 YOLANDA AVE
RESEDA, 91335
PHONE: (818) 609-9629

SCALE MODEL CO.
4613 WEST ROSECRANS AVE
HAWTHORNE, 90250
PHONE: (213) 679-1435

SCENERY WEST
1126 N CITRUS AVE
HOLLYWOOD, 90038
PHONE: (213) 467-7495
FAX: (213) 467-1622

SHARP, DAVID B. PRODUCTIONS
12743 DARYL AVE
GRANADA HILLS, 91344
PHONE: (818) 366-2169

MONTGOMERY SHOOK
23664 CLOVER TRAIL
CALABASAS, 91302
PHONE: (818) 887-9750

SHORT, ROBERT, PRODUCTIONS, INC
4228 GLENCOE AVE
MARINA DEL REY, 90291
PHONE: (213) 306-6842

STARLIGHT EFFECTS
923 N LOUISE ST., SUITE C
GLENDALE, 91207
PHONE: (818) 765-7739/760-3392
FAX: (818) 760-6701

STETSON VISUAL SERVICES INC
5200 WEST 83RD ST
LOS ANGELES, 90045
PHONE: (213) 645-8822
FAX: (213) 645-5363

STIPES PRODUCTIONS INC, DAVID
10665 VANOWEN ST
BURBANK, 91505
PHONE: (818) 753-9093
FAX: (818) 760-6701

TED SUSSMAN/ THE PURE IMAGINATION CO.
7940 YOLANDA AVE
RESEDA, 91335
PHONE: (818) 609-9629

WONDER WORKS, INC/ BRICK PRICE
7231 REMMET AVE
CANOGA PARK, 91303
PHONE: (818) 992-8811
FAX: (818) 347-4330

YOUNG, GENE, EFFECTS
517 WEST WINDSOR ST
GLENDALE, 91204
PHONE: (818) 848-7471/243-8593

PROPS

A & A SPECIAL EFFECTS
7021 HAYVENHURST AVE.
VAN NUYS, 91406
PHONE: (818) 782-6558

AAA BILLIARDS, INC
6326 LAUREL CANYON BLVD
NORTH HOLLYWOOD, 91606
PHONE: (818) 762-2040

ABC/IMAGE DESIGN
34 14TH ST
HERMOSA BEACH, 90254
PHONE: (213) 379-0374

AIRLINE FILM & TV PROMOTION
13246 WEIDNER ST
PACOIMA, 91331
PHONE: (818) 899-1151

ALFONSOS BREAKAWAY GLASS
8070 SAN FERNANDO RD
SUN VALLEY, 91352
PHONE: (818) 768-7402

ALL AMERICAN BEVERAGE SERVICE
18228 PARTHENIA ST
NORTHRIDGE, 91325
PHONE: (818) 989-3171

ANIMAL MAKERS
85 DANDELION COURT
THOUSAND OAKS, 91320
PHONE: (805) 499-9779
FAX: (805) 499-3454

AQUARIUM STOCK COMPANY
8070 BEVERLY BLVD
LOS ANGELES, 90048
PHONE: (213) 653-8930

ART DECO LAS PROP HOUSE
1025 N SYCAMORE AVE
LOS ANGELES, 90038
PHONE: (213) 462-5474

ATI LIMITED
640 PAULA AVE
GLENDALE, 91201
PHONE: (818) 240-5020

AUDIO PROP HOUSE, THE
P.O. BOX 2046
MALIBU, 90265
PHONE: (213) 456-8823

BAPTY & CO. LTD
703 HARROW RD
LONDON NW10 5NY
PHONE: 081-969 6671
TELEX: 939099 BAPTY G
FAX: 081-960 1106

BETHANIS APPLIANCE STORE
433 N GLENOAKS
BURBANK, 91502
PHONE: (818) 842-6191

BILL BAKER PRODUCTIONS
4154 CARAGENA DR. SUITE A
SAN DIEGO, 92115
PHONE: (619) 284-2400

BISCHOFFS TAXIDERMY
449 S SAN FERNANDO BLVD
BURBANK, 91502
PHONE: (818) 843-7561

BROOK FURNITURE RENTAL
3281 WILSHIRE BLVD
LOS ANGELES, 90010
PHONE: (213) 382-8262

BURBANK STUDIOS, THE
4000 WARNER BLVD
BURBANK, 91522
PHONE: (818) 954-2171

CAMERA READY CARS
1577 PLACENTIA AVE
NEWPORT BEACH, 92663
PHONE: (714) 645-3100

CHARISMA DESIGN STUDIO INC
15333 RAYEN ST
SEPULVEDA, 91343
PHONE: (818) 891-8617

CINDERELLA CARRIAGE CO.
PHONE: (619) 239-8080

COLBY POSTER PRINTING CO
1332 W 12TH PL
LOS ANGELES, 90015
PHONE: (213) 747-5108

COTTAGE SHOPS
7922 W 3RD ST
LOS ANGELES, 90048
PHONE: (213) 658-6066

COUNTRY COUNTRY CO
7123 MELROSE AVE
LOS ANGELES, 90046
PHONE: (213) 937-3566

CULVER STUDIOS PROP DEPARTMENT, THE
9336 W WASHINGTON BLVD
CULVER CITY, 90230
PHONE: (213) 202-3350

D & H ENTERPRISES
10725 E RUSH ST.
SOUTH EL MONTE 91733
PHONE: (818) 443-5912

DAMIAN CANVAS WORKS
4111 LINCOLN BLVD #405
MARINA DEL REY, 90292
PHONE: (213) 822-2343

DAVE'S DISPLAY WORLD
1306 KETTNER BLVD
SAN DIEGO, 92101
PHONE: (619) 232-3097

DECADES
6666 SANTA MONICA BLVD
LOS ANGELES, 90038
PHONE: (213) 464-0696

DIANE & RUDYS FOUNTS./ STATUES
19130 VENTURA BLVD
TARZANA, 91356
PHONE: (818) 343-4321

DICKENS, TED
2418 ARTESIA BLVD
REDONDO BEACH, 90278
PHONE: (213) 538-1389

DOZAR OFFICE FURNISHINGS
2656 S WESTERN AVE
LOS ANGELES, 90018
PHONE: (213) 732-6173

ELLIS MERCANTILE CO.
169 NORTH LA BREA
HOLLYWOOD, 90036
PHONE: (213) 933-7334

ENGINEERED STORAGE SYSTEMS
15034 E PROCTOR
INDUSTRY, 91746
PHONE: (818) 961-0961

EVANS RENTS FURNITURE
14140 VENTURA
SHERMAN OAKS,
PHONE: (818) 907-5496
8668 WILSHIRE BLVD
BEVERLY HILLS, 90212
PHONE: (213) 855-1148

FIXTURES BY HOWIE
901 SOUTH MAIN ST
LOS ANGELES, 90015
PHONE: (213) 627-5952

FRAMING COMPANY, THE
10674 RIVERSIDE DR
TOLUCA LAKE, 91602
PHONE: (818) 766-7127

HANASSAAB ORIENTAL RUGS
8687 MELROSE AVE #186
LOS ANGELES, 90069
PHONE: (213) 657-3674

HAND PROP ROOM, INC. THE
5700 VENICE BLVD
LOS ANGELES, 90019
PHONE: (213) 931-1534
FAX: (213) 931-2145

HEMISPHERE
1426 MONTANA AVE. #9
SANTA MONICA, 90403
PHONE: (213) 458-6853

HOLLYWOOD CENTRAL PROPS
7333 RADFORD AVE
NORTH HOLLYWOOD, 91605
PHONE: (818) 765-1923

HOLLYWOOD PIANO RENTAL CO
1647 N HIGHLAND AVE
HOLLYWOOD, 90028
PHONE: (213) 462-2329

HOLLYWOOD STUDIO GALLERY
1035 N CAHUENGA BLVD
HOLLYWOOD, 90038
PHONE: (213) 462-1116

HOLTZMAN OFFICE FURNITURE
2155 E 6TH ST
LOS ANGELES, 90023
PHONE: (213) 749-7021

HOUSE OF MUZZLE LOADING
1019 E PALMER AVE
GLENDALE, 91205
PHONE: (818) 241-0455

HOUSE OF PROPS
1117 NORTH GOWER
HOLLYWOOD, 90038
PHONE: (213) 463-3166

IMPERIAL DISPLAY
1117 NORTH GOWER
HOLLYWOOD, 90038
PHONE: (213) 735-1011

INDEPENDENT DESIGN/ FABRICATION
801 SOUTH MAIN STREET
BURBANK, 91506
PHONE: (818) 845-3848

INDEPENDENT STUDIO SERVICES
11907 WICKS
SUN VALLEY, 91352
PHONE: (818) 768-5711

IRWIN PRODUCTIONS
6211 YARROW DR., SUITE B
CARLSBAD, 92009
PHONE: (619) 931-1103

JAY'S DISPLAYS
730 F STREET
SAN DIEGO, 92101
PHONE: (619) 233-3151

JAZZ FURNITURE & LIGHTING
8687 MELROSE AVE #G178
LOS ANGELES, 90069
PHONE: (213) 652-2015

KITCHEN PROP HOUSE
900 N CITRUS AVE
HOLLYWOOD, 90038
PHONE: (213) 475-3023

LANDMARK ENTERTAINMENT GROUP
11044 WEDDINGTON ST
NORTH HOLLYWOOD, 91601
PHONE: (818) 753-6777

LEATHER & TREASURES
7571 MELROSE AVE
LOS ANGELES, 90046
PHONE: (213) 655-7541

M.E.L.
7110 LAUREL CANYON BLVD
NORTH HOLLYWOOD, 91605
PHONE: (818) 982-1483
FAX: (818) 982-5712

MICHAEL'S CLASSIC WICKER
8532 MELROSE AVE
LOS ANGELES, 90069
PHONE: (213) 659-1121

MINDS EYE PRODUCTION SERVICE
767 NORTHPOINT
SAN FRANCISCO, 04109
PHONE: (415) 441-4578
FAX: (415) 441-2853

MINIATURE ESTATES
1451 S ROBERTSON BL
LOS ANGELES, 90035
PHONE: (213) 552-2200

MODERN PROPS
4063 REDWOOD AVE
LOS ANGELES, 90066
PHONE: (213) 306-1400

MOVIE ARMS MANAGEMENT
LOS ANGELES,
PHONE: (213) 456-6843

MUSIC CENTER
5616 SANTA MONICA BL
HOLLYWOOD, 90038
PHONE: (213) 469-8143

OMEGA/CP2
5857 SANTA MONICA BL
HOLLYWOOD, 90038
PHONE: (213) 466-8201

PACIFIC MEDICAL PROPS
1112 S VICTORY BL
BURBANK, 91502
PHONE: (818) 567-4800

PARAMEDICAL EQUIPMENT
1710 STANDARD
GLENDALE, 91221
PHONE: (818) 240-8250

PERIOD PROPS
235 W OLIVE AVE
BURBANK, 91502
PHONE: (818) 848-7767

PRACTICAL PROPS
11100 MAGNOLIA BL
NORTH HOLLYWOOD, 91601
PHONE: (818) 980-3198

PRI MEDICAL PROPS
1706 STANDARD AVE
GLENDALE, 91201
PHONE: (818) 240-8250

PROP. SERVICES WEST
915 NORTH CITRUS AVE
HOLLYWOOD, 90038
PHONE: (213) 461-3371

RAPHAEL STUDIOS, INC
7763 MELROSE AVE
LOS ANGELES, 90046
PHONE: (213) 653-5952

RAY PRODUCTS
P.O. BOX 1087
EL MONTE, 91734
PHONE: (818) 579-4250

LYNDA RECHT PRODUCTIONS PLUS
PHONE: (619) 226-0371

REFRIGERATION EQUIPMENT CO
2251 VENICE BL
LOS ANGELES, 90006
PHONE: (213) 732-0123

RICHARD'S CARPETS INC
1446 S ROBERTSON BL
LOS ANGELES, 90035
PHONE: (213) 273-1464

ROSCHU'
7100 FAIR AVE
NORTH HOLLYWOOD, 91605
PHONE: (818) 503-9392

SCHOOL DAYS EQUIPMENT CO.
PHONE: (213) 223-3474

SCHWABE BOOKS & IMPORTS
14472 AMHERST
MOORPARK, 93021
PHONE: (805) 529-4297

SEPANEK, TERRY
P.O. BOX 6593
BURBANK, 91510
PHONE: (818) 849-4370

SHIPS TRADER
21235 SAN MIGUEL ST
WOODLAND HILLS, 91364
PHONE: (818) 884-9008

SPELLMAN DESK COMPANY
6159 SANTA MONICA BL
HOLLYWOOD, 90038
PHONE: (213) 467-0628

STARLIGHT EFX
10665 VANOWEN ST
BURBANK, 91505
PHONE: (818) 760-3392

STEMBRIDGE GUN RENTALS
431 MAGNOLIA AVE
GLENDALE, 91204
PHONE: (818) 246-4333

STUDIO LINEN RENTALS
915 NORTH CITRUS AVE
HOLLYWOOD, 90030
PHONE: (213) 461-3371

TALLMANTZ AVIATION INC
ORANGE COUNTY AIRPORT
SANTA ANA, 92707
PHONE: (213) 629-2770 /
(714) 545-1193

THERMAL
19431 BUSINESS CENTER DR #41
NORTHRIDGE, 91324
PHONE: (818) 701-7983

TRAFFIC CONTROL SERVICE INC
1881 BETMOR LN
ANAHEIM, 92805
PHONE: (714) 937-0422

TRAFTON & ASSOCIATES
11101 CALABASH AVE
FONTANA, 92335
PHONE: (714) 357-7130

TROPICS, THE
7056 SANTA MONICA BL
LOS ANGELES, 90038
PHONE: (213) 469-1682

TWENTIETH CENTURY PROPS
1237 N VINE ST
HOLLYWOOD, 90038
PHONE: (213) 463-9306

UNIVERSAL STUDIOS PROPERTY
100 UNIVERSAL CITY PLAZA
UNIVERSAL CITY, 91608
PHONE: (818) 777-2784

WEEKS, MIKE
2997 RIKKARD DR
THOUSAND OAKS, 91362
PHONE: (805) 492-8303

WERTZ BROTHERS FURNITURE
11879 SANTA MONICA BL
LOS ANGELES, 90025
PHONE: (213) 477-4251

WONDERWORKS, INC
7231 REMMET AVE
CANOGA PARK, 91303
PHONE: (818)992-8811

WOODYS ELECTRICAL PROPS
9165 SAN FERNANDO BL
SUN VALLEY 91352
PHONE: (818) 768-6637

YERKES CIRCUS PRODUCTIONS
17721 ROSCOE BL
NORTHRIDGE, 91325
PHONE: (213) 462-2301

PROPS - DESIGN AND BUILD

ACTESON PAPER CO.
7633 HASKELL AVE
VAN NUYS, 91406
PHONE: (818) 785-7471

ACTION JETS F/X
6546 HOLLYWOOD BLVD #201
HOLLYWOOD, 90028
PHONE: (213) 464-8381

THE ALEON COLLECTION
PHONE: (619) 340-0628

ALL ART SERVICES
5870 GREEN VALLEY CIRCLE
CULVER CITY, 90230
PHONE: (213) 410-1893

ANIMAL MAKERS
85 DANDELION COURT
THOUSAND OAKS, 91320
PHONE: (805) 499-9779

APOGEE PRODUCTIONS, INC
6842 VALJEAN AVE
VAN NUYS, 91406
PHONE: (818) 989-5757
FAX: (818) 781-6671

ART & MAGIC
7338 VARNA, #5
NORTH HOLLYWOOD, 91605
PHONE: (818) 865-0662
FAX: (818) 765-3965

ART FORMS
1450 E SIXTH ST
LOS ANGELES, 90021
PHONE: (213) 626-2230
FAX: (213) 626-5124

ARTEFFEX/DANN O'QUINN
5419 CLEON ST
NORTH HOLLYWOOD, 91601
PHONE: (818) 506-5358
FAX: (818) 781-6671

ARTISAN RESTORATION
8743 W. WASHINGTON BLVD
CULVER CITY, 90232
PHONE: (213) 559-8182

ASSOCIATED MODEL SERVICES
1044 PIONEER WAY SUITE G
EL CAJON, 92020
PHONE: (619) 447-1445

BOB BAKER PRODUCTIONS
3401 PASADENA AVE
LOS ANGELES, 90031
PHONE: (213) 227-1337
FAX: (213) 227-1724

BISCHOFF'S TAXIDERMY
449 S. SAN FERNANDO BLVD
BURBANK, 91502
PHONE: (818) 843-7561

BLUE LINE SCENERY & DESIGN, INC
5218 VINELAND AVE
NORTH HOLLYWOOD, 91601
PHONE: (818) 508-1913
FAX: (818) 508-1907

BRUEHL, DONALD
7527 COLDWATER CYN AVE
NORTH HOLLYWOOD, 91605
PHONE: (818) 982-5356

BUENA VISTA STUDIOS
500 SOUTH BUENA VISTA ST
BURBANK, 91521
PHONE: (818) 560-5295
FAX: (818) 841-8328

CALIFORNIA COUNTRY TREES
74-885 JONI DR., SUITE 2
PALM DESERT, 92260
PHONE: (619) 341-7884/
(800) 872-1889
FAX: (619) 341-3429

CENTRAL PROPERTIES
514 WEST 49TH ST
NEW YORK, NY, 10019
PHONE: (212) 265-7767

CINNABAR
1040 NORTH LAS PALMAS
LOS ANGELES, 90038
PHONE: (213) 462-3737
FAX: (213) 462-0515

CIRCLE K PRODUCTS
20814 SOUTH NORMANDIE AVE
TORRANCE, 90502
PHONE: (213) 320-4218

CONCEPT DESIGN
374 NORTH WILTON PL
LOS ANGELES, 90004
PHONE: (213) 856-0717/718-1934
FAX: (213) 856-0716

CONTINENTAL SCENERY, INC
1022 N. LA BREA AVE
HOLLYWOOD, 90038
PHONE: (213) 461-4139
FAX: (213) 461-1424

CREATURE SHOP
15424 CABRITO RD., SUITE 7
VAN NUYS, 91406
PHONE: (818) 989-0220
FAX: (818) 989-0419

DENNIS CURTIN STUDIOS, INC
446 N. LA BREA AVE
LOS ANGELES, 90036
PHONE: (213) 936-1131
FAX: (213) 936-2215

CUSTOM CHARACTERS, INC
820 THOMPSON AVE., SUITE 6
VAN NUYS, 91402
PHONE: (818) 785-1248/993-3756

DAMIAN CANVAS WORKS
4111 LINCOLN BLVD
MARINA DEL REY, 90292
PHONE: (213) 822-2343

DEROUCHEY URETHANE FOAM
15881 VIEWPOINT DR
RIVERSIDE, 92504
PHONE: (714) 780-1912

DESIGN WORKS
P.O. BOX 65582
LOS ANGELES, 90065
PHONE: (213) 285-7989

DILIGENT DWARVES EFFECTS LAB
7119 LAUREL CANYON BLVD.,
SUITE 4
NORTH HOLLYWOOD, 91605
PHONE: (818) 503-9273/458
FAX: (818) 503-9459

DREAMATION
10021 1/2 AVE
CHATSWORTH, 91311
PHONE: (818) 341-2131
FAX: (818) 718-2862

THE EDGE
26516 GOLDEN VALLEY RD.,
UNIT 210
SAUGUS, 91351
PHONE: (805) 251-7236

ELLIS MERCANTILE COMPANY
169 NORTH LA BREA AVE
LOS ANGELES, 90036
PHONE: (213) 933-7334
FAX: (213) 930-1268

JOE ENGLISH
PHONE: (213) 463-6445

ENVIRONMENTAL ARTS/ KARL WEST
29446 TRAILWAY LANE
AGOURA HILLS, 91301
PHONE: (818) 991-8861
FAX: (818) 991-2162

FILMTRIX, INC
11054 CHANDLER BLVD
NORTH HOLLYWOOD, 91601
PHONE: (818) 980-3700
FAX: (818) 980-3703

FIORITTO, LARRY, SPECIAL EFFECTS SERVICES
1067 EAST ORANGE GROVE
BURBANK, 91501
PHONE: (818) 954-9828

FIRST TAKE
8070 SAN FERNANDO RD
SUN VALLEY, 91352
PHONE: (818) 767-8261
FAX: (818) 767-6964

FOAM-TEC, INC
1100 EILINITA AVE
GLENDALE, 91208
PHONE: (818) 248-3692

GARDNER, JOHN, COMPANY
2080 LAURA AVE
HUNTINGTON PARK, 90255
PHONE: (213) 623-3028

GOLDEN WEST BILLIARDS
21260 DEERING COURT
CANOGA PARK, 91304
PHONE: (818) 888-2300

GREENBERG, MIMI
348 SOUTH ORANGE DR
LOS ANGELES, 90036
PHONE: (213) 936-1131

NORMAN GRIFFIN STUDIO
919 ISABEL ST
BURBANK, 91506
PHONE: (818) 840-8913

HAMILTON FX
9570 WEST WASHINGTON BLVD
LOS ANGELES, 90016
PHONE: (213) 840-4793
FAX: (213) 857-9321

THE HAND PROP ROOM, INC
5700 VENICE BLVD
LOS ANGELES, 90019
PHONE: (213) 931-1534
FAX: (213) 931-2145

HOLLYWOOD BREAKAWAY
15125 CALIFA ST., SUITE B
VAN NUYS, 91411
PHONE: (818) 781-0621

HOT PROPS / KRIS NAGLE
7504 FOUNTAIN AVE
WEST HOLLYWOOD, 90046
PHONE: (213) 876-6549/340-1990
FAX: (213) 874-3607

CLARK HUNTER & ASSOCIATES
1945 N HOOVER ST
LOS ANGELES, 90027
PHONE: (213) 666-9816

I.D.F. STUDIO SCENERY
801 S MAIN ST
BURBANK, 91506
PHONE: (818) 845-3848
FAX: (818) 841-1572

IMAGE CREATORS
2712 6TH ST
SANTA MONICA, 90405
PHONE: (213) 392-3583

IMAGINE THAT
28220 AVENUE CROCKER,
SUITE 400
VALENCIA, 91355
PHONE: (805) 294-0061

**INDEPENDENT STUDIO
SERVICES, INC**
11907 WICKS ST
SUN VALLEY, 91352
PHONE: (818) 764-0840/768-5711
FAX: (818) 768-6320

INDUSTRIAL ARTISTS
815 WESTERN AVE., SUITE 4
GLENDALE, 91201
PHONE: (818) 244-4100
FAX: (818) 244-9339

INTER VIDEO/TRITRONICS, INC
733 NORTH VICTORY BLVD
BURBANK, 91502
PHONE: (818) 3633/569-4000

IWASAKI IMAGES
19330 VAN NESS AVE
TORRANCE, 90501
PHONE: (213) 328-7121
FAX: (213) 618-0876

**DAVID D. JOHNSON
PRODUCTIONS**
210 PASADENA AVE
SOUTH PASADENA, 91030
PHONE: (818) 441-4869

JOHNSON, RAY, STUDIO
5434 DENNY AVE
NORTH HOLLYWOOD, 91601
PHONE: (818) 508-7348
FAX: (818) 508-8156

ROBERT JOYCE STUDIO
7017 CANOGA AVE., SUITE E
CANOGA PARK, 91303
PHONE: (818) 716-9464

**KNOTT LIMITED SPECIAL
EFFECTS**
6919 TREASURE TRAIL
LOS ANGELES, 90068
PHONE: (213) 876-9724
FAX: (213) 876-2356

**LEXINGTON SCENERY & PROPS,
INC**
13005 SATICOY ST
NORTH HOLLYWOOD, 91605
PHONE: (818) 765-0443/ (213)
469-4372
FAX: (818) 765-7082

LUCASARTS ATTRACTIONS
P.O. BOX 2459
SAN RAFAEL, 94912
PHONE: (415) 258-2000
FAX: (415) 453-7891

M.E.L., INC
7110 LAUREL CANYON BLVD.,
SUITE E
NORTH HOLLYWOOD, 91605
PHONE: (818) 982-1483

**MAGICRAFT OR MAGICRAFT
DESIGN & FABRICATION
GROUP**
1218 S. GERHART AVE
LOS ANGELES, 90022
PHONE: (213) 724-2279
FAX: (213) 724-2693

**MARTY MARIAS PROPS &
SPECIAL EFFECTS**
6345 FOUNTAIN AVE
HOLLYWOOD, 90028
PHONE: (213) 962-2032/ (818)
784-6522
FAX: (213) 856-4971

JOELLE MCGONAGLE
15118 WEDDINGTON ST
VAN NUYS, 91411
PHONE: (818) 986-5704

STEPHEN J. MCHALE
PHONE: (818) 785-5683/593-
1229
FAX: (818) 782-6978

**MCINTIRE, WILLIAM,
ENTERPRISES**
P.O. BOX 4244
PORTLAND, OR, 97208
PHONE: (503) 286-4193

JOANNE MCPHERSON STUDIO
PHONE: (213) 393-5660

**MEININGER/DESIGN: BRAZIL
STUDIO**
4522 BRAZIL ST
LOS ANGELES, 90039
PHONE: (818) 500-8098
FAX: (818) 500-1467

MICHAEL'S CLASSIC WICKER
8532 MELROSE AVE
LOS ANGELES, 90069
PHONE: (213) 659-1121

MISTY IMAGES
11800 KITTRIDGE ST., #43
NORTH HOLLYWOOD, 91606
PHONE: (818) 753-9610

MODELWERKES
28206 TAMBORA DR
CANYON COUNTRY, 91351
PHONE: (805) 298-0627

MODERN PROPS
4063 REDWOOD AVE
LOS ANGELES, 90066
PHONE: (213) 306-1400
FAX: (213) 822-5992

MOTION PICTURE MODELS
15357 LEADWELL ST
VAN NUYS, 91406
PHONE: (818) 780-5827

NEON SHOP
13026 SATICOY ST UNIT 28
NORTH HOLLYWOOD, 91605
PHONE: (818) 764-7181

NEOTEK
211 WEST PALM AVE
BURBANK, 91502
PHONE: (818) 840-8225/849-1502
FAX: (818) 840-8301

NIGHTS OF NEON, INC
7442 VARNA AVE
NORTH HOLLYWOOD, 91605
PHONE: (818) 982-3592

NORCOSTCO INC
5867 LANKERSHIM BLVD
NORTH HOLLYWOOD, 91601
PHONE: (818) 760-2911

ORIOL PRODUCTIONS
175 N SYCAMORE ST
LOS ANGELES, 90036
PHONE: (213) 933-1812

PACIFIC ART & LIGHT
15831 ROMAR ST
SEPULVEDA, 91343
PHONE: (818) 895-7662
FAX: (818) 891-6411

PANACHE
836 WEST EL CAMINO
SUNNYVALE, 94087
PHONE: (408) 730-9923

**PATINO, STEVE S.P.F.X.
STUDIOS**
6740-A BRIGHT AVE
WHITTIER, 90601
PHONE: (213) 696-2441

PAUL'S WEST
641 NORTH WESTERN AVE
LOS ANGELES, 90004
PHONE: (213) 462-0758

PENNINGTON, DONALD
7641 JELLICO AVE
NORTHRIDGE, 91325
PHONE: (818) 705-2956

POST, DON STUDIOS, INC
8211 LANKERSHIM BLVD
NORTH HOLLYWOOD, 91605
PHONE: (818) 768-0811

P.R.I. MEDICAL PROPS
1706 STANDARD AVE
GLENDALE, 91201
PHONE: (818) 240-8250

PROP-ART
PHONE: (213) 871-2320

PROP FX
6850 VINELAND AVE., SUITE E
NORTH HOLLYWOOD, 91605
PHONE: (818) 762-1225

PROP MASTERS, INC
420 SOUTH FIRST ST
BURBANK, 91502
PHONE: (818) 846-3915
FAX: (818) 846-1278

**PURE IMAGINATION
COMPANY, THE**
7940 YOLANDA AVE
RESEDA, 91335
PHONE: (818) 609-9629

QUANTUM ROBOTICS
5629 MESMER AVE
CULVER CITY, 90230
PHONE: (213) 391-4484

RAMBALDI ENTERPRISES
18409 BRYANT ST
NORTHRIDGE, 91325
PHONE: (818) 701-5868

REEL EFX, INC
5300 MELROSE AVE., SUITE 201D
HOLLYWOOD, 90038
PHONE: (213) 963-4500
FAX: (213) 960-4577

RICK'S CUSTOM PROPS
3106 RESERVOIR DR
SIMI VALLEY, 93065
PHONE: (805) 581-5808
FAX: (805) 522-4050

RICK'S STUNT CARS
7754 DEERING AVE., SUITE 1
CANOGA PARK, 91304
PHONE: (818) 702-0740

ROBERTS STUDIO DESIGN
8700 RINCON AVE
SUN VALLEY, 91352
PHONE: (818) 504-0742

ROSCHU
7100 FAIR AVE
NORTH HOLLYWOOD, 91605
PHONE: (818) 503-9392/
(213) 469-2749

**SAND SCULPTORS
INTERNATIONAL**
425 VIA ANITA
REDONDO BEACH, 90277
PHONE: (213) 378-5559/4522

SCENERY WEST
1126 N CITRUS AVE
HOLLYWOOD, 90038
PHONE: (213) 467-7435
FAX: (213) 467-1622

SCENIC EXPRESS
3025 FLETCHER DR
LOS ANGELES, 92376
PHONE: (213) 254-4351

THE SCULPTURE STUDIO
8210 LANKERSHIM BLVD.,
SUITE 14
NORTH HOLLYWOOD, 91605
PHONE: (818) 7868-2880

SERRURIER & ASSOCIATES
61 WEST MOUNTAIN ST
PASADENA, 91103
PHONE: (818) 798-0951
FAX: (818) 798-3040

MONTGOMERY SHOOK
23664 CLOVER TRAIL
CALABASAS, 91302
PHONE: (818) 887-9750

**SHORT, ROBERT, PRODUC-
TIONS INC**
4228 GLENCOE AVE
MARINA DEL REY, 90291
PHONE: (213) 306-6842

SILVESTRI STUDIOS
1733 WEST CORDOVA ST
LOS ANGELES, 90007
PHONE: (213) 735-1481

SOLTER PLASTICS
12016 W. PICO BLVD
WEST LOS ANGELES 90064
PHONE: (213) 473-5155

STARLIGHT EFFECTS
923 N. LOUISE ST., SUITE C
GLENDALE, 91207
PHONE: (818) 765-7739/760-3392
FAX: (818) 760-6701

**STETSON VISUAL SERVICES,
INC**
5200 W. 83RD ST
LOS ANGELES, 90045
PHONE: (213) 645-8822
FAX: (213) 645-5363

STUDIO LEATHER
PHONE (818) 764-0620

**ED SUSSMAN/ THE PURE
IMAGINATION CO**
7940 YOLANDA AVE
RESEDA, 91335
PHONE: (818) 609-9629

**TASTEFULLY YOURS PROP
FOOD**
7848 SEPULVEDA BLVD., SUITE B
VAN NUYS, 91405
PHONE: (818) 901-1507

URBANO, TONY, PRODUCTIONS
11925 GOSHEN AVE., SUITE D
LOS ANGELES, 90049
PHONE: (213) 826-7214/392-5365

WEISS, IRVIN
5203 YOLANDA AVE
TARZANA, 91356
PHONE: (818) 344-4136

WILDFIRE, INC., VIOLET LIGHT TECHNOLOGY
152 SOUTH LASKY DR
BEVERLY HILLS, 90212
PHONE: (213) 858-0229

WONDER WORKS, INC/ BRICK PRICE
7231 REMMET AVE
CANOGA PARK, 91303
PHONE: (818) 992-8811
FAX: (818) 347-4330

YOUNG, GENE, EFFECTS
517 WEST WINDSOR ST
GLENDALE,91204
PHONE: (818) 848-7471/243-8593

PROPS: HIRE AND SALE

20TH CENTURY PROPS
1237 VINE ST
HOLLYWOOD,90038
PHONE: (213) 463-9306

AAA BILLIARDS, INC
6326 LAUREL CYN BLVD
NORTH HOLLYWOOD, 91606
PHONE: (818) 762-2040

A & P INDUSTRIAL RENTALS
777 E. GAGE AVE
LOS ANGELES, 90001
PHONE: (213) 231-1145

AARON-SCOTT
3232 SANTA MONICA BLVD
SANTA MONICA, 90404
PHONE: (213) 829-4441

ABBEY/FOSTER
1551 N VERMONT AVE
HOLLYWOOD, 90027
PHONE: (213) 666-7470
2200 WILSHIRE BLVD
SANTA MONICA, 90403
PHONE: (213) 826-7895

ABRAHAM RUG GALLERY
525 N LA CIENEGA BLVD
LOS ANGELES, 90048
PHONE: (213) 652-6520/
(800) 222-7847

ACCENT PARTY LINENS
270 N CANON DR., SUITE 1328
BEVERLY HILLS, 90210
PHONE: (213) 273-8191

ACME DISPLAY FIXTURE CO
1057 S OLIVE ST
LOS ANGELES,90015
PHONE: (213) 749-9191

ACTION WATER SPORTS
2110 HARBOR BLVD
COSTA MESA, 92627
PHONE: (714) 645-2062

AERO PARTS INTERNATIONAL
9625 W. SIERRA HWY.
AGUA DULCE, 91350
PHONE: (805) 268-8123
FAX: (805) 268-8125

AIR DIMENSIONAL DESIGN, INC
1305 MAIN ST
VENICE, 90291
PHONE: (213) 399-2030

ALL CARE MEDICAL
3606 WEST MAGNOLIA BLVD
BURBANK, 91505
PHONE: (818) 848-4451

AL'S STUDIO RENTALS
6025 HOLLYWOOD BLVD
HOLLYWOOD, 90028
PHONE: (818) 845-8071

AMERICAN MILITARY MUSEUM/ HERITAGE PARK
1918 NORTH ROSEMEAD BLVD
EL MONTE, 91733
PHONE: (818) 442-1776

ANIMAL MAKERS
85 DANDELION COURT
THOUSAND OAKS, 91320
PHONE: (805) 499-9779
FAX: (805) 499-3454

ANTIQUARIAN TRADERS
650 N. LA PEER DR
LOS ANGELES, 90069
PHONE (213) 289-0345
8483 MELROSE AVE.
WEST HOLLYWOOD, 90069
PHONE: (213) 658-6394

ANTIQUARIAN TRADERS WAREHOUSE
4851 S. ALAMEDA ST
LOS ANGELES, 90058
PHONE: (213) 627-2144

APEX ELECTRONICS
8909 SAN FERNANDO RD
SUN VALLEY, 91352
PHONE: (213) 875-1308/
(818) 767-7202
FAX: (818) 767-1341

AQUARIUM SALES AND SERVICES
1837 FLOWER ST
GLENDALE, 91201
PHONE: (818) 409-1730

AQUARIUM STOCK COMPANY
8070 BEVERLY BLVD
LOS ANGELES, 90048
PHONE: (213) 653-8930

ART DECO L.A.
1025 NORTH SYCAMORE AVE
LOS ANGELES, 90038
PHONE: (213) 462-5474
FAX: (213) 462-5056

ARTE DE MEXICO
5356 RIVERTON AVE
NORTH HOLLYWOOD, 91601
PHONE: (818) 769-5090
FAX: (818) 769-9425

AT & T
4444 RIVERSIDE DR., SUITE 110
BURBANK, 91505
PHONE: (818) 841-5801/954-6944

A.T.I. PRODUCTIONS
640 PAULA AVE
GLENDALE, 91201
PHONE: (818) 502-9965/
240-5020

ATTIC FANATIC
18612 VENTURA BLVD
TARZANA, 91356
PHONE: (818) 343-7315

THE AUDIO PROP HOUSE
20058 PACIFIC COAST HWY.
MALIBU, 90265
PHONE: (213) 456-8823

AVADON, DAVID, MAGICIAN
3414 CENTINELA AVE
LOS ANGELES, 90066
PHONE: (213) 397-5539

BARTON'S HORSE DRAWN CARRIAGES
518 FAIRVIEW AVE
ARCADIA, 91007
PHONE: (818) 447-6693

BEVERLY PACKING
645 NORTH FAIRFAX
LOS ANGELES, 90036
PHONE: (213) 658-8365
FAX: (213) 658-5815

BISCHOFF'S TAXIDERMY
449 S SAN FERNANDO BLVD
BURBANK, 91502
PHONE: (818) 843-7561

BLEAU-BUSH CO. INC.
3225 WEST WASHINGTON BLVD
LOS ANGELES, 90018
PHONE: (213) 735-1561

BOSES COLLECTION
8300 SUNSET BLVD., SUITE 400
LOS ANGELES, 90069
PHONE: (213) 650-6484
FAX: (213) 650-4589

BOWEN & COMPANY
2940 MAIN ST
SANTA MONICA, 90405
PHONE: (213) 392-3057

BREUNERS
3281 WILSHIRE BLVD
LOS ANGELES, 90010
PHONE (213) 382-8262

BRINGAS BROTHER MUSIC CO.
608 1/2 EAST SEVENTH ST
LOS ANGELES, 90021
PHONE: (213) 622-6300

MEL BROWN FURNITURE
5840 S. FIGUEROA ST
LOS ANGELES, 90003
PHONE: (213) 778-4444

BUSS CARSON BLEACHERS
7905 LLOYD AVE
NORTH HOLLYWOOD, 91605
PHONE: (818) 780-1735

C & T FARM
11078 MCBROOM ST
SUNLAND, 91040
PHONE: (818) 767-0668

CADILLAC JACK
6911 MELROSE AVE
LOS ANGELES, 90038
PHONE: (213) 931-8864

CALIFORNIA ATTRACTIONS, LTD
7023 CANOGA AVE
CANOGA PARK, 91303
PHONE: (818) 999-6255

MARGARET CAVIGGA QUILT COLLECTION
8648 MELROSE AVE
LOS ANGELES, 90069
PHONE: (213) 659-3020/3

CENTER THEATRE GROUP COSTUME SHOP
3301 EAST 14TH ST
LOS ANGELES, 90023
PHONE: (213) 267-1230

CINEMAFLOAT
1624 WEST OCEAN FRONT WALD
NEWPORT BEACH, 92663
PHONE: (714) 675-8888
FAX: (714) 673-1531

CIVILIZATION
8921 VENICE BLVD
LOS ANGELES, 90034
PHONE: (213) 202-8883

COAST KITES, INC
15953 MINNESOTA AVE
PARAMOUNT, 90723
PHONE: (213) 634-3630

COLBY POSTER PRINTING CO
1332 W. 12TH PL
LOS ANGELES, 90015
PHONE: (213) 747-5108

COMPUTER RENTAL CENTER
975 NORTH MICHILINDA AVE
PASADENA, 91107
PHONE: (818) 351-5310/
(213) 231-6784
FAX: (818) 351-7428

CONCEPT DESIGN
374 NORTH WILTON PL
LOS ANGELES, 90004
PHONE: (213) 856-0717/
718-1934
FAX: (213) 856-0716

COPYRITE OFFICE MACHINES
347 SOUTH ROBERTSON BLVD
BEVERLY HILLS, 90211
PHONE: (213) 652-8900
FAX: (213) 652-9565

COTTAGE SHOPS, INC
7922 WEST THIRD ST
WEST HOLLYWOOD, 90048
PHONE: (213) 658-6066
FAX: (213) 658-6305

COUNTRY COUNTRY CO.
7123 MELROSE AVE
LOS ANGELES, 90046
PHONE: (213) 937-3566

CREST OFFICE FURNITURE
100 W. 17TH ST
LOS ANGELES, 90015
PHONE: (213) 749-9425

CRYSTALARIUM
646 W. THIRD ST
LOS ANGELES, 90048
PHONE: (213) 932-1114

CUSTOM SERVICES COMPANY
14516 ARMINTA ST
VAN NUYS, 91402
PHONE: (818) 785-1248/
993-3756

D & H ENTERPRISES
10725 E. RUSH ST
SOUTH EL MONTE, 91733
PHONE: (818) 443-5912

KEN DAY'S ORIENTAL & NAVAJO RUGS
4996 MELROSE AVE
LOS ANGELES, 90029
PHONE: (213) 469-3621

DECADES
6666 SANTA MONICA BLVD
HOLLYWOOD, 90025
PHONE: (213) 464-0696

DELTA GROUP WEAPONS RENTAL, INC
415 PARK AVE
SAN FERNANDO, 91340
PHONE: (818) 898-3200

DEWAYNE BROTHERS
16520 DIVER AVE
CANYON COUNTRY, 91351
PHONE: (805) 251-4342

DIANE & RUDY'S
19130 VENTURA BLVD
TARZANA, 91356
PHONE: (818) 343-4321

DISPLAY & DESIGN DEPOT, INC
1507 ESSEX ST
LOS ANGELES, 90021
PHONE: (213) 748-8991
FAX: (213) 748-8994

DIVA
8818 BEVERLY BLVD
LOS ANGELES, 90048
PHONE: (213) 274-0650/278-3191
FAX: (213) 274-7189

DONATO GYM EQUIPMENT COMPANY
103 NORTH MACLAY AVE
SAN FERNANDO, 91340
PHONE: (818) 365-3177/361-7722

DOZAR OFFICE FURNISHINGS
2656 S. WESTERN AVE
LOS ANGELES, 90018
PHONE: (213) 732-6173

DUNN, BOB, ANIMAL RENTALS
16001 YARNELL ST
SYLMAR, 91342
PHONE: (818) 896-0394

EARL HAYS PRESS
10707 SHERMAN WAY
SUN VALLEY, 91352
PHONE: (818) 765-0700

ELLIS MERCANTILE CO
169 NORTH LA BREA AVE
HOLLYWOOD, 90036
PHONE: (213) 933-7334
FAX: (213) 930-1268

ENCINO PATIO & BABY
17555 VENTURA BLVD
ENCINO, 91316
PHONE: (818) 986-1074

CATHY ENDFIELD
PHONE: (818) 783-4357

ENGINEERED STORAGE SYSTEMS
15034 E. PROCTOR ST
CITY OF INDUSTRY, 91746
PHONE: (818) 961-0961

EQUATOR ANTIQUES
160 N LA BREA AVE
LOS ANGELES, 90036
PHONE: (213) 933-6535

EVER-WEAR TIRE PRODUCTS CO
7435 S ALAMEDA ST
LOS ANGELES, 90001
PHONE: (213) 582-4259

EXCLUSIVE ORIENTAL RUGS, INC
8583 MELROSE AVE
WEST HOLLYWOOD, 90046
PHONE: (213) 858-7847

EXPRESS FURNITURE RENTAL
9033 WILSHIRE BLVD.,
SUITE 100
BEVERLY HILLS, 90211
PHONE: (213) 275-8282
FAX: (213) 275-1631
15020 VENTURA BLVD
SHERMAN OAKS, 91402
PHONE: (818) 905-6600
FAX: (88) 905-6670

EYES ON MAIN
3110 MAIN ST., SUITE 108
SANTA MONICA, 90405
PHONE: (213) 399-3302

F.C. ENTERPRISES
3901 MEDFORD ST
LOS ANGELES, 90063
PHONE: (213) 262-5476

FAMILY AMUSEMENT CORPORATION
876 NORTH VERMONT AVE
LOS ANGELES, 90029
PHONE: (213) 660-8180/ (800) 262-6467

FANTASY LITES
7126 MELROSE AVE
HOLLYWOOD, 90046
PHONE: (213) 933-7244

FIRST STREET FURNITURE
1123 NORTH BRONSON AVE
HOLLYWOOD, 90038
PHONE: (213) 462-6306

FITNESS STORE
17639 CHATSWORTH ST
GRANADA HILLS, 91344
PHONE: (818) 831-1300

FLAX, INC
10852 LINDBROOK DR
LOS ANGELES, 90045
PHONE: (213) 208-3529

FLAX, INC
8801 SOUTH SEPULVEDA BLVD
LOS ANGELES, 90045
PHONE: (213) 641-7995

FLORENTINE ART STUDIO
2221 POTRERO AVE
SOUTH EL MONTE, 91733
PHONE: (818) 443-8873

GALAXY
5411 SHEILA ST
LOS ANGELES, 90040
PHONE: (213) 728-3980

GANTON MICRO COMPUTER RENTALS
1201 SOUTH FLOWER ST
BURBANK, 91502
PHONE: (818) 842-6866/ (213) 785-9319

GEARY'S
351 NORTH BEVERLY DR
BEVERY HILLS, 90210
PHONE: (213) 273-4741

GINA B SHOWROOM
8714 SANTA MONICA BLVD
LOS ANGELES, 90069
PHONE: (213) 652-4488/8437
FAX: (213) 657-4180

GLOBE DRUM & BARREL CO
1149 E. EASTERN AVE
LOS ANGELES, 90022
PHONE: (213) 263-2132

GRAND AMERICAN FARE
32017 LIVE OAK CANYON RD
REDLANDS, 92373
PHONE: (714) 795-4903

H.B. VENDING
7352 RADFORD AVE
NORTH HOLLYWOOD, 91605
PHONE: (818) 932-4698/ (800) 350-3363

H.B. HALICKI PRODUCTIONS
17902 SOUTH VERMONT AVE
GARDENA, 90247
PHONE: (213) 327-1744

HANASSAB IMPORTS DISTINCTIVE RUGS
8687 MELROSE AVE
LOS ANGELES, 90069
PHONE: (213) 657-3674
FAX: (213) 659-4146

HAND PROP ROOM, INC
5700 VENICE BLVD
LOS ANGELES, 90019
PHONE: (213) 931-3671
FAX: (213) 931-2145

HARRY
148 S LA BREA AVE
LOS ANGELES, 90036
PHONE: (213) 938-3344
FAX: (213) 936-8939
8639 VENICE BLVD
LOS ANGELES, 90034
PHONE: (213) 559-7863

HARVEY'S & TROPICAL SUN RATTAN
7367 MELROSE AVE
LOS ANGELES, 90046
PHONE: (213) 852-1271

HAVE PHONE WILL TRAVEL
5255 BELLINGHAM #115
NORTH HOLLYWOOD, 91607
PHONE: (818) 506-8422

EARL HAYS PRESS
10707 SHERMAN WAY
SUN VALLEY, 91352
PHONE: (818) 765-0700
FAX: (818) 765-5245

HAYWIRE
5247 MELROSE AVE
LOS ANGELES, 90038
PHONE: (213) 466-6676

HEMISPHERE
1426 MONTANA AVE., SUITE 9
SANTA MONICA, 90403
PHONE: (213) 458-6853

HERITAGE UNFINISHED FURNITURE
2516 W. HELLMAN AVE
ALHAMBRA, 91803
PHONE: (818) 281-1052

THE HIGH WHEELERS, INC
109 S HIDALGO AVE
ALHAMBRA, 91801
PHONE: (818) 576-8648/ 288-9431

HISTORY FOR HIRE
7103 FAIR AVE
NORTH HOLLYWOOD, 91605
PHONE: (818) 765-7767
FAX: (818) 765-7871

HOLLYWOOD CENTRAL PROPS
7333 RADFORD AVE
NORTH HOLLYWOOD, 91605
PHONE: (818) 765-1923

HOLLYWOOD PIANO RENTAL COMPANY
1647 N HIGHLAND AVE
HOLLYWOOD, 90028
PHONE: (213) 462-2329

HOLLYWOOD STUDIO GALLERY
1035 CAHUENGA BLVD
HOLLYWOOD, 90038
PHONE: (213) 462-1116

HOLTZMAN OFFICE INTERIOR ENVIRONMENTS
2155 E SEVENTH ST
LOS ANGELES, 90023
PHONE: (213) 266-5700/ (800) 437-4545
FAX: (213) 261-5692

HORIZON SHOWROOM OF CONTEMPORARY FURNITURE
8600 WEST PICO BLVD
LOS ANGELES, 90035
PHONE: (213) 655-8800

HOUSE OF BRIENZA
7922 MELROSE AVE
LOS ANGELES, 90046
PHONE: (213) 655-2654

HOUSE OF PROPS, INC
1117 NORTH GOWER ST
HOLLYWOOD, 90038
PHONE: (213) 463-3166

HUME'S SALES & RENTALS
1024 WEST BURBANK BLVD
BURBANK, 91506
PHONE: (213) 849-1614

IMPERIAL DISPLAY
3410 WEST WASHINGTON BLVD
LOS ANGELES, 90018
PHONE: (213) 735-1011

INDEPENDENT STUDIO SERVICES, INC
11907 WICKS ST
SUN VALLEY, 91352
PHONE: (818) 764-0840/ 768-5711
FAX: (818) 768-6320

W.G. INGALLS CO. LTD
5202 INDUSTRY AVE
PICO RIVERA, 90560
PHONE: (213) 948-4410/ (800) 826-4554
FAX: (213) 801-9178

INTER VIDEO/TRITRONICS, INC
733 N VICTORY BLVD
BURBANK, 91502
PHONE: (818) 569-4000

INTERNATIONAL HOUSE OF MUSIC
344 SOUTH BROADWAY
LOS ANGELES, 90013
PHONE: (213) 628-9161

INTERNATIONAL TERRACOTTA
690 N ROBERTSON BLVD
WEST HOLLYWOOD, 90069
PHONE: (23) 657-3752

IWASAKI IMAGES
20460 GRAMERCY PL
TORRANCE, 90501
PHONE: (213) 328-7121

JAZZ FURNITURE & LIGHTING
8687 MELROSE AVE., SUITE G178
LOS ANGELES, 90069
PHONE: (213) 652-2015

RAY JOHNSON STUDIO
5425 DENNY AVE
NORTH HOLLYWOOD, 91601
PHONE: (818) 508-7348

RAMON JUNCAL, INC
3446 W. FIRST ST
LOS ANGELES, 90004
PHONE: (213) 385-0474
FAX: (213) 385-4557

JUST 4 FUN
22444 CARDIFF DR
SAUGUS, 91350
PHONE: (818) 894-9560

KID EQUIPMENT/SCHOOL DAYS
2525 MEDFORD ST
LOS ANGELES, 90033
PHONE: (213) 221-1023

L.A. EYEWORKS
7407 MELROSE ST
LOS ANGELES, 90046
PHONE: (213) 653-8255

L.A. SIGNS & GRAPHICS
3421 SAN FERNANDO RD.,
SUITE G
LOS ANGELES, 90065
PHONE: (213) 257-3955/931-9137
FAX: (213) 257-3545
25135 ANZA DR
VALENCIA, 91335
PHONE: (805) 295-0588
FAX: (805) 295-5711

LENNIE MARVIN ENTERPRISES INC
1105 HOLLYWOOD WAY
BURBANK, 91505
PHONE: (818) 841-5882/ (800) 451-9007
FAX: (818) 841-2896

LEXINGTON
13005 SATICOY ST
NORTH HOLLYWOOD, 91605
PHONE: (818) 765-0443
FAX: (818) 765-7082

LUCKY ENTERTAINMENT
10271 ALMAYO AVE., SUITE 101
LOS ANGELES, 90064
PHONE: (213) 277-9666

LUNDIN FARM
27506 NORTH OAK SPRINGS CYN RD
CANYON COUNTRY, 91351
PHONE: (805) 252-6410

MADE IN AMERICAN GALLERY
10653 WEST PICO BLVD
LOS ANGELES, 90064
PHONE: (213) 470-3366

MAGIC BALLOONS
8945 WEST PICO BLVD
LOS ANGELES, 90035
PHONE: (213) 273-5501

MARGARET CAVIGGA QUILT COLLECTION
8648 MELROSE AVE
LOS ANGELES, 90069
PHONE: (213) 659-3020

MARKET FIXTURES UNLIMITED, INC
13235 WOODRUFF AVE
DOWNEY, 90242
PHONE: (213) 803-5553/6658
FAX: (213) 803-6650

BOB MARRIOTT'S FLYFISHING STORE
2700 W. ORANGETHORPE AVE
FULLERTON, 92633
PHONE: (714) 525-1827
FAX: (714) 525-5783

MCINTIRE, WILLIAM. ENTERPRISES
P.O. BOX 4244
PORTLAND, OR, 97208
PHONE: (503) 286-4193

MCMULLEN'S JAPANESE ANTIQUES
146 N ROBERTSON BLVD
LOS ANGELES, 90048
PHONE: (213) 652-9492

MICHAEL'S CLASSIC WICKER
8552 MELROSE AVE
LOS ANGELES, 90069
PHONE: (213) 659-1121

MODERN PROPS
4063 REDWOOD AVE
LOS ANGELES, 90066
PHONE: (213) 306-1400
FAX: (213) 822-5992

MODERN TIMES
338 N LA BREA AVE
LOS ANGELES, 90036
PHONE: (213) 930-1150

MONARCHS GYMNASTICS
5331 DERRY AVE., SUITE D-H
AGOURA HILLS, 91301
PHONE: (818) 889-3634/ (805) 497-1011

MOTION PICTURE MARINE, INC
616 VENICE BLVD
MARINA DEL REY, 90291
PHONE: (213) 822-1100

MR. POOL/P.M. SALES
18441 VANOWEN ST
RESEDA, 91335
PHONE: (818) 345-1528
FAX: (818) 345-0292

MURREY INTERNATIONAL
407 WEST ROSECRANS AVE
GARDENA, 91335
PHONE: (213) 770-3644
FAX: (213) 217-0504

MUSIC CENTER
5616 SANTA MONICA BLVD
HOLLYWOOD, 90038
PHONE: (213) 469-8143

NAUTICAL DECOR
222 BROAD ST
WILMINGTON, 90744
PHONE: (213) 823-5505

NEIMAN SEWING MACHINE CO. INC
1810 S MAIN ST
LOS ANGELES, 90015
PHONE: (213) 747-2345

NEOTEK
211 WEST PALM AVE
BURBANK, 91502
PHONE: (818) 840-8225/ (213) 849-1502
FAX: (818) 840-8301

NICK'S NEW & USED MERCHANDISE
9607 IMPERIAL HWY
DOWNEY, 90242
PHONE: (213) 803-4140

NICKELODEON MEMORABILIA SHOP
13826 VENTURA BLVD
SHERMAN OAKS, 90023
PHONE: (818) 981-5325/ 341-9628

NIGHTS OF NEON, INC
7442 VARNA AVE
NORTH HOLLYWOOD, 91605
PHONE: (818) 982-3592
FAX: (818) 503-1090

NORTON SALES
7429 LAUREL CANYON BLVD
NORTH HOLLYWOOD, 91605
PHONE: (818) 983-1941/ (213) 877-0107

NOVOCOM INC
6314 SANTA MONICA BLVD
HOLLYWOOD, 90038
PHONE: (213) 461-3688
FAX: (213) 462-3505

OAKWOOD FOUNTAINS & RESTORATIONS
1124 NOWITA PL
VENICE, 90291
PHONE: (213) 399-8256

OBJECTS
7221 BEVERLY BLVD
LOS ANGELES, 90069
PHONE: (213) 933-4333

OFF THE WALL
7325 MELROSE AVE
LOS ANGELES, 90046
PHONE: (213) 930-1185
FAX: (213) 930-1595

OLIVER PEOPLES
8642 SUNSET BLVD
LOS ANGELES, 90069
PHONE: (213) 657-2553

OMEGA/CINEMA PROPS
5857 SANTA MONICA BLVD
LOS ANGELES, 90038
PHONE: (213) 466-8201
FAX: (213) 461-3643

ON TRACK
820 THOMPSON AVE., SUITE 17
GLENDALE, 91201
PHONE: (818) 956-1803

P.R. FITNESS
8211 MELROSE AVE., SUITE 200
LOS ANGELES, 90046
PHONE: (213) 655-9941

PACIFIC MEDICAL RENTALS, INC
1112 SOUTH VICTORY BLVD
BURBANK, 91502
PHONE: (818) 567-4800
FAX: (818) 567-488

PARAMOUNT STUDIOS
5555 MELROSE AVE
HOLLYWOOD, 90038
PHONE: (213) 468-5000

PAUL'S WEST
641 NORTH WESTERN AVE
LOS ANGELES, 90004
PHONE: (23) 462-0758

PEDAL PUSHER
122 23RD ST
NEWPORT BEACH, 92663
PHONE: (714) 675-2570

PEGGS COMPANY INC
4851 S FELSPAR ST
RIVERSIDE, 92509
PHONE: (714) 360-9170

PHYSICO, INC
16554 ARMINTA
VAN NUYS, 91325
PHONE: (818) 988-0225

PHYSICO FITNESS SUPERSTORES
BEVERLY CONNECTION, 100 N. LA CIENEGA BLVD
LOS ANGELES, 90048
PHONE: (213) 657-5605

PICO RENTS
6035 WEST PICO BLVD
LOS ANGELES, 90035
PHONE: (213) 275-9431

THE PINE MINE
7974 MELROSE AVE
LOS ANGELES, 90046
PHONE: (213) 653-9726/ 852-1939

PRI MEDICAL STUDIO RENTALS
1706 STANDARD AVE
GLENDALE, 91201
PHONE: (818) 240-8250

PRODUCTION SERVICES HOLLYWOOD
8033 SUNSET BLVD., SUITE 5035
LOS ANGELES, 90046
PHONE: (213) 858-7610

PROP MASTERS, INC
420 SOUTH FIRST ST
BURBANK, 91502
PHONE: (818) 846-3915
FAX: (818) 846-1278

PROP SERVICES WEST, INC
915 NORTH CITRUS AVE
HOLLYWOOD, 90038
PHONE: (213) 461-3371

PROPFX/BARRY CONNER
6850 VINELAND AVE., SUITE E
NORTH HOLLYWOOD, 91605
PHONE: (818) 762-1225

QUALITY EQUIPMENT
711 N LA BREA AVE
INGLEWOOD, 90302
PHONE: (213) 677-7600

R.L. SPEAR
5510 SATSUMA AVE
NORTH HOLLYWOOD, 91601
PHONE: (213) 877-5533/ (818) 980-0266

R.W.B. PARTY PROPS, INC
128 SOUTH CYPRESS ST
ORANGE, 92666
PHONE: (714) 538-8629/8020
FAX: (714) 538-5764

RAPHAEL STUDIOS INC
7763 MELROSE AVE
LOS ANGELES, 90046
PHONE: (213) 653-5952

RAMON JUNCAL, INC
3446 WEST FIRST ST
LOS ANGELES, 90004
PHONE: (213) 385-0474

RATTON INTERIORS
5543 SATSUMA AVE
NORTH HOLLYWOOD, 91601
PHONE: (818) 766-4343/ (213) 852-0672
FAX: (213) 653-9031

RC VINTAGE
1644 N CHEROKEE AVE
HOLLYWOOD, 90028
PHONE: (213) 462-4510

REFLECTIONS - ICE RELATED ENTERPRISES
4918 ESCOBEDO DR
WOODLAND HILLS, 91364
PHONE: (818) 883-6223

RENT FROM RICKARDS
33173 W. MULHOLLAND DR
MALIBU,'90265
PHONE: (818) 889-1447

RICHELIEU COLLECTION
420 N LA BREA AVE
LOS ANGELES, 90036
PHONE: (213) 931-1855
FAX: (213) 931-0978

RICHARD MULLIGAN
8471 MELROSE AVE
WEST HOLLYWOOD, 90069
PHONE: (213) 653-0204

RICK ENTERPRISES
5350 STROHM SVE., SUITE 9
NORTH HOLLYWOOD, 91601
PHONE: (818) 762-2934
FAX: (818) 763-7931

RITUALS
756 N LA CIENEGA BLVD
LOS ANGELES, 90069
PHONE: (213) 854-0848

ROBERTS RENTS FURNITURE
8719 WILSHIRE BLVD
BEVERLY HILLS, 90211
PHONE: (213) 659-7300

THE ROBOT COMPANY
881 W. 18TH ST
COSTA MESA, 92627
PHONE: (714) 722-0890

ROSCHU
7100 FAIR AVE
NORTH HOLLYWOOD, 91605
PHONE: (818) 503-9392/ (213) 469-2749

RUSSAIR LTD
PHONE: (818) 985-9308

SATELLITE CITY
4920 TOPANGA CYN BLVD
WOODLAND HILLS, 91364
PHONE: (818) 710-9348

SCHAEFER'S AMBULANCE
4627 BEVERLY BLVD
LOS ANGELES, 90004
PHONE: (213) 469-1473

SCHOOL SERVICE CO.
647 SOUTH LA BREA AVE
LOS ANGELES, 90036
PHONE: (213) 933-5691

FRED SEGAL
8100 MELROSE AVE
WEST HOLLYWOOD, 90046
PHONE: (213) 651-1298

SHIP'S TRADER
21235 SAN MIGUEL ST
WOODLAND HILLS, 91364
PHONE: (818) 884-9088

SHOW CARPET SPECIALTIES
15131 CLARK AVE
CITY OF INDUSTRY, 91744
PHONE: (818) 336-7469

SIGNSMITH SIGN COMPANY
1827 VICTORY BLVD,. SUITE D
GLENDALE, 91201
PHONE: (818) 241-9412

SPACE OF NIPPON
146 N ROBERTSON BLVD
LOS ANGELES, 90048
PHONE: (213) 657-7317
FAX: (213) 477-2751

SPECIAL EFFECTS UNLIMITED INC
752 N CAHUENGA BLVD
HOLLYWOOD,90038
PHONE: (213) 466-3361

SPELLMAN DESK CO.
6159 SANTA MONICA BLVD
HOLLYWOOD, 90038
PHONE: (213) 467-0628/8874
FAX: (213) 467-2518

STANDARD ENGINEERING COMPANY
115 EAST 23RD ST
LOS ANGELES, 90011
PHONE: (213) 759-1366

STEMBRIDGE GUN RENTALS
431 MAGNOLIA AVE
GLENDALE, 91204
PHONE: (818) 246-4333

STUDIO PROP RENTAL & REPAIR
7527 COLDWATER CYN AVE
NORTH HOLLYWOOD, 91605
PHONE: (818) 982-5356

SUPERIOR SIGNAL SERVICE
7224 SCOUT AVE
BELL GARDENS, 90201
PHONE: (213) 927-4488
FAX: (213) 928-9689

TAVERN SERVICE INC
PHONE: (818) 989-3171/ 349-8337
FAX: (818) 886-1646

THANKS FOR THE MEMORIES
8319 MELROSE AVE
LOS ANGELES, 90069
PHONE: (213) 852-9407

THERMAL
19431-41 BUSINESS CENTER DR
NORTHRIDGE, 91324
PHONE: (818) 701-7983

TOKORO INC
310 N. ROBERTSON BLVD
LOS ANGELES, 90048
PHONE: (213) 657-3806
FAX: (213) 657-5238

TOPS
23410 CIVIC CENTER WAY
MALIBU, 90265
PHONE: (213) 456-8677

TOWARDS 2000
5302 VINELAND AVE
NORTH HOLLYWOOD, 91601
PHONE: (818) 769-5622
FAX: (818) 769-5699

THE TRAIN SHACK
1030 N HOLLYWOOD WAY
BURBANK, 91505
PHONE: (818) 842-3330
FAX: (818) 842-4562

TRI-ESS SCIENCES, INC
1020 W. CHESTNUT ST
BURBANK,91506
PHONE: (213) 245-7685/ (800) 274-6910

ULTRA CARE
110 HILL ST
HERMOSA BEACH, 90254
PHONE: (213) 374-6535

UNIVERSAL FURNITURE CO
11274 VENTURA BLVD
STUDIO CITY, 91604
PHONE: (818) 762-9088

VECTREX CORP
1731 BERKELEY ST
SANTA MONICA, 90404
PHONE: (213) 828-5533

THE VICTORIAN ROSE
3421 W. MAGNOLIA BLVD
BURBANK, 91505
PHONE: (818) 842-3201

VIDE-U
612 N SEPULVEDA BLVD
LOS ANGELES, 90049
PHONE: (213) 657-4385

VIDEO PLAYBACK SERVICES
11684 VENTURA BLVD., SUITE 805
STUDIO CITY, 91604
PHONE: (818) 240-0017

**VINTAGE CYCLERY OF
PASADENA**
　PHONE: (818) 440-1730
VISION CONNECTION
　6106 SUNSET BLVD
　HOLLYWOOD, 90028
　PHONE: (213) 464-4477
WANNA BUY A WATCH?
　7410 MELROSE AVE
　LOS ANGELES, 90046
　PHONE: (213) 653-0467
WERTZ BROTHERS
　210 N WESTERN AVE
　LOS ANGELES, 90004
　PHONE: (213) 462-3155
　14550 VICTORY BLVD
　VAN NUYS, 91401
　PHONE: (818) 997-7951
　11879 SANTA MONICA BLVD
　WEST LOS ANGELES, 90025
　PHONE: (213) 477-4251
WILDER PLACE
　79751/2 MELROSE AVE
　LOS ANGELES, 90046
　PHONE: (213) 665-9072
WOODY'S BICYCLE WORLD
　3157 LOS FELIZ BLVD
　LOS ANGELES, 90039
　PHONE: (213) 661-6665
WOODY'S ELECTRICAL PROPS
　9165 SAN FERNANDO RD
　SUN VALLEY, 91352
　PHONE: (818) 768-6637

PROP SUPPLIES

**ACHIEVEMENT BADGES &
RIBBON AWARDS**
　1518 W. SEVENTH ST
　LOS ANGELES, 90017
　PHONE: (213) 483-7981
**ALFONSO'S BREAKAWAY
GLASS**
　8070 SAN FERNANDO RD
　SUN VALLEY, 91352
　PHONE: (818) 768-7402
　FAX: (818) 767-6969
APEX ELECTRONICS
　8909 SAN FERNANDO RD
　SUN VALLEY, 91352
　PHONE: (213) 875-1308/
　(818) 767-7502
　FAX: (818) 767-1341
AQUARIUM STOCK COMPANY
　8070 BEVERLY BLD
　LOS ANGELES, 90048
　PHONE: (213) 653-8930
BERGER SPECIALTY CO
　413 E. EIGHTH ST
　LOS ANGELES, 90014
　PHONE: (213) 627-8783
C.C. BROWN'S
　7007 HOLLYWOOD BLVD
　HOLLYWOOD, 90028
　PHONE: (213) 462-9262/ 464-
　9726
**CALIFORNIA GRAPHIC
SYSTEMS**
　7262 BELLAIRE AVE
　NORTH HOLLYWOOD, 91605
　PHONE: (818) 765-5715/
　(800) 227-8745
　FAX: (818) 764-1907
CANE & BASKET SUPPLY CO
　1283 COCHRAN AVE
　LOS ANGELES, 90019
　PHONE: (213) 939-9644
CANVAS SPECIALTIES/LA
　7344 E BANDINI BLVD
　LOS ANGELES, 90040
　PHONE: (213) 723-8311
CAPTIVE SEA
　8687 MELROSE AVE.
　PACIFIC DESIGN CENTER
　SHOWROOM 109
　LOS ANGELES, 90069
　PHONE: (213) 657-3232

CHIME CITY
　7615 CRENSHAW BLVD
　LOS ANGELES, 90043
　PHONE: (213) 751-1153
CIRCLE K PRODUCTS
　20814 S. NORMANDIE AVE
　TORRANCE, 90502
　PHONE: (213) 320-4218
CONWIN CARBONIC CO.
　4510 SPERRY ST
　LOS ANGELES, 90039
　PHONE: (213) 245-2842/
　(818) 246-9233
COSTA'S FEED BARN
　10159 SUNLAND BLVD.
　SUNLAND, 91040
　PHONE: (818) 352-5577
CREATIVE EFFECTS, INC
　760 ARROYO AVE
　SAN FERNANDO, 91340
　PHONE: (818) 365-0655
　FAX: (818) 365-0651
DAVIS DENTAL SUPPLY
　13060 SATICOY ST
　NORTH HOLLYWOOD, 91605
　PHONE: (818) 765-4994/ (800)
　842-4203
　FAX: (818) 765-6508
**DECORATIVE PAPER
PRODUCTS**
　2481 LILLYVALE AVE
　LOS ANGELES 90032
　PHONE: (213) 223-2676
ELECTRONIC CITY
　4001 W. BURBANK BLVD
　BURBANK, 91505
　PHONE: (818) 842-5275
**ELLIS MERCANTILE STUDIO
SUPPLY STORE**
　169 N. LA BREA AVE
　LOS ANGELES, 90036
　PHONE: (213) 933-7334
　FAX: (213) 930-1268
FIRST TAKE
　8070 SAN FERNANDO RD
　SUN VALLEY, 91352
　PHONE: (818) 767-8261
　FAX: (818) 767-6964
FISH & TACKLE UNLIMITED
　121 N. VICTORY BLVD
　BURBANK, 91502
　PHONE: (818) 845-1000
FLORA SET, INC
　1021 N MCCADDEN PL
　HOLLYWOOD, 90038
　PHONE: (213) 465-9487
FOAM MART
　628 N VICTORY BLVD
　BURBANK, 91502
　PHONE: (818) 848-3626/ (800)
　640-3626
GLOBE DRUM & BARREL CO
　1149 S. EASTERN AVE
　LOS ANGELES, 90022
　PHONE: (213) 263-2132
H.G. MCGARY & CO
　2900 E. 11TH ST
　LOS ANGELES, 90023
　PHONE: (213) 266-4131
THE HAND PROP ROOM, INC
　5700 VENICE BLVD
　LOS ANGELES, 90019
　PHONE: (213) 931-1534
　FAX: (213) 931-2145
JEWELER'S EMPORIUM
　6013 HOLLYWOOD BLVD
　HOLLYWOOD, 90028
　PHONE: (213) 463-4855
JOHN'S PIPE SHOP
　6765 HOLLYWOOD BLVD
　HOLLYWOOD, 90028
　PHONE: (213) 462-3013/461-
　9400

L.A. FLORAL SUPPLIES
　818 S. WALL ST
　LOS ANGELE, 90014
　PHONE: (213) 622-1700
LUCKY ENTERTAINMENT
　10271 ALMAYO AVE., SUITE 10
　LOS ANGELES, 90064
　PHONE: (213) 277-9666
MAGNET SALES & MGG. CO
　11248 PLAYA CT
　CULVER CITY, 90230
　PHONE: (213) 391-7213
OCEANIC ARTS
　12414 E. WHITTIER BLVD
　WHITTIER, 90602
　PHONE: (213) 698-6960
　FAX: (213) 945-0868
ORIENTAL GIFTS
　10770 W. WASHINGTON BLVD
　CULVER CITY, 90232
　PHONE: (213) 839-4741/9328
OUR LADY'S GIFT SHOP
　6665 SUNSET BLVD
　HOLLYWOOD, 90028
　PHONE: (231) 462-0611
PEGGS COMPANY, INC
　4851 S. FELSPAR ST
　RIVERSIDE, 92509
　PHONE: (714) 360-9170
　FAX: (714) 360-9186
RAINBOW CRAFTS
　531 N HOLLYWOOD WAY
　BURBANK, 91505
　PHONE: (818) 846-1135
RAINBOW FEATHER CO
　813 S. VICTORY BLVD
　BURBANK, 91502
　PHONE: (818) 842-3107
**SHANNON LUMINOUS
MATERIALS, INC**
　304 N. TOWNSEND ST., SUITE A
　SANTA ANA, 92703
　PHONE: (714) 550-9931
FRANK STEIN NOVELTY CO.
　1969 S. LOS ANGELES ST
　LOS ANGELES, 90011
　PHONE: (213) 747-9585
STUDIO SPECIALTIES
　3013 GILROY ST
　LOS ANGELES, 90039
　PHONE: (213) 662-3031
　FAX: (213) 662-0004
**SUPERIOR CHEMICAL
PRODUCTS**
　2035 E. VERNON BLVD
　VERNON, 90058
　PHONE: (213) 232-3521
**TASTEFULLY YOURS PROP
FOOD**
　7848 SEPULVEDA BLVD., SUITE B
　VAN NUYS, 91405
　PHONE: (818) 901-1507
WE WRAP
　4371 WOODMAN AVE
　SHERMAN OAKS, 91423
　PHONE: (818) 905-8427
THE WHOLESALE SUPPLY CO
　1005 LILLIAN WAY
　LOS ANGELES 90038
　PHONE: (213) 467-4194
　FAX: (818) 848-3521

PYROTECHNICIANS

ACORD, CALVIN J.
　(805) 497-8412
ALBAIN, RICHARD F.
　(213) 782-6558
ALBIEZ, PETER H.
　(818) 846-4664
ALDRIDGE, WM. R.
　(213) 368-6120
AMBORN, STANLEY A.
　(213) 368-5681
ANDERSON, IRA JR.
　(503) 332-1841 [OREGON]

ANDERSON, LARZ
　(213) 399-3214
ARBOGAST, ROY
　(805) 251-1701
AUER, GREG M.
　(714) 627-3333
BAUR, TASSILO
　(818) 841-1271
BEAUCHAMP, JAMES
　(818) 997-6866
BEETZ, RAYMOND
　(818) 353-8321
BELYEU, JON
　(213) 892-8256
BENTLEY, GARY F.
　(805) 253-1985
BLITSTEIN, DAVID M.
　(213) 991-4237
BORDEAUX, JOHN
　(213) 493-4549
BRESIN, MARTIN
　(213) 987-1177
BROWNFIELD, RICHARD
　(213) 245-4715
BURDETTE, WM. G., JR.
　S(213) 956-5100
CAMOMILE, JAMES
　(805) 522-2650
CANGEMI, DANNY
　(318) 367-2554
CARLUCCI, LOUIS
　(818) 353-1723
CAVANAUGH, LARRY
　(805) 532-7055
CEGLIA, FRANK C.
　(818) 848-2614
CHESNEY, PETER M.
　(818) 843-6497
COONEY, KAM
　(818) 886-2233
COOPER, LOUIS ROY
　(213) 986-7641
COPLEN, THEODORE
　(213) 663-8748
CORY, PHILIP C.
　(805) 259-4382
CRAMER, FRED J
　(818) 705-1267
CRAWFORD, GARY L.
　(213) 214-1773
CRUM, EUGENE
　(818) 889-2982
CURTIS, WILLIAM G.
　(213) 352-7495
DAWSON, ROBERT N.
　(213) 340-4282
DEL GENIO, MICHAEL
　(213) 848-8671
DELGENIO, THOMAS F.
　(213) 848-8671
DIGAETANO, JOSEPH III
　(818) 845-7037
DION, DENNIS
　(805) 251-1770
DION, WALTER C.
　(213) 881-2748
DISARRO, ALFRED S.
　(213) 360-5127
DIXON, ARTHUR L.
　(213) 892-5725
DOANE, WILLIAM B.
　(805) 495-8765
DOLAN, CHARLES E.
　(213) 894-7041
DOWNEY, ROY
　(818) 763-4975
EDMONSON, MICHAEL E.
　(805) 255-5407
EGGETT, JOHN D.
　(714) 538-2713
ESTES, KENNETH E.
　(305) 268-0818
EVANS, ANDREW C.
　(213) 841-9149
FAGGARD, JACK
　(213) 762-4753

FARIA, GUY EDWARD
(818) 366-7164
FIORITTO, LAWRENCE R.
(818) 954-9828
FISHER, THOMAS L.
(805) 497-0656
FRAZEE, LOGAN Z.
(714) 677-6733
FRAZEE, TERRY
(714) 676-2242
FRAZIER, JOHN
(213) 353-7370
FREDBURG, JAMES K.
(805) 268-1012
FUENTES, LARRY L.
(805) 251-0686
GALICH, STEVE
(818) 353-1490
GASPAR, CHARLES G.
(213) 862-2088
GEHR, ROCKY A.
(213) 506-6110
GIRSKIS, JOHN
(213) 666-0633
GRAY, JOHN E.
(805) 492-4475
GRIGG, DEWEY
(818) 504-0449
GROZEA, COSTEL B.
(213) 855-0764
HAINES, PAUL H. JR.
(818) 704-1554
HAKIAN, JOSHUA
(818) 769-3936
HALL, ALLEN L.
(805) 492-6207
HANSEN, ROGER
(805) 495-1634
HARRISON, WILLIAM D.
(818) 363-3104
HART, JAMES M.
(818) 360-8578
HERNANDEZ, ABRAHAM
(213) 994-3026
HERNANDEZ, SAVAS K.
(213) 839-3132
HESSEY, CHARLES W.
(805) 584-9342
HESSEY, RUSSELL B.
(213) 763-8671
HICKERSON, PAUL H.
(805) 245-1011
HILL, RICHARD L.
(818) 353-9663
HOHMAN, ROBERT J.
(213) 255-5038
HOLADAY, PHILO
(818) 705-6299
INCORVAIA, CARL J.
(213) 363-3832
JAMES, JOHN
(213) 893-0130
JARVIS, JEFF
(213) 998-3052
JOHNSON, RICHARD E.
(213) 893-2927
KING, GARY
(213) 215-3379
KLINGER, WILLIAM A., JR.
(818) 365-5923
KNOTT, ROBERT L.
(213) 876-9724
KOERNER, HARRY
(818) 846-5022
KUROYAMA, BRUCE Y.
(213) 822-7419
LANDERER, GREGORY C.
(818) 894-0994
LANNUTTI, ALBERT
(213) 455-2431
LANTIEPI, MICHAEL
(213) 246-1951
LOMBARDI, JOSEPH JOHN
(213) 784-3031
MARTIN, DALE
(213) 343-8916

MATTOX, WM. B.
(805) 259-2072
MCCARTHY, KEVIN F.
(818) 360-4553
MCCARTHY, ROBERT E.
(818) 368-5084
MCLEOD, JOHN A.
(415) 388-0114
MEINARDUS, MICHAEL
(818) 352-0375
MERTZ, THOMAS R.
(805) 269-1115
MILLAR, HENRY E. JR.
(503) 946-1402 [OREGON]
MOEHNKE, THOEDORE E.
(415) 924-7789
MONAK, GARY
(818) 715-9366
MORAN, TIMOTHY
(818) 760-2042
MORRIS, THAINE
(213) 641-7332
MUNOZ, FRANK
(818) 841-8311
MYATT, BILLY L.
(805) 268-0813
MYERS, DONALD JR.
(213) 474-5581
NEWKIRK, DALE
(714) 943-3151
NEWMAN, TOM
(213) 391-5188
NICHOLS, PAUL B.
(213) 241-0537
OBERG, RAYMOND WARREN
(213) 342-3945
PARKS, STAN
(213) 363-5436
PEPIOT, KENNETH
(805) 268-1067
PETERSEN, DENNIS K.
(818) 899-9640
PETERSON, ROBERT L.
(213) 845-9510
PEYSER, JOHN R.
(213) 884-7731
PIKE, JOHN KEVIN
(213) 760-8020
POPE, FRANK
(213) 769-3693
POWER, DONALD FRANK
(213) 767-4996
PRITCHETT, DARRELL
(213) 481-8262
PURCELL, STEVEN R.
(805) 583-3121
RATLIFF, RICHARD
(213) 837-1412
REPECKA, GINTARAS
(818) 363-8346
RHEAUME, DELL
(213) 346-7720
ROBERTS, JOHN L.
(213) 709-4855
ROBLES, JOHN BRUCE
(213) 379-7838
RYLANDER, ERIC
(805) 648-5636
SCHIRMER, WILLIAM
(213) 843-2509
SIMMONS, DAVID L.
(805) 296-9620
SMITH, NEIL
(818) 353-4040
SOLIS, LEO L.
(818) 706-2572
SPEED, KENNETH
(805) 984-5654
VANZEEBROECK, BRUNO
(818) 991-6069
VISKOCIL, JOSEPH J.
(213) 246-0577
WARD, THOMAS R.
(805) 252-3086
WENGER, CLIFFORD
(805) 495-2654

WILLIAMS, JERRY D.
(818) 895-2109
WOOD, MICHAEL
(805) 252-3023
WRIGHT, ALTON ASHBY JR.
(213) 540-5009
ZAMORA, GEORGE M.
(818) 992-1838
ZARRUS, JEFF
(213) 784-7340

PYROTECHNICS

A & A SPECIAL EFFECTS
7021 HAVENHURST ST
VAN NUYS, 91406
PHONE: (818) 782-0635
FAX: (818) 782-0635
APOGEE PRODUCTIONS INC
6842 VALJEAN AVE
VAN NUYS, 91406
PHONE: (818) 989-5757
APOLLO EFFECTS LTD
13105 SATICOY ST
NORTH HOLLYWOOD, 91605
PHONE: (818) 982-9398
ASTRO PYROTECHNICS
13449 EXCELSIOR DR
NORWALK, 91650
PHONE: (714) 521-1424/
(213) 921-6418
**GARY F. BENTLEY SPECIAL
EFFECTS SYSTEMS**
26846 OAK AVE., SUITE J
CANYON COUNTRY, 91350
PHONE: (805) 251-1333/
523-4002
FAX: (805) 251-6619
BOSS FILM CORPORATION
13335 MAXELLA AVE
MARINA DEL REY, 90292
PHONE: (213) 305-8576
**BUENA VISTA STUDIOS -
SPECIAL EFFECTS
DEPARTMENT**
500 SOUTH BUENA VISTA ST
BURBANK, 91521
PHONE: (818) 560-5244
CARLUCCI, LOU
11022 CARDAMINE DR
TUJUNGA, 91042
PHONE: (818) 353-1723
KAM COONEY
PHONE: (818) 886-2233
FAX: (818) 886-6652
CREATIVE EFFECTS,INC
760 ARROYO AVE
SAN FERNANDO, 91340
PHONE: (818) 365-0655
FAX: (818) 365-0651
**DE LA MARE ENGINEERING,
INC**
1910 FIRST ST
SAN FERNANDO, 91340
PHONE: (818) 365-9208
FANTASY II FILM EFFECTS
504 SOUTH VARNEY ST
BURBANK, 91502
PHONE: (818) 843-1413
FILMTRIX, INC
11054 CHANDLER BLVD
NORTH HOLLYWOOD, 91601
PHONE: (818) 980-3700
**FIORITTO, LARRY SPECIAL
EFFECTS SERVICES**
1067 EAST ORANGE GROVE
BURBANK, 9151
PHONE: (818) 954-9828
**KNOTT LIMITED SPECIAL
EFFECTS**
6919 TREASURE TRAIL
LOS ANGELES, 90068
PHONE: (213) 876-8724
FAX: (213) 876-2356
LIGHTING EFFECTS, INC
PHONE: (213) 256-3604

MARIAS, MARTY
13562 VALLEY VISTA BLVD
SHERMAN OAKS, 91423
PHONE: (818) 784-6522/
(213) 565-6392
FAX: (213) 856-4971
MARTIN, DALE L.
PHONE: (805) 255-7002/15
**MCINTIRE, WILLIAM,
ENTERPRISES**
P.O. BOX 4244
PORTLAND, OR , 97208
PHONE: (503) 286-419
PLAYERS SPECIAL EFFECTS
11028 PASO ROBLES AVE.
GRANADA HILLS CA.91344
PHONE: 818- 360-4558
PYRO-SPECTACULARS, INC
3196 NORTH LOCUST AVE
RIALTO, 92376
PHONE: (714) 874-1644
SCENIC EXPRESS
3025 FLETCHER DR
LOS ANGELES, 92376
PHONE: (213) 254-4351
FAX: (213) 254-4411
SPECIAL EFFECTS SYSTEMS
26846 OAK AVE UNIT J
CANYON COUNTRY, 91351-2473
PHONE: (805) 251-1333
FAX: (805) 251-6619
ROGER GEORGE RENTALS, INC
14525 1/2 BESSEMER ST
VAN NUYS, 91411
PHONE: (818) 994-3049/
762-6478
FAX: (818) 993-943
**SPECIAL EFFECTS UNLIMITED,
INC**
752 NORTH CAHUENGA BLVD
HOLLYWOOD, 90038
PHONE: (213) 466-3361
TRI-ESS SCIENCES INC
1020 WEST CHESTNUT ST
BURBANK, 91506
PHONE: (800) 274-6910/
(213) 245-7685
FAX: (818) 848-3521
VISKOCIL, JOE
PHONE: (818) 246-0577
YLS PRODUCTIONS
PHONE: (213) 430-2890
WIZARDS INC.
18333 LAHEY ST
NORTH RIDGE CA. 91326.
PHONE: 818-368-5084
FAX 818-368-5084

SPECIAL ITEMS

*AIRPLANE PROPS FOR WIND
MACHINES*
AVIATION WAREHOUSE
HCR 10, BOX 120
ADELANTO 92301
PHONE: (619) 388-4215
NATIONAL AIRCRAFT & SALES
3170 CHERRY AVE.
LONG BEACH 90807
PHONE: (213) 426-8300
TIME-AVIATION SERVICES
9255 SAN FERNANDO RD.
SUN VALLEY 91352
PHONE: (213) 875-0650

AMMUNITION,BLANKS,WEAPONS
ELLIS MERCANTILE COMPANY
169 N. LA BREA AVE.
LOS ANGELES 90036
PHONE: (213) 933-7334
HAND PROP ROOM
5700 VENICE BLVD
LOS ANGELES 90019
PHONE: (213) 931-1534

STEMBRIDGE GUN RENTAL
431 MAGNOLIA BLVD
GLENDALE 91204
PHONE: (818) 246-4333

ARMY-NAVY-STORES & SUPPLIES

AA SURPLUS SALES
1700 EAST OLYMPIC BLVD
LOS ANGELES 90021
PHONE: (213) 680-1610

ARMY NAVY STORE
131 EAST 6TH ST
LOS ANGELES 90014
PHONE: (213) 623-3141

HISTORY FOR HIRE
7103 FAIR AVE
NORTH HOLLYWOOD 91605
PHONE: (818) 765-7767

THE SUPPLY SERGEANT
6664 HOLLYWOOD BLVD
HOLLYWOOD 90028
PHONE: (213) 463-4730

ART SUPPLIES

AARON BROTHERS HOLLYWOOD
716 N. LA BREA AVE
HOLLYWOOD 90038
PHONE: (213)930-2611

ALEXANDER STATIONERS
1531 NORTH CAHUENGA BLVD
HOLLYWOOD 90028
PHONE :(213)464-1151

BULLETIN & DIRECT BOARD CO.
2656 S. GRAND AVE
LOS ANGELES'92605
PHONE: (213)328-1147

CARTER-SEXTON
5308 LAUREL CANYON
NORTH HOLLYWOOD'91607
PHONE:(818)877-5050

CLINTON ART SUPPLIES
160 SOUTH LA BREA AVE
LOS ANGELES90036
PHONE: (213)936-8166

H.G. DANIELS
2543 WEST 6TH ST
LOS ANGELES90057

MICHAELS ART & ENG. SUPPLIES
1518 NORTH HIGHLAND AVE
HOLLYWOOD90028
PHONE: (213)466-5295

PAS GRAPHICS
1292 E. COLORADO BLVD
PASADENA91106
PHONE: (213)681-0615

THE ART STORE
7200 BEVERLY BLVD
LOS ANGELES90036
PHONE: (213)933-9284

WORLD SUPPLY INC
3425 WEST CAHUENGA
HOLLYWOOD90068
PHONE: (213)851-1350

AUTOMOBILES

ANTIQUE AND CLASSIC CAR RENT
611 1/2 WEST VERNON BLVD
LOS ANGELES 90037
PHONE: (213) 232-7211

FRANK DEVANEY FORD
3701 W. OAK ST., BLDG. 3
BURBANK 91505
PHONE: (818) 954-6944

GALPIN FORD
15505 ROSCOE BLVD
SEPULVEDA 91343
PHONE: (818) 787-3800

PICTURE VEHICLE UNLIMITED
5518 VINELAND AVE
NORTH HOLLYWOOD 91601
PHONE: (818) 766-2200

RENT A WRECK
12333 WEST PICO
WEST LOS ANGELES 90064
PHONE: (213) 478-0676

SPECIALTY PICTURE CARS
10158 CANOGA BLVD
CHATSWORTH 91311
PHONE: (818) 998-5765

BALLOONS

AIR DEMENTIONAL DESIGN CO.
10573 W. PICO BLVD.,
SUITE 101
LOS ANGELES 90064
PHONE: (213) 479-6623

BALLON FACTORY
8766 HOLLOWAY DR.
WEST HOLLYWOOD 90069
PHONE: (213) 275-0007

BALLOON ART BY TREB
2010 S. EASTWOOD
SANTA ANA 92705
PHONE: (213) 836-3121

CALIFORNIA ATTRACTIONS LIMITED
7023 CANOGA AVE., SUITE E
CANOGA PARK 91303
PHONE: (818) 999-6255

CONWIN CARBONIC CO.
4510 SPERRY
LOS ANGELES 90039
PHONE: (213) 246-9233

HOLLYWOOD TOYS & COSTUMES INC.
6265 HOLLYWOOD BLVD
HOLLYWOOD 90028
PHONE: (213) 465-3119

L.A. BALLOONS
P.O. BOX 3033
NORTH HOLLYWOOD 91609
PHONE: (818) 764-5772

MAX BALLOONS
11950 VOSE ST
NORTH HOLLYWOOD 91605
PHONE: (818) 764-5772

STAR CELEBRITY LOOK-ALIKES
11023 FRUITLAND DR. #2
STUDIO CITY 91604
PHONE: (213) 876-7866

OCEANIC ARTS
6540 MAGNOLIA
WHITTIER 90601
PHONE: (213) 699-0911

BOOTHS, GAME

CALIFORNIA ATTRACTIONS LIMITED
7023 CANOGA AVE. SUITE E
CANOGA PARK91303
PHONE: (818)999-6255

STAR CELEBRITY LOOK-ALIKES
11023 FRUITLAND DR. #2
STUDIO CITY91606
PHONE: (213) 876-7866

BOTTLES, PLASTIC WATER

ARROWHEAD DRINKING WATER
1544 EAST WASHINGTON BLVD
LOS ANGELES 90021
PHONE: (213) 741-0011

SPARKLETTS DRINKING WATER
4500 YORK BLVD
LOS ANGELES 90041
PHONE: (213) 258-4400

BOWS & ARROWS/PROPS

CULVER STUDIOS PROP DEPT
9339 W. WASHINGTON BLVD
CULVER CITY 90232
PHONE: (213) 202-3350
FAX: (213) 202-3272

ELLIS MERCANTILE COMPANY
169 N. LA BREA AVE.
LOS ANGELES 90036
PHONE: (213) 933-7334
FAX: (213) 930-1268

HAND PROP ROOM
5700 VENICE BLVD
LOS ANGELES 90019
PHONE: (213) 931-1534

HISTORY FOR HIRE
7103 FAIR AVE.
NORTH HOLLYWOOD 91605
PHONE: (818) 765-7767

BOXES FOR STUNTS

ANDERSON WINN PAPER CO.
1101 WEST CHESTNUT ST.
BURBANK 91506
PHONE: (213) 849-3381/ (818)
842-4814

ATLAS BOX CO.
9926 SAN FERNANDO RD
PACOIMA 91331
PHONE: (818) 897-0711

CAL STATE STORE FIXTURES
14210 VAN NUYS BLVD
PACOIMA 91331
PHONE: (818) 899-0686

TRADE PAPER CO.
4501 SOUTH SANTA FE AVE
LOS ANGELES 90058
PHONE: (213) 582-7411

BREAKAWAYS

AA SPECIAL EFFECTS
7017 HAYVENHURST BLVD
VAN NUYS 91406
PHONE: (818) 782-6558

ALFONSOS BREAKAWAY GLASS
8070 SAN FERNANDO RD.
SUN VALLEY 91352
PHONE: (818) 768-7402

BURBANK STUDIOS
4000 WARNER BLVD
BURBANK 91522
PHONE: (818) 954-6000

CINEMA SECRETS INC
4400 RIVERSIDE DR.
BURBANK 91505
PHONE: (818) 846-0579

CULVER STUDIOS PROP DEPT
9339 W. WASHINGTON BLVD
CULVER CITY 90232
PHONE: (213) 202-3350
FAX: (213) 202-3272

ELLIS MERCANTILE CO.
169 N. LA BREA AVE.
LOS ANGELES 90036
PHONE: (213) 933-7334
FAX: (213) 930-1268

FIRST TAKE PRODUCTIONS, INC
8070 SAN FERNANDO RD.
SUN VALLEY 91352
PHONE: (818) 767-8261

HAND PROP ROOM
5700 VENICE BLVD
LOS ANGELES 90019
PHONE: (213) 931-1534

HOLLYWOOD BREAKAWAYS
15125 CALIFA ST.
VAN NUYS CA.91411.
PHONE (818) 781-06

OLESEN
1535 IVAR AVE
HOLLYWOOD 90028
PHONE: (213) 461-4631

SPECIAL EFFECTS UNLIMITED
752 N. CAHUENGA BLVD
HOLLYWOOD 90038
PHONE: (213) 466-3361

UNIVERSAL STUDIOS
100 UNIVERSAL CITY PLAZA
UNIVERSAL CITY 91608
PHONE: (818) 777-1656

BUBBLE MACHINES

ACEY DECY EQUIPMENT CO.
5420 VINELAND AVE.
NORTH HOLLYWOOD 91601
PHONE: (818) 766-9445
FAX: (818) 756-4758

CINNABAR
1040 NORTH LAS PALMAS
HOLLYWOOD 90038
PHONE: (213) 462-3737

CALIFORNIA ATTRACTIONS LIMITED
7023 CANOGA AVE. SUITE E
CANOGA PARK 91303
PHONE: (818) 999-62

AA SPECIAL EFFECTS
7017 HAVENHURST BLVD.
VAN NUYS 91406
PHONE (818) 782-6558

SPECIAL EFFECTS UNLIMITED
752 N. CAHUENGA BLVD.
HOLLYWOOD 90038
PHONE.(213) 466-3361

CANDLES

MOSKATELS
733 SOUTH SAN JULIAN
LOS ANGELES 90014
PHONE: (213) 589-4330

STATS
120 SOUTH RAYMOND AVE
PASADENA 91105
PHONE: (818) 795-9308

CARNIVAL SHOOTING GALLERIES

CALIFORNIA ATTRACTIONS LIMITED
7023 CANOGA AVE. SUITE E
CANOGA PARK91303
PHONE: (818)999-6255

CINEMA SECRETS INC.
4400 RIVERSIDE DR.
BURBANK91505
PHONE: (818)846-0579

FRANK STEIN NOVELTY CO.
1969 SOUTH LOS ANGELES ST
LOS ANGELES90011
PHONE: (213)747-9585

LENNIE MARVIN ENTRPRISES
1105 NORTH HOLLYWOOD
WAY
BURBANK91505
PHONE: (818)841-5882

LUCKY ENTERTAINMENT
10271 ALMAYO AVE. #1
LOS ANGELES90064
PHONE: (213)277-9666

PAGEANTRY PRODUCTIONS
11122 WRIGHT RD.
LYNWOOD90262
PHONE: (213)632-5600
FAX: (213) 632-3304

RC VINTAGE STUDIO RENTALS
1644 CHEROKEE AVE
HOLLYWOOD90028
PHONE: (213)462-4510

CASKETS/COFFINS

CINEMA SECRETS INC
4400 RIVERSIDE DR.
BURBANK91505
PHONE: (818)846-0579

ELLIS MERCANTILE COMPANY
169 NORTH LA BREA AVE
LOS ANGELES90036
PHONE: (213)933-7334
FAX: (213) 930-1268

FIRST STREET FURNITURE
1123 NORTH BRONSON AVE
HOLLYWOOD90038
PHONE: (213)462-6306

HAND PROP ROOM
5700 VENICE BLVD
LOS ANGELES90019
PHONE: (213)931-1534

HISTORY FOR HIRE
7103 FAIR AVE.
NORTH HOLLYWOOD 91605
PHONE: (818) 765-7767

CHEMICAL SPECIAL EFFECTS

ROGER GEORGE RENTALS, INC.
145 1/2 BESSEMER ST.
VAN NUYS 91411
PHONE (818) 994-3049/
762-6478
FAX: (818) 994-9432

TRE-ESS SCIENCES, INC
1020 W. CHESTNUT ST
BURBANK 91506
PHONE: (213) 245-7685/
(800) 274-6910
FAX: (818) 848-3521

COSTUMES

ABC WARDROBE
4151 PROSPECT AVE
HOLLYWOOD 90027
PHONE: (213) 557-5328

ADELES OF HOLLYWOOD
5034 HOLLYWOOD BLVD
HOLLYWOOD 90027
PHONE: (213) 663-2231

BILL HARGATE COSTUMES
1111 NORTH FORMOSA AVE
WEST HOLLYWOOD 90046
PHONE: (213) 876-4432

CALIF. NORCOSTCO-COSTUME CO.
5867 LANKERSHIM BLVD
NORTH HOLLYWOOD 91601
PHONE: (818) 760-2911

CALIFORNIA MILLINERY SUPPLY CO.
621 S. SPRING ST.
LOS ANGELES 90014
PHONE: (213) 622-8746

CAPEZIO DANCE SHOE CO.
1777 NORTH VINE
HOLLYWOOD 90028
PHONE: (213) 465-3744

CINEMA SECRETS INC
4400 RIVERSIDE DR
BURBANK '91505
PHONE: (818) 846-0579

COSTUMES RENTAL CORP
7007 LANKERSHIM BLVD
NORTH HOLLYWOOD 91605
PHONE: (818) 765-8877

ELIZABETH COURTNEY
8636 MELROSE AVE
LOS ANGELES 90069
PHONE: (213) 657-4361

FIDDLERS COSTUMES & DESIGNS
6600 LEXINGTON AVE
LOS ANGELES 90038
PHONE: (213) 464-2525

G.L.I.
SUNSET GOWER STUDIOS
SUITE 221
HOLLYWOOD 90028
PHONE: (213) 962-5491

HOLLYWOOD FANCY FEATHER CO.
12140 SHERMAN WAY
NORTH HOLLYWOOD 91605
PHONE: (818) 765-1767

HOLLYWOOD SOUTH WEST
4918 VINELAND AVE
NORTH HOLLYWOOD 91601
PHONE: (818) 753-9050

INTERNATIONAL COSTUME INC
1423 MARCELINA
TORRANCE 90501
PHONE: (213) 320-6392

MAKE BELIEVE INC
3229 PICO BLVD
SANTA MONICA 90405
PHONE: (213) 829-7165

PALACE COSTUME CO
835 NORTH FAIRFAX AVE
LOS ANGELES 90046
PHONE: (213) 651-5458

THE STUDIO WARDROBE DEPT
P.O. BOX 3158
VAN NUYS 91407
PHONE: (818) 781-4267

TUXEDO CENTER
7360 SUNSET BLVD
HOLLYWOOD 90046
PHONE: (213) 874-4200

UNIVERSAL STUDIOS
100 UNIVERSAL CITY PLAZA
UNIVERSAL CITY 91608
PHONE: (818) 777-2722

WESTERN COSTUME CO
5335 MELROSE AVE
HOLLYWOOD 90038
PHONE: (213) 469-1451

CREDIT-CRAWLS & TITLES

PACIFIC TITLE AND ART
6350 SANTA MONICA BLVD
LOS ANGELES, 90038
PHONE: (213) 464-0121

TITLE HOUSE
738 NORTH CAHUENGA BLVD
HOLLYWOOD, 90028
PHONE: (213) 469-8171

DUMMIES

HISTORY FOR HIRE
7103 FAIR AVE
NORTH HOLLYWOOD 91605
PHONE: (818) 765-7767
FAX: (818) 765-7871

DUNK TANK

CALIFORNIA ATTRACTIONS LIMITED
7023 CANOGA AVE. SUITE E
CANOGA PARK 91303
PHONE: (818) 999-6255

LUCKY ENTERTAINMENT
10271 ALMAYO AVE. #101
LOS ANGELES 90064
PHONE: (213) 277-9666

ELECTRIC CHAIR PROPS

OMEGA/CINEMA PROPS
5857 SANTA MONICA BLVD
HOLLYWOOD 90038
PHONE: (213) 466-8201

ELECTRICAL SUPPLIES

ACEY DECEY EQUIPMENT CO.
5420 VINELAND AVE.
NORTH HOLLYWOOD, 91601
PHONE: (818) 766-9445

ALL PHASE
2101 EMPIRE AVE
BURBANK, 91504
PHONE: (818) 843-4111

ELECTRIC SWITCHES, INC
2478 FLETCHER DR
LOS ANGELES, 90039
PHONE: (213) 664-1141

PACIFIC STATE ELECTRIC
757 EAST WASHINGTON BLVD
LOS ANGELES, 90021
PHONE: (213) 749-7881

ELECTRONIC EQUIPMENT

ACEY DECEY EQUIPMENT CO.
5420 VINELAND AVE
NORTH HOLLYWOOD, 91601
PHONE: (818) 766-9445

C.P. TWO
5755 SANTA MONICA BLVD
LOS ANGELES, 90038
PHONE: (213) 466-8201

ELLIS MERCANTILE COMPANY
169 N. LA BREA AVE
LOS ANGELES, 90036
PHONE: (213) 933-7334
FAX: (213) 930-1268

HISTORY FOR HIRE
7103 FAIR AVE.
NORTH HOLLYWOOD, 91605
PHONE: (818) 765-7767
FAX: (818) 765-7871

MODERN PROPS
4063 REDWOOD AVE
LOS ANGELES, 90066
PHONE: (213) 306-1400

PACIFIC MEDICAL RENTALS
1112 S. VICTORY BLVD
BURBANK, 91502
PHONE: (818) 567-4885

SUN-SETS, INC
7600 WHEATLAND AVE
SUN VALLEY, 91352
PHONE: (818) 767-0900

WOODYS
9165 SAN FERNANDO RD
SUN VALLEY, 91352
PHONE: (818) 768-6637

ELECTRONIC SPECIAL EFFECTS

FIRST TAKE PRODUCTIONS, INC
8070 SAN FERNANDO RD
SUN VALLEY, 91352
PHONE: (818) 767-8261

VISTA ELECTRONICS, INC
3019 ANDRITA ST
LOS ANGELES, 90065
PHONE: (213) 258-8864

EXPENDABLE SUPPLIES

ACEY DECEY EQUIPMENT CO
5420 VINELAND AVE
NORTH HOLLYWOOD 91601
PHONE: (818) 766-9445
FAX: (818) 766-4758

AUTOMATED STUDIO LIGHTING
545 RODIER DR
GLENDALE 91201
PHONE: (818) 500-1646

CINEMA SECRETS INC
4400 RIVERSIDE DR
BURBANK 91505
PHONE: (818) 846-0579

ELLIS MERCANTILE COMPANY
169 N. LA BREA AVE
LOS ANGELES 90036
PHONE: (213) 933-7334

EXPENDABLE SUPPLIES
7848 N. SAN FERNANDO RD.
SUN VALLEY 91352
PHONE: (818) 768-8018

EXPENDABLE SUPPLY STORE INC
1316 N. WESTERN AVE.
HOLLYWOOD 90027
PHONE: (213) 849-1326

KEYLITE P.S.I.
11200 SHERMAN WAY
SUN VALLEY 91352
PHONE: (818) 503-0900

PACIFIC MEDICAL RENTALS
1112 S. VICTORY BLVD
BURBANK 91502
PHONE: (818) 567-4885

ROSCO
1135 N. HIGHLAND AVE
LOS ANGELES 90038
PHONE: (213) 462-2233

STUDIO SPECIALTIES LTD
3013 GILROY ST
LOS ANGELES 90039
PHONE: (213) 662-3031

THEATRE VISTION, INC
5426 FAIR AVE
NORTH HOLLYWOOD 91601
PHONE: (818) 769-0928

FABRICS

ACME UPHOLSTERY SUPPLY & FABRIC CO.
4820 SOUTH VERMONT
LOS ANGELES 90037
PHONE: (213) 750-1191

BEVERLY HILLS SILKS & WOOLENS
417 NORTH CANON DR
BEVERLY HILLS 90210
PHONE: (213) 272-2565

CALIFORNIA FLAMEPROOFING
170 NORTH HALSTEAD ST
PASADENA 91107
PHONE: (213) 681-6773

CLOTH WORLD
7260 NORTH ROSEMEAD
SAN GABRIEL 91775
PHONE: (818) 287-0475

DAZIANS INC
165 SOUTH ROBERTSON BLVD
BEVERLY HILLS 90211
PHONE: (213) 657-8900

DECORATORS WAREHOUSE
1627 1/2 S. SAN PEDRO S
LOS ANGELES 90015
PHONE: (213) 749-7605

DOWNTOWN YARDAGE
5536 SANTA MONICA BLVD
LOS ANGELES 90038
PHONE: (213) 464-7089

ERES
10305 SANTA MONICA BLD
LOS ANGELES 90025
PHONE: (213) 273-1142
LOS ANGELES 90015

HAL DAVID INTERIORS INC
1415 SOUTH LA CIENEGA BLVD
LOS ANGELES 90035
PHONE: (213) 652-1470

HOUSE OF FABRIC - GLENDALE
185 NORTH ORANGE BLVD
GLENDALE 91205
PHONE: (818) 246-7342

INTERNATIONAL SILK & WOOLENS
8347 BEVERLY BLVD
LOS ANGELES 90048
PHONE: (213) 653-6453

OLESEN
1535 IVAR AVE
HOLLYWOOD 90028
PHONE: (213) 461-4631

PINDLER & PINDLER ARCH. DESIGN
145 NORTH ROBERTSON BLVD
LOS ANGELES 90048
PHONE: (213) 274-9841

THE ASTRUP CO
13137 ARTIC CIRCLE
SANTA FE SPRINGS 90670
PHONE: (213) 921-8884

THE LEFT BANK FABRIC CO
8354 WEST 3RD ST
LOS ANGELES 90048
PHONE: (213) 655-7289

FANS AND BLOWERS

BROWN & GOLD LIGHTING
176 NORTH LA BREA AVE
LOS ANGELES 90036
PHONE: (213) 933-7149

ELLIS MERCANTILE COMPANY
169 N. LA BREA AVE
LOS ANGELES 90036
PHONE: (213) 933-7334

HOLLYWOOD CENTRAL PROPS
525 W. ELK AVE
GLENDALE 91204
PHONE: (213) 240-4504

LENNIE MARVIN ENTERPRISES
1105 N. HOLLYWOOD WAY
BURBANK 91505
PHONE: (818) 841-5882

OLDE TYME CEILING FAN CO
22743 VENTURA BLVD
WOODLAND HILLS 91364
PHONE: (818) 888-8176
ROGER GEORGE RENTALS, INC
145 1/2 BESSEMER ST.
VAN NUYS 91411
PHONE: (818) 994-3049/
762-6478
FAX: (818) 994-9432
SPECIAL EFFECTS UNLIMITED
752 N. CAHUENGA BLVD
HOLLYWOOD 90038
PHONE: (213) 466-3361
TRI-ESS SCIENCES, INC
1020 W. CHESTNUT ST
BURBANK 91506
PHONE: (213) 245-7685/ (800)
274-6910
FAX: (818) 848-3521

FARM EQUIPMENT

FIRST STREET FURNITURE
1123 N. BRONSON AVE
HOLLYWOOD ,90038
PHONE:213)462-6306
HISTORY FOR HIRE
7103 FAIR AVE
NORTH HOLLYWOOD,91605
PHONE: (818)765-7767
FAX:(818) 765-7871

FEATHERS

AMCAN FEATHER
8338 BEVERLY BLVD
HOLLYWOOD 90048
PHONE: (213) 653-1508
**CALIFORNIA MILLINERY
SUPPLY CO**
621 S. SPRING ST
LOS ANGELES 90014
PHONE: (213) 622-8746
MOSKATELS
733 SOUTH SAN JULIAN
LOS ANGELES 90014
PHONE: (213) 689-4830
RAINBOW FEATHER CO
813 S. VICTORY BLVD
BURBANK 91502
PHONE: (818) 842-3107
STATS
120 SOUTH RAYMOND AVE
PASADENA 91105
PHONE: (213) 795-9308

FIRE EXTINGUISHERS

FIRE EQUIPMENT COMPANY
8717 VENICE BLVD
LOS ANGELES 90034
PHONE: (213) 870-5700
FIREMASTER
2684 LACY ST
LOS ANGELES 90031
PHONE: (213) 225-6666

FIREARMS

AA SPECIAL EFFECTS
7017 HAYVENHURST BLVD
VAN NUYS 91406
PHONE: (818) 782-6558
CINEMA SECRETS INC
4400 RIVERSIDE DR
BURBANK 91505
PHONE: (818) 846-0579
ELLIS MERCANTILE COMPANY
169 N. LA BREA AVE
LOS ANGELES 90036
PHONE: (213) 933-7334
FAX: (213) 930-1268
**FIRST TAKE PRODUCTIONS,
INC**
8070 SAN FERNANDO RD
SUN VALLEY 91352
PHONE: (818) 767-8261

HAND PROP ROOM
5700 VENICE BLVD
LOS ANGELES 90019
PHONE: (213) 931-1534
SPECIAL EFFECTS UNLIMITED
752 N. CAHUENGA BLVD
HOLLYWOOD 90038
PHONE: (213) 466-3361
STEMBRIDGE GUN RENTALS
431 MAGNOLIA BLVD
GLENDALE 91204
PHONE: (818) 246-4333

FIREPLACE EQUIPMENT

AISCO & FIREPLACE SHOP
1122 SOUTH CENTRAL
GLENDALE 91204
PHONE: (213) 245-5826
ELLIS MERCANTILE COMPANY
169 N. LA BREA AVE
LOS ANGELES 90036
PHONE: (213) 933-7334
FAX: (213) 930-1268
HISTORY FOR HIRE
7103 FAIR AVE
NORTH HOLLYWOOD 91605
PHONE: (818) 765-7767
FAX: (818) 765-7871
HOLLYWOOD CENTRAL PROPS
525 W. ELK AVE
GLENDALE 91204
PHONE: (818) 240-4504
OMEGA/CINEMA PROPS
5857 SANTA MONICA BLVD
HOLLYWOOD 90038
PHONE: (213) 466-8201
PAGEANTRY PRODUCTIONS
11122 WRIGHT RD
LYNWOOD 90262
PHONE: (213) 632-5600
PAYKEL
1443 LINCOLN BLVD
SANTA MONICA 90401
PHONE: (213) 394-1441
PROP SERVICES WEST
915 N. CITRUS AVE.
HOLLYWOOD 90038
PHONE: (213) 461-3371
THEATRE VISION, INC
5426 FAIR AVE
NORTH HOLLYWOOD 91601
PHONE: (818) 769-0928
WILSHIRE FIREPLACE
8636 WILSHIRE BLVD
BEVERLY HILLS 90211
PHONE: (213) 657-7176

FIREWOOD

HOLLYWOOD FIREWOOD
946 N. FORMOSA AVE
HOLLYWOOD 90046
PHONE: (213) 876-4393
**WALTER ALLEN PLANT
RENTALS**
4996 MELROSE AVE
HOLLYWOOD 90029
PHONE: (213) 469-3621
WHITTS FIREWOOD
2356 SOUTH SEPULVEDA BLVD
WEST LOS ANGELES 90064
PHONE: (213) 478-2630

FISH NETS

OCEANIC ARTS
6540 MAGNOLIA
WHITTIER 90601
PHONE: (213) 699-0911

FLAMEPROOFING

CALIFORNIA FLAMEPROOFING
170 NORTH HALSTEAD ST
PASADENA 91107
PHONE: (213) 681-6773

FABRIC FLAMEPROOFING CO.
835 MILFORD ST
GLENDALE 91203
PHONE: (213) 245-1701
OLESEN
1535 IVAR AVE
HOLLYWOOD 90028
PHONE: (213) 461-4631
R.L. GROSH AND SONS SCENIC
4114 SUNSET BLVD
HOLLYWOOD 90029
PHONE: (213) 662-1134
ROGER GEORGE RENTALS, INC
145 1/2 BESSEMER ST.
VAN NUYS 91411
PHONE: (818) 994-3049/
762-6478
FAX: (818) 994-9432
SHOWBIZ ENTERPRISES
15541 LANARK ST
VAN NUYS 91406
PHONE: (818) 989-5005
FAX: (818) 989-6006
SPECIAL EFFECTS UNLIMITED
752 N. CAHUENGA BLVD
HOLLYWOOD 90038
PHONE: (213) 466-3361
THEATRE VISION INC
5426 FAIR AVE
NORTH HOLLYWOOD 91601
PHONE: (818) 769-0928
TRI-ESS SCIENCES, INC
1020 W. CHESTNUT ST
BURBANK 91506
PHONE: (213) 245-7685/
(800) 274-6910
FAX: (818) 848-3521

FOAM BOARDS

ACEY DECY EQUIPMENT CO
5420 VINELAND AVE
NORTH HOLLYWOOD, 91601
PHONE: (818) 766-9445
FAX: (818) 766-4758

FOAM RUBBER

**FOAM N HOME UPHOLSTRY
SUPPLY**
409 NORTH ALAMEDA
COMPTON, 90220
PHONE: (213) 636-6377
**FIRST TAKE PRODUCTIONS,
INC**
8070 SAN FERNANDO RD
SUN VALLEY, 91352
PHONE: (818) 767-8261
WILSHIRE FOAM
1240 E. 230TH ST
CARSON, 90745
PHONE: (213) 775-3761

FOAM URETHANE

CINNABAR
1040 NORTH LAS PALMAS
HOLLYWOOD, 90038
PHONE: (213) 462-3737
FOAM TECH
PHONE: (818) 248-3692

FOG MACHINES

ACEY DECY EQUIPMENT CO.
5420 VINELAND AVE
NORTH HOLLYWOOD 91601
PHONE: (818) 766-9445
**AUTOMATED STUDIOS
LIGHTING**
545 RODIER DR
GLENDALE 91201
PHONE: (818) 500-1646
CINNABAR
1040 NORTH LAS PALMAS
HOLLYWOOD 90038
PHONE: (213) 462-3737

ROGER GEORGE RENTALS, INC
145 1/2 BESSEMER ST
VAN NUYS 91411
PHONE: (818) 994-3049/
762-6478
FAX: (818) 994-9432
SPECIAL EFFECTS UNLIMITED
752 N CAHUENGA BLVD
HOLLYWOOD 90038
PHONE: (213) 466-3361
TRI-ESS SCIENCES, INC
1020 W. CHESTNUT ST
BURBANK 91506
PHONE: (213) 245-7635/
(800) 274-6910
FAX: (818) 848-3521
THEATRE VISION, INC
5426 FAIR AVE
NORTH HOLLYWOOD 91601
PHONE: (818) 769-0928

FOUNTAINS

ROSCHU
6514 SANTA MONICA BLVD
HOLLYWOOD 90038
PHONE: (213) 469-2749

FURNITURE, PADS

CANVAS SPECIALITY
7344 EAST BANDINI BLVD
LOS ANGELES, 90040
PHONE: (213) 723-8311
ELLIS MERCANTILE COMPANY
169 N. LA BREA AVE
LOS ANGELES 90036
PHONE: (213) 933-7344
FAX: (213) 930-1268
**NEW HAVEN MOVING
EQUIPMENT**
1513 PALOMA ST
LOS ANGELES, 90036
PHONE: (213) 749-8181
U-HAUL
4550 HOLLYWOOD BLVD
HOLLYWOOD, 90027
PHONE: (213) 666-7326

GLASS

A TO Z GLASS
5821 EAST BEVERLY BLVD
LOS ANGELES 90022
PHONE: (213) 723-3449
EZER & STONE GLASS CO
1005 NORTH LA BREA AVE
LOS ANGELES 90038
PHONE: (213) 851-2600
FELDMAN GLASS CO
2925 MAIN ST
SANTA MONICA 90405
PHONE: (213) 396-2643
ROUNTREE GLASS CO
625 W COLORADO ST
GLENDALE 91204
PHONE: (818) 246-1785

GLASS, STAINED/LEADED

GLASS STUDIO ONE
3645 10 AVE. ST
LOS ANGELES 91008
PHONE (213) 731-1169

GLITTER

CINEMA SECRETS INC
4400 RIVERSIDE DR
BURBANK 91505
PHONE: (818) 846-0579
MANN BROS. PAINTS
757 NORTH LA BREA AVE
LOS ANGELES 90038
PHONE: (213) 936-5158
MOSKATELS
733 SOUTH SAN JULIAN
LOS ANGELES 90014
PHONE: (213) 689-4830

ROGER GEORGE RENTALS, INC
145 1/2 BESSEMER ST
VAN NUYS 91411
PHONE: (818) 994-3049/
762-6478
FAX: (818) 994-9432

SPECIAL EFFECTS UNLIMITED
752 N CAHUENGA BLVD
HOLLYWOOD 90038
PHONE: (213) 466-3361

STATS
120 SOUTH RAYMOND AVE
PASADENA 91105
PHONE: (818) 795-9308

STUDIO SPECIALTIES LTD
3013 GILROY ST
LOS ANGELES 90039
PHONE: (213) 662-3031

TRI-ESS SCIENCES, INC
1020 W. CHESTNUT ST
BURBANK 91506
PHONE: (213) 245-7685/
(800) 274-6910
FAX: (818) 848-3521

VINE-AMERICAN PARTY STORE
5969 MELROSE AVE
LOS ANGELES 90038
PHONE: (213) 467-7124

GRASS MATS

STUDIO SPECIALTIES LTD
3013 GILROY ST
LOS ANGELES 90039
PHONE: (213) 662-3031

WALTER ALLEN PLANT RENTALS
4996 MELROSE AVE
HOLLYWOOD 90029
PHONE: (213) 469-3621

OCEANIC ARTS
6540 MAGNOLIA
WHITTIER 90601
PHONE: (213) 699-0911

GUILLOTINES

ELLIS MERCANTILE COMPANY
169 N. LA BREA AVE.
LOS ANGELES, 90036
PHONE: (213) 933-7334
FAX: (213) 930-1278

HISTORY FOR HIRE
7103 FAIR AVE
NORTH HOLLYWOOD, 91605
PHONE: (818) 765-7767
FAX: (818) 765-7871

HAND TRUCKS (PERIOD)

HISTORY FOR HIRE
7103 FAIR AVE
NORTH HOLLYWOOD, 91605
PHONE: (818) 765-7767
FAX: (818) 765-7871

HARDWARE SUPPLIES

ANAWALT LUMBER CO
641 NORTH ROBERTSON
WEST HOLLYWOOD 90069
PHONE: (213) 652-6202

BUILDERS EMPORIUM
4650 WEST PICO BLVD
LOS ANGELES 90019
PHONE: (213) 935-1151

BUILDERS EMPORIUM
5525 WEST SUNSET BLVD
HOLLYWOOD 90027
PHONE: (213) 462-1120

BUILDERS EMPORIUM
5960 NORTH SEPULVEDA BLVD
VAN NUYS 91411
PHONE: (818) 785-8601

CALIFORNIA HARDWARE
13085 EAST TEMPLE
CITY OF INDUSTRY 91746
PHONE: (818) 369-9411

DECO BRASS
18919 VENTURA BLVD
TARZANA 91356
PHONE: (818) 345-5481

JONES LUMBER AND HARDWARE
10711 S. ALAMEDA
LINWOOD 90262
PHONE: (213) 564-6656

KOONTZ HARDWARE
650 N. LA PEER
WEST HOLLYWOOD 90069
PHONE: (213) 652-0184

OMEGA/CINEMA PROPS
5857 SANTA MONICA BLVD
HOLLYWOOD 90038
PHONE: (213) 466-8201
FAX: (213) 461-3643

PARAMOUNT STUDIOS
5555 MELROSE AVE
HOLLYWOOD 90038
PHONE: (213) 468-5000

ROMPAGE
5916 HOLLYWOOD BLVD
HOLLYWOOD 90027
PHONE: (213) 467-2129

THE OLD PASADENA BRASSORIE
100 N. FAIR OAKS
PASADENA 91103
PHONE (818) 793-2883

VALLEY SASH AND DOOR
14829 OXNARD ST
VAN NUYS 91411
PHONE: (818) 785-8628

HEADSTONES

ALS STUDIO RENTALS
6025 HOLLYWOOD BLVD
CULVER CITY 90028
PHONE: (213)469-8155

BURBANK STUDIOS
4000 WARNER BLVD
BURBANK 91522
PHONE: (818)954-6000

UNIVERSAL STUDIOS
100 UNIVERSAL CITY PLAZA
UNIVERSAL CITY 91608
PHONE: (818)777-2784

WALTER ALLEN PLANT RENTALS
4996 MELROSE AVE
HOLLYWOOD 90029
PHONE: (213)469-3621

HEATERS, OUTDOOR

R.P. BROOKS & SONS INC
1933 BROADWAY #1145
LOS ANGELES 90007
PHONE: (213) 749-7374

ROGER GEORGE RENTALS, INC
145 1/2 BESSEMER ST
VAN NUYS 91411
PHONE: (818) 994-3049/762-6478
FAX: (818) 994-9432

SPECIAL EFFECTS UNLIMITED
752 N CAHUENGA BLVD
HOLLYWOOD 90038
PHONE: (213) 466-3361

STAR CELEBRITY LOOK-ALIKES
8022 HOLLYWOOD BLVD
LOS ANGELES 90046
PHONE: (213) 654-3500

TRI-ESS SCIENCES, INC
1020 W. CHESTNUT ST
BURBANK 91506
PHONE: (213) 245-7685/ (800) 274-6910
FAX: (818) 848-3521

HELIUM

CONWIN CARBONIC CO
4510 SPERRY
LOS ANGELES 90039
PHONE: (818) 246-9233

MAX BALLOONS
11950 VOSE ST
NORTH HOLLYWOOD 91605
PHONE: (213) 654-3500

HORNS

ACEY DECY EQUIPMENT CO
5420 VINELAND AVE
NORTH HOLLYWOOD, 91601
PHONE: (818) 766-9445

ELLIS MERCANTILE COMPANY
169 N. LA BREA AVE
LOS ANGELES, 90036
PHONE: (213) 933-7334
FAX: (213) 930-1268

HOT GLUE SYSTEMS

STUDIO SPECIALTIES LTD
3013 GILROY ST
LOS ANGELES 90039
PHONE: (213) 662-3031

ICE SKATING RINK

REFLECTIONS
4918 ESCOBEDO DR
WOODLAND HILLS, 91364
PHONE: (818) 883-6223

INSTRUMENT CASES

A&S CASE CO.INC.
1111 N. GORDON ST
HOLLYWOOD, 90038
PHONE: (213) 466-6181

JANITORIAL SUPPLIES (PROP)

ALS STUDIO RENTALS
6025 HOLLYWOOD BLVD
HOLLYWOOD, 90026
PHONE: (213) 469-8155
FAX: (213) 468-8150

C.P.TWO
5755 SANTA MONICA BLVD
LOS ANGELES, 90038
PHONE: (213) 466-8201
FAX: (213) 461-3643

JUGGLERS

CALIFORNIA ATTRACTIONS LIMITED
7023 CANOGA AVE. SUITE E
CANOGA PARK, 91303
PHONE: (818) 999-6255

STAR CELEBRITY LOOK-ALIKES
11023 FRUITLAND DR. #2
STUDIO CITY, 91604
PHONE: (213) 876-7866

JUNGLE DRESSING

OCEANIC ARTS
6540 MAGNOLIA
WHITTIER, 90601
PHONE: (213) 699-0911

JUNK

CLEVELAND WRECKING CO
3170 EAST WASHINGTON BLVD
LOS ANGELES, 90023
PHONE: (213) 269-0633

LOS ANGELES WRECKING CO
810 E. 9TH ST
LOS ANGELES, 90021
PHONE: (213) 622-5135

SCAVENGERS PARADISE
4360 TUJUNGA
STUDIO CITY, 91604
PHONE: (213) 877-7945

KEYS/LOCKS

DOOR KEYPER
1105 N. ALLEN AVE
PASADENA, 91104
PHONE: (818) 794-6940/(213) 684-5076

LOS FELIZ LOCK & KEY
1856 NORTH VERMONT BLVD
HOLLYWOOD, 90027
PHONE: (213) 663-8351

N.Y.LOCK & KEY
1040 NORTH WESTERN AVE
LOS ANGELES, 90029
PHONE: (213) 469-2183

KNIVES

HISTORY FOR HIRE
7103 FAIR AVE
NORTH HOLLYWOOD 91605
PHONE: (818) 765-7767
FAX: (818) 765-7871

LABELS

ALEXANDER STATIONERS
1531 NORTH CAHUENGA BLVD.
HOLLYWOOD, 90028
PHONE: (213) 464-1151

EARL HAYS PRESS
10707 SHERMAN WAY
SUN VALLEY, 91352
PHONE: (818) 765-0700

HAND PROP ROOM
5700 VENICE BLVD
LOS ANGELES, 90019
PHONE: (213) 931-1534

LAB EQUIPMENT

HOLLYWOOD CENTRAL PROPS
525 W. ELK AVE
GLENDALE 91204
PHONE: (818) 240-4504

ROGER GEORGE RENTALS, INC
145 1/2 BESSEMER ST
VAN NUYS 91411
PHONE: (818) 994-3049/
762-6478
FAX: (818) 994-9432

SPECIAL EFFECTS UNLIMITED
752 N CAHUENGA BLVD
HOLLYWOOD 90038
PHONE: (213) 466-3361

TRI-ESS SCIENCES, INC
1020 W. CHESTNUT ST
BURBANK 91506
PHONE: (213) 245-7685/
(800) 274-6910
FAX: (818) 848-3521

LADDERS

A. PALMER SCAFFOLD & EQUIPMENT
4600 BRAZIL ST
LOS ANGELES, 90039
PHONE: (213) 245-3127

ACEY DECY EQUIPMENT CO
5420 VINELAND AVE
NORTH HOLLYWOOD, 91601

LADDERS INDUSTRIES
1819 BARRANCA ST
LOS ANGELES, 90031

SUNSET LADDER CO
4421 FOUNTAIN AVE
LOS ANGELES, 90029
PHONE: (213) 665-5877

THEATRE VISION, INC
5426 FAIR AVE
NORTH HOLLYWOOD, 91601
PHONE: (818) 769-0928

LAMINATING

HAND PROP ROOM
5700 VENICE BLVD
LOS ANGELES, 90019
PHONE: (213) 931-1534

PERMA PLAQUE CORPORATION
7251 VARNA AVE
NORTH HOLLYWOOD, 91605
PHONE: (818) 764-3100

LAMP POSTS

ACEY DECY EQUIPMENT CO
5420 VINELAND AVE
NORTH HOLLYWOOD, 91601
PHONE: (818) 766-9445
FAX: (818) 766-4758

FIRST STREET FURNITURE
1123 N. BRONSON AVE
HOLLYWOOD, 90038
PHONE: (213) 462-6306

HOLLYWOOD CENTRAL PROPS
525 W. ELK AVE
GLENDALE, 91204
PHONE: (818) 240-4504

PAGEANTRY PRODUCTIONS
11122 WRIGHT RD
LYNWOOD, 90262
PHONE: (213) 632-5600

PROP SERVICES WEST
915 N. CITRUS AVE
HOLLYWOOD, 90038
PHONE: (213) 461-3371

ROSCHU
6514 SANTA MONICA BLVD
HOLLYWOOD, 90038
PHONE: (213) 469-2749

THEATRE VISION,INC
5426 FAIR AVE
NORTH HOLLYWOOD, 91601
PHONE: (818) 769-0928

LANTERNS

HISTORY FOR HIRE
7103 FAIR AVE
NORTH HOLLYWOOD, 91605
PHONE: (818) 765-7767
FAX: (818) 765-7871

LEATHER

MACPHERSON LEATHER CO
420 S. SAN PEDRO ST
LOS ANGELES 90013
PHONE: (213) 626-4831

PACIFIC HIDE & LEATHER CO
1400 S. BROADWAY
LOS ANGELES 90061
PHONE: (213) 321-6730

LEAVES

WALTER ALLEN PLANT RENTALS
4996 MELROSE AVE
LOS ANGELES, 90029
PHONE: (213) 469-3621

LICENSE PLATES

EARL HAYS PRESS
10707 SHERMAN WAY
SUN VALLEY, 91352
PHONE: (818) 765-0700

ELLIS MERCANTILE COMPANY
169 N. LA BREA AVE
LOS ANGELES, 90036
PHONE: (213) 933-7334
FAX: (213) 931-1268

F.C. ENTERPRISES
3901 MEDFORD ST
LOS ANGELES, 90063
PHONE: (213) 262-5476

HAND PROP ROOM
5700 VENICE BLVD
LOS ANGELES, 90019
PHONE: (213) 931-1534

HISTORY FOR HIRE
7103 FAIR AVE
NORTH HOLLYWOOD, 91605
PHONE: (818) 765-7767
FAX: (818) 765-7871

MAGICIANS

CALIFORNIA ATTRACTIONS LIMITED
7023 CANOGA AVE. SUITE E
CANOGA PARK, 91303
PHONE: (818) 999-6255

MAGIC ASTLE
7001 FRANKLIN AVE
HOLLYWOOD, 90028
PHONE: (213) 851-3313

STAR CELEBRITY LOOK-ALIKES
11023 FRUITLAND DR.#2
STUDIO CITY, 91604
PHONE: (213) 876-7866

MAGICIANS SUPPLIES

CINEMA SECRETS INC
4400 RIVERSIDE DR
BURBANK, 91505
PHONE: (818) 846-0579

HOLLYWOOD MAGIC INC
6614 HOLLYWOOD BLVD
HOLLYWOOD, 90028
PHONE: (213) 464-5610

MAGIC EMPORIUM
19401 PARTHENIA
NORTHRIDGE, 91324
PHONE: (818) 344-2525

MAGICRAFT
1218 S. GERHART AVE
LOS ANGELES, 90022
PHONE: (213) 724-2279

OWEN MAGIC
734 NORTH MCKEEVER
AZUSA, 91702
PHONE: (818) 969-4519

MAKEUP

BEN NYE CO
11571 SANTA MONICA BLVD
LOS ANGELES 90025
PHONE: (213) 477-0443

CALIF. COSTUME-NORCOSTCO CO
1842 PUENTE AVE
BALDWIN PARK 91706
PHONE: (818) 960-4711

CINEMA SECRETS INC
4400 RIVERSIDE AVE
BURBANK 91505
PHONE: (818) 846-0579

COLUMBIA STAGE & SCREEN COSMETICS
1440 N. GOWER
HOLLYWOOD 90028
PHONE: (213) 464-7555

FRENDS BEAUTY SUPPLY
5270 LAUREL CANYON BLVD
NORTH HOLLYWOOD 91607
PHONE: (818) 769-3834

HOFFMANS INTERNATIONAL
8126 SANTA MONICA BLVD
WEST HOLLYWOOD 90046
PHONE: (213) 654-6840

JOE BLASCO MAKE-UP CENTER
1708 HILLHURST AVE
LOS ANGELES 90046
PHONE: (213) 467-4949

OLESEN
1535 IVAR AVE
HOLLYWOOD 90028
PHONE: (213) 461-4631

SPERLING BEAUTY SUPPLY
13639 VAN OWEN ST
VAN NUYS 91405
PHONE: (818) 781-6300

MANNEQUINS

C.P. TWO
5755 SANTA MONICA BLVD
LOS ANGELES 90038
PHONE: (213) 466-8201
FAX: (213) 461-3643

CAL STATE STORE FIXTURES
14210 VAN NUYS BLVD
PACOIMA 91331
PHONE: (818) 899-0686

DECTER MANNEQUIN
1118 EAST 8TH ST
LOS ANGELES 90021
PHONE: (213) 627-9842

FIXTURES BY HOWIE
901 SOUTH MAIN ST
LOS ANGELES 90015
PHONE: (213) 627-5953

HISTORY FOR HIRE
7103 FAIR AVE
NORTH HOLLYWOOD 91605
PHONE: (818) 765-7767
FAX: (818) 765-7871

IMPERIAL DISPLAY
3410 WEST WASHINGTON BLVD
LOS ANGELES 90018
PHONE: (213) 735-1011

MIKES STORE FIXTURES
770 SOUTH SAN PEDRO ST
LOS ANGELES 90014
PHONE: (213) 622-21

SILVESTRI STUDIO, INC
1733 CORDOVA ST
LOS ANGELES 90007
PHONE: (213) 735-1481

MASKS

CINEMA SECRETS INC
4400 RIVERSIDE DR
BURBANK 91505
PHONE: (818) 846-0579

DON POST STUDIOS
8211 LANKERSHIM BLVD
NORTH HOLLYWOOD 91405
PHONE: (818) 768-0811

ELLIS MERCANTILE COMPANY
169 N. LA BREA AVE
LOS ANGELES 90036
PHONE: (213) 933-7334
FAX: (213) 930-1268

HAND PROP ROOM
5700 VENICE BLVD
LOS ANGELES 90019
PHONE: (213) 931-1534

HOLLYWOOD CENTRAL PROPS
525 W. ELK AVE
GLENDALE 91204
PHONE: (818) 240-4504

HOLLYWOOD MAGIC INC
6614 HOLLYWOOD BLVD
HOLLYWOOD 90028
PHONE: (213) 464-5610

HOLLYWOOD STUDIO GALLERY
1035 CAHUENGA BLVD
HOLLYWOOD 90038
PHONE: (213) 462-1116

HOLLYWOOD TOYS & COSTUMES INC
6562 HOLLYWOOD BLVD
HOLLYWOOD 90028
PHONE: (213) 465-3119

MAGIC EMPORIUM
19401 PARTHENIA
NORTHRIDGE 91324
PHONE: (818) 344-2525

OMEGA/CINEMA PROPS
5857 SANTA MONICA BLVD
HOLLYWOOD 90038
PHONE: (213) 466-8201
FAX: (213) 461-3643

OCEANIC ARTS
6540 MAGNOLIA
WHITTIER 90601
PHONE: (213) 699-0911

MINIATURES

BOB BAKER MARIONETTE THEATER
1345 WEST 1ST ST
LOS ANGELES 90026
PHONE: (213) 250-9995

DOLL & MINIATURE SHOP
661 S. LA BREA AVE
LOS ANGELES 90036
PHONE: (213) 934-6229

FIRST TAKE PRODUCTIONS, INC
3070 SAN FERNANDO RD
SUN VALLEY 91352
PHONE: (818) 767-8251

MINIATURE ESTATES
1451 SOUTH ROBERTSON BLVD
LOS ANGELES 90035
PHONE: (213) 552-2200

MINIATURE TOWNE U.S.A.
17624 SHERMAN WAY
VAN NUYS 91406
PHONE: (818) 996-3330

MIRROR BALLS

ACEY DECY EQUIPMENT CO
5420 VINELAND AVE
NORTH HOLLYWOOD 91601
PHONE: (818) 766-9445
FAX: (818) 766-4758

LENNIE MARVIN ENTERPRISES
1105 N. HOLLYWOOD WAY
BURBANK 91505
PHONE: (818) 841-5882

PAGEANTRY PRODUCTIONS
11122 WRIGHT RD
LYNWOOD 90626
PHONE: (213) 632-5600
FAX: (213) 632-3304

PRODUCTION LIGHTING SYSTEMS INC
ONE WEST ALAMEDA
BURBANK 91502
PHONE: (818) 845-1200

ROSCHU
6514 SANTA MONICA BLVD
HOLLYWOOD 90038
PHONE: (213) 469-2749

STUDIO SPECIALTIES LTD
3013 GILROY ST
LOS ANGELES 90039
PHONE: (213) 662-3031

THEATRE VISION, INC
5426 FAIR AVE
NORTH HOLLYWOOD 91601
PHONE: (818) 769-0928

MIRRORS

C.P. TWO
5755 SANTA MONICA BLVD
LOS ANGELES 90038
PHONE: (213) 466-8201
FAX: (213) 461-3643

PROP SERVICES WEST
915 N. CITRUS AVE.
HOLLYWOOD 90038
PHONE: (213) 461-3371

MIRRORS, FUN HOUSE

LENNIE MARVIN ENTERPRISES
1105 N. HOLLYWOOD WAY
BURBANK, 91505
PHONE: (818) 841-5882

LUCKY ENTERTAINMENT
10271 ALMAYO AVE. #101
LOS ANGELES, 90064
PHONE: (213) 277-9666

MORTICIANS WAX

CALIFORNIA COSTUME-NORCOSTCO CO
1842 PUENTE AVE
BALDWIN PARK 91706
PHONE: (818) 960-4711

CINEMA SECRETS INC
4400 RIVERSIDE DR
BURBANK 91505
PHONE: (818) 846-0579

COLUMBIA STAGE & SCREEN COSMETICS
1440 N. GOWER
HOLLYWOOD 90028
PHONE: (213) 464-7555

NATURO DIST.
4250 EAST WASHINGTON BLVD
LOS ANGELES 90023
PHONE: (213) 268-7291

MOTORCYCLE RENTALS

ANDERSON RENTALS INC
18432 VAN OWEN
RESEDA, 91335
PHONE: (818) 881-9088

BILL ROBERTSON & SONS, INC
6525 SANTA MONICA BLVD
HOLLYWOOD, 90038
PHONE: (213) 466-7191

BURBANK KAWASAKI
1329 NORTH HOLLYWOOD WAY
BURBANK, 91505
PHONE: (818) 848-6627

HOLLYWOOD KAWASAKI
1339 NORTH HIGHLAND AVE.
HOLLYWOOD, 90028
PHONE: (213) 466-8451

I. MARTIN IMPORTS
8330 BEVERLY BLVD
LOS ANGELES, 90048
PHONE: (213) 653-6900

MOVIE SCREENS

AMETRON RENTS
1200 NORTH VINE ST
HOLLYWOOD 90038
PHONE: (213) 466-4321

BACKGROUND ENGINEERING
1213 FLOWER ST
GLENDALE 91201
PHONE: (818) 465-4161

MUSICAL INSTRUMENTS

C.P. TWO
5755 SANTA MONICA BLVD
LOS ANGELES,
90038
PHONE: (213) 466-8201
FAX: (213) 461-3643

DAVID ABELL
8162 BEVERLY BLVD
LOS ANGELES, 90048
PHONE: (213) 651-3060

ELLIS MERCANTILE COMPANY
169 N. LA BREA AVE
LOS ANGELES, 90036
PHONE: (213) 933-7334
FAX: (213) 930-1268

HAND PROP ROOM
5700 VENICE BLVD
LOS ANGELES, 90019
PHONE: (213) 931-1534

HISTORY FOR HIRE
7103 FAIR AVE
NORTH HOLLYWOOD, 91605
PHONE: (818) 765-7767
FAX: (818) 765-7871

HOLLYWOOD PIANO RENTAL
1647 NORTH HIGHLAND AVE
HOLLYWOOD, 90028
PHONE: (818) 462-2329

LEEDS/REH. STUDIO
11131 WEDDINGTON
NORTH HOLLYWOOD, 91601
PHONE: (818) (80-7774

MUSIC CENTER
5616 SANTA MONICA BLVD
HOLLYWOOD, 90038
PHONE: (213) 469-8143

MUSICIANS SUPPLY SHOP
11732 WEST PICO BLVD
WEST LOS ANGELES, 90064
PHONE: (213) 478-7836

STUDIO INSTRUMENT RENTAL
6048 SUNSET BLVD
HOLLYWOOD, 90028
PHONE: (213) 466-3417

VALLEY PIANO
933 WEST OLIVE
BURBANK, 915606
PHONE: (818) 849-1666

MYLAR

MANN BROS. PAINTS
757 NORTH LA BREA AVE
LOS ANGELES 90038
PHONE: (213) 936-5168

SHOWBIZ ENTERPRISES
15541 LANARK ST
VAN NUYS 91406
PHONE: (818) 989-5005

STUDIO SPECIALTIES LTD
3013 GILROY ST
LOS ANGELES 90039
PHONE: (213) 662-3031

NEON

AMERICAN NEON & GRAPHICS
5542 SATSUMAN WAY
NORTH HOLLYWOOD 91601
PHONE: (818) 875-1815

CUSTOM NEON
2210 SOUTH LA BREA AVE
LOS ANGELES 90016
PHONE: (213) 937-6366

NIGHTS OF NEON
7442 VARNA AVE
NORTH HOLLYWOOD 91605
PHONE: (818) 982-3592

RC VINTAGE STUDIO RENTALS
1644 CHEROKEE AVE
HOLLYWOOD 90028
PHONE: (213) 462-4510

SPECIAL EFFECTS UNLIMITED
752 N. CAHUENGA BLVD
HOLLYWOOD 90038
PHONE: (213) 466-3361

TWENTY FIRST CENTURY NEON
2644 ARMSTRONG BLVD
LOS ANGELES 92372
PHONE: (213) 662-6666

NOVELTY ITEMS

HOLLYWOOD MAGIC INC
6614 HOLLYWOOD BLVD
HOLLYWOOD, 90028
PHONE: (213) 464-5610

NIGHTS OF NEON
7442 VARNA AVE
NORTH HOLLYWOOD, 91605
PHONE: (818) 982-3592
FAX: (818) 503-1090

OIL DRUMS

C.P. TWO
5755 SANTA MONICA BLVD
LOS ANGELES 90038
PHONE: (213) 466-8201
FAX: (213) 461-3643

ELLIS MERCANTILE COMPANY
169 N. LA BREA AVE
LOS ANGELES 90036
PHONE: (213) 933-7334
FAX: (213) 930-1268

FIRST STREET FURNITURE
1123 N. BRONSON AVE
HOLLYWOOD 90038
PHONE: (213) 462-6306

HISTORY FOR HIRE
7103 FAIR AVE
NORTH HOLLYWOOD 91605
PHONE: (818) 765-7767
FAX: (818) 765-7871

OVERSIZED PROPS

CALIFORNIA ATTRACTIONS LIMITED
7023 CANOGA AVE. SUITE E
CANOGA PARK 91303
PHONE: (818) 999-6255

ROSCHU
6514 SANTA MONICA BLVD
HOLLYWOOD 90038
PHONE: (213) 469-2749

FIRST TAKE PRODUCTIONS, INC
8070 SAN FERNANDO RD
SUN VALLEY 91352
PHONE: (818) 767-8261

PAINTING SUPPLIES

BUILDERS EMPORIUM
5525 SUNSET BLVD
HOLLYWOOD 90028
PHONE: (213) 462-1120

DUNN-EDWARDS PAINT (HOLLYWOOD)
960 NORTH HIGHLAND AVE
HOLLYWOOD 90046
PHONE: (213) 464-4157

FULLER-OBRIEN PAINTS
6313 PACIFIC BLVD
HUNTINGTON PARK 90255
PHONE: (213) 588-4277

HOLLYWOOD PAINT
8016 MELROSE AVE
HOLLYWOOD 90046
PHONE: (213) 852-0402

MANN BROS. PAINTS
757 NORTH LA BREA AVE
LOS ANGELES 90038
PHONE: (213) 936-5168

OLESEN
1535 IVAR AVE
HOLLYWOOD 90028
PHONE: (213) 461-4631
FAX: (213) 464-0444

ROSCO
1135 N. HIGHLAND AVE
LOS ANGELES 90038
PHONE: (213) 462-2233

SHANNON LUMINOUS MATERIALS CO.
304 A TOWNSEND ST
SANTA ANA 92703
PHONE: (714) 550-9931

SINCLAIR PAINTS
12203 VENTURA BLVD
STUDIOS CITY 91604
PHONE: (818) 984-1516

SINCLAIR PAINTS
5600 SANTA MONICA BLVD
HOLLYWOOD 90038
PHONE: (213) 465-7161

STANDARD BRANDS
861 NORTH VINE ST
HOLLYWOOD 90038
PHONE: (213) 465-0271

WESTSIDE PAINT
5706 WEST PICO BLVD
LOS ANGELES 90035

PAPER, SEAMLESS

ACEY DECY EQUIPMENT CO.
5420 VINELAND AVE
NORTH HOLLYWOOD, 91601
PHONE: (818) 766-9445
FAX: (818) 766-4758

BOB GAMBLE
5170 SANTA MONICA BLVD
LOS ANGELES, 90029
PHONE: (213) 663-9251

FREISTYLE PHOTO
5124 W. SUNSET BLVD
LOS ANGELES, 90027
PHONE: (213) 660-3460

PAN PACIFIC CAMERA CENTER
825 NORTH LA BREA AVE
LOS ANGELES, 90018
PHONE: (213) 933-5888

STUDIO SPECIAL TIES LTD
3013 GILROY ST
LOS ANGELES, 90039
PHONE: (213) 662-3031

THEATRE VISION, INC
5426 FAIR AVE
NORTH HOLLYWOOD, 91601
PHONE: (818) 769-0928

PHOTO BLOW-UPS

AARON BROTHERS-LA CIENEGA
330 NORTH LA CIENEGA
LOS ANGELES, 90048
PHONE: (213) 657-7588

B&R GRAPHICS
1132 VINE ST
HOLLYWOOD, 90046
PHONE: (213) 466-6179

COLOR HOUSE
1919 EMPIRE BLVD
BURBANK, 91504
PHONE: (213) 649-3255

CREATIVE DISPLAY
3716 WEST JEFFERSON
LOS ANGELES, 90016
PHONE: (213) 731-0897

DUPLICATE PHOTO LAB
1522 NORTH HIGHLAND AVE
HOLLYWOOD, 90028
PHONE: (213) 466-7544

HOLLYWOOD STUDIO GALLERY
1035 CAHUENGA BLVD
HOLLYWOOD, 90038
PHONE: (213) 462-1116

NEWELL COLOR LAB
221 N. WESTMORELAND
LOS ANGELES, 90004
PHONE: (213) 380-2980

PACIFIC STUDIOS INC
8315 MELROSE AVE
LOS ANGELES, 90069
PHONE: (213) 653-3093

PINBALL MACHINES

C. A. ROBINSON
2891 WEST PICO BLVD
WEST LOS ANGELES, 90006
PHONE: (213) 735-3001

HISTORY FOR HIRE
7103 FAIR AVE
NORTH HOLLYWOOD, 91605
PHONE: (818) 765-7767
FAX: (818) 765-7871

LENNIE MARVIN ENTERPRISES
1105 N,. HOLLYWOOD WAY
BURBANK, 91505
PHONE: (818) 841-5881

LUCKY ENTERTAINMENT
10271 ALMAYO AVE. #101
LOS ANGELES, 90064
PHONE: (213) 277-9666

STAR CELEBRITY LOOK-ALIKES
11023 FRUITLAND DR. #2
STUDIO CITY, 91604
PHONE: (213) 876-7866

PLASTIC ICE CUBES

HAND PROP ROOM
5700 VENICE BLVD
LOS ANGELES 90019
PHONE: (213) 931-1534

PROP SERVICES WEST
915 N. CITRUS AVE
HOLLYWOOD 90038
PHONE: (213) 461-3371

STUDIO SPECIALTIES LTD
3013 GILROY ST
LOS ANGELES 90039
PHONE: (213) 662-3031

PLASTIC SUPPLIES

CADILLAC PLASTICS
11255 VANOWEN ST
NORTH HOLLYWOOD 91605
PHONE: (818) 980-0840

CINEMA SECRETS INC
4400 RIVERSIDE DR
BURBANK 91505
PHONE: (818) 846-0579

LEED PLASTICS
793 E. PICO BLVD
LOS ANGELES 90021
PHONE: (213) 746-5984

PLASTIC MART
2101 PICO BLVD
SANTA MONICA 90405
PHONE: (213) 451-1701

REGAL PLASTIC SUPPLY
965 ARTESIA BLVD
CARSON 90746
PHONE: (213) 538-5860

STUDIO SPECIALTIES LTD
3013 GILROY ST
LOS ANGELES 90039
PHONE: (213) 662-3031

PLATFORMS/PORTABLE STAGES

CINNABAR
1040 NORTH LAS PALMAS
HOLLYWOOD 90038
PHONE: (213) 462-3737

COMBINED ENTERTAINMENT SERVICES INC
3019 ANDRITA ST
LOS ANGELES 90065
PHONE: (213) 258-5055
FAX: (213) 258-6950

MIKE BROWN GRANDSTANDS, INC
2800 E. HUNTINGTON DR
DUARTE 91010
PHONE: (818) 357-1161

RUSSELL & RUSSELL
3226 E. WASHINGTON BLVD
LOS ANGELES 90023
PHONE: (213) 262-1161

SCENIC EXPRESS
3025 FLETCHER DR
LOS ANGELES 90035
PHONE: (213) 254-4351
FAX: (213) 254-4411

STAGE MECHANIX
1634 EAST 23RD ST
LOS ANGELES 90011
PHONE: (213) 232-3663

THEATRE VISION, INC
5426 FAIR AVE
NORTH HOLLYWOOD 91601
PHONE: (818) 769-0928

PLAYER PIANOS

HOLLYWOOD CENTRAL PROPS
525 W. ELK AVE
GLENDALE 91204
PHONE: (818) 240-4504

LENNIE MARVIN ENTERPRISES
1105 N. HOLLYWOOD WAY
BURBANK 91505
PHONE: (818) 841-5882

PLUMBING FIXTURES

COAST FIXTURES
424 SOUTH LOS ANGELES ST
LOS ANGELES, 90013
PHONE: (213) 687-4411

FIRST STREET FURNITURE
1123 N. BRONSON AVE
HOLLYWOOD, 90038
PHONE: (213) 462-6306

SNYDER-DIAMOND
1399 OLYMPIC BLVD
SANTA MONICA, 90404
PHONE: (213) 870-6667

WESTERN BEVERLY PLUMBING
538 NORTH WESTERN AVE
LOS ANGELES, 90004
PHONE: (213) 462-7281

PRODUCT PLACEMENT

G.L.I.
SUNSET GOWER STUDIOS
SUITE 221
HOLLYWOOD, 90028
PHONE: (213) 962-5491

PROJECTION ROOMS

UNIVERSAL STUDIOS
100 UNIVERSAL CITY PLAZA
UNIVERSAL CITY, 91608
PHONE: (818) 777-1331

PROJECTORS

HISTORY FOR HIRE
7103 FAIR AVE
NORTH HOLLYWOOD, 91605
PHONE: (818) 765-7767
FAX: (818) 765-7871

PROP HOUSES

ALS STUDIO RENTALS
6025 HOLLYWOOD BLVD
HOLLYWOOD, 90028
PHONE: (213) 469-8155
FAX: (213) 469-8150

ARTE DE MEXICO
5356 RIVERTON AVE
NORTH HOLLYWOOD, 91601
PHONE: (818) 769-5090

BURBANK STUDIOS
4000 WARNER BLVD
BURBANK, 91522
PHONE: (818) 954-6000

C.P. TWO
5755 SANTA MONICA BLVD
LOS ANGELES, 90038
PHONE: (213) 466-8201
FAX: (213) 461-3643

CULVER STUDIOS PROP DEPT
9339 W. WASHINGTON BLVD
CULVER CITY, 90232
PHONE: (213) 202-3350
FAX: (213) 202-3272

CUSTOM NEON
2210 SOUTH LA BREA AVE
LOS ANGELES, 90016
PHONE: (213) 937-6366

DECADES
6666 SANTA MONICA BLVD
HOLLYWOOD, 90038
PHONE: (213) 464-0696

ELLIS MERCANTILE COMPANY
169 N. LA BREA AVE
LOS ANGELES, 90036
PHONE: (213) 933-7334
FAX: (213) 930-1268

FIRST STREET FURNITURE
1123 N. BRONSON AVE
HOLLYWOOD, 90038
PHONE: (213) 462-6306

HAND PROP ROOM
5700 VENICE BLVD
LOS ANGELES, 90019
PHONE: (213) 931-1534

HISTORY FOR HIRE
7103 FAIR AVE
NORTH HOLLYWOOD, 91605
PHONE: (818) 765-7767
FAX: (818) 765-7871

HOLLYWOOD CENTRAL PROPS
525 W. ELK AVE
GLENDALE, 91204
PHONE: (818) 240-4505

HOLLYWOOD SOUTHWEST
4918 VINELAND AVE
NORTH HOLLYWOOD, 91601
PHONE: (818) 753-9050

HOLLYWOOD STUDIO GALLERY
1035 CAHUENGA BLVD
HOLLYWOOD, 90038
PHONE: (213) 462-1116

HOUSE OF PROPS
1117 N. GOWER ST
LOS ANGELES, 90038
PHONE: (213) 463-3166

INDEPENDENT STUDIO SERVICE INC
11907 WICKS ST
SUN VALLEY, 91352
PHONE: (818) 768-5711

LENNIE MARVIN ENTERPRISES
1105 N. HOLLYWOOD WAY
BURBANK, 91505
PHONE: (818) 841-5882

MODERN PROPS
4063 REDWOOD AVE
LOS ANGELES, 90066
PHONE: (213) 306-1400

NIGHTS OF NEON
7442 VARNA AVE
NORTH HOLLYWOOD, 91605
PHONE: (818) 982-3592

OCEANIC ARTS
6540 MAGNOLIA
WHITTIER, 90601
PHONE: (213) 699-0911

OMEGA/CINEMA PROPS
5857 SANTA MONICA BLVD
HOLLYWOOD, 90038
PHONE: (213) 466-820L
FAX: (213) 461-3643

PACIFIC MEDICAL RENTALS
1112 S. VICTORY BLVD
BURBANK, 91503
PHONE: (818) 567-4885

PAGEANTRY PRODUCTIONS
11122 WRIGHT RD
LYNWOOD, 90262
PHONE: (213) 632-5600

PARAMOUNT STUDIOS
5555 MELROSE AVE
HOLLYWOOD, 90038
PHONE: (213) 468-5000

PROP SERVICES WEST
915 N. CITRUS AVE
HOLLYWOOD, 90038
PHONE: (213) 461-3371

RC VINTAGE STUDIO RENTALS
1644 CHEROKEE AVE
HOLLYWOOD, 90028
PHONE: (213) 462-4510

ROSCHU
6514 SANTA MONICA BLVD
HOLLYWOOD, 90038
PHONE: (213) 469-2749

SPELLMAN DESK
6159 SANTA MONICA BLVD
HOLLYWOOD, 90038
PHONE: (213) 467-0628

TR TRADING CO.
15604 S. BROADWAY
GARDENA, 90248
PHONE: (213) 329-9242
FAX: (213) 329-0789

UNIVERSAL STUDIOS
100 UNIVERSAL CITY PLAZA
UNIVERSAL CITY, 91608
PHONE: (818) 777-2784

WALT DISNEY PRODUCTIONS
500 SOUTH BUENA VISTA ST
BURBANK, 91521
PHONE: (818) 560-1000 EXT. 1321

WALTER ALLEN PLANT RENTALS
4996 MELROSE AVE
HOLLYWOOD, 90029
PHONE: (213) 469-3621

WOODYS ELECTRONIC PROPS
9167 SAN FERNANDO RD
SUN VALLEY, 91352
PHONE: (818) 768-6637

PUPPETS

PUPPET WORKS
1680 VINE ST. SUITE 604
HOLLYWOOD 90028
PHONE: (213) 461-1415

RENE & HIS ARTISTS PRODUCTIONS
707-A MAIN ST
BURBANK 91506
PHONE: (213) 849-4115
FAX: (818) 848-4861

STAR CELEBRITY LOOK-ALIKES
11023 FRUITLAND DR. #2
STUDIO CITY 91604
PHONE: (213) 876-7866

RADIATORS

ELLIS MERCANTILE COMPANY
169 N. LA BREA AVE
LOS ANGELES, 90036
PHONE: (213) 933-7334
FAX: (213) 930-1268

FIRST STREET FURNITURE
1123 N. BRONSON AVE
HOLLYWOOD, 90038
PHONE: (213) 462-6306

HISTORY FOR HIRE
7103 FAIR AVE
NORTH HOLLYWOOD, 91605
PHONE: (818) 765-7767
FAX: (818) 765-7871

HOLLYWOOD CENTRAL PROPS
525 W. ELK AVE
GLENDALE, 91204
PHONE: (818) 240-4504

PROP SERVICES WEST
915 N. CITRUS AVE
HOLLYWOOD, 90038
PHONE: (213) 461-3371

WEST COAST THEATRE SUPPLY
4671 ALGER ST
LOS ANGELES, 90039
PHONE: (818) 240-8797

RIGGING EQUIPMENT

AMERICAN SCENERY
18555 EDDY ST
NORTHRIDGE, 91324
PHONE: (818) 886-1585

F.W. WARD CO
2319 NORTH TROY AVE. #6
SOUTH EL MONTE, 91733
PHONE: (213) 686-0480/ (818) 443-5465

MIKE BROWN GRANDSTANDS INC.
2800 E. HUNTINGTON DR
DUARTE, 91010
PHONE: (818) 357-1161

OLESEN
1535 IVAR AVE
HOLLYWOOD, 90028
PHONE: (213) 461-4631
FAX: (213) 464-0444

THEATRE VISION, INC
5426 FAIR AVE
NORTH HOLLYWOOD, 91601
PHONE: (818) 769-0928

TRIANGLE SCENERY & LIGHTING CO.
1215 BATES
LOS ANGELES, 90029
PHONE: (213) 662-8129

TRU-ROLL CORPORATION
622 SONORA BLVD
GLENDALE, 91201
PHONE: (818) 243-9567

WEST COAST THEATRE SUPPLY
4671 ALGER ST
LOS ANGELES, 90039
PHONE: (818) 240-8797

ROBOTS

CINEMA SECRETS INC
4400 RIVERSIDE DR
BURBANK 91505
PHONE: (818) 846-0579

ROCKS

ELLIS MERCANTILE COMPANY
169 N. LA BREA AVE
LOS ANGELES, 90036
PHONE: (213) 933-7334
FAX: (213) 930-1268

HAND PROP ROOM
5700 VENICE BLVD
LOS ANGELES, 90019
PHONE: (213) 931-1534

PAGEANTRY PRODUCTIONS
11122 WRIGHT RD
LYNWOOD, 90262
PHONE: (213) 632-5600
FAX: (213) 632-3304

STUDIO SPECIALTIES LTD
3013 GILROY ST
LOS ANGELES, 90039
PHONE: (213) 662-3031

WALTER ALLEN PLANT RENTALS
4996 MELROSE AVE
HOLLYWOOD, 90029
PHONE: (213) 469-3621

SCENERY CONSTRUCTION
AMERICAN SCENERY
18555 EDDY ST
NORTHRIDGE, 91324
PHONE: (818) 886-1585
CINNABAR
1040 NORTH LAS PALMAS
HOLLYWOOD, 90038
PHONE: (213) 462-0515
COMBINED ENTERTAINMENT SERVICES INC
3019 ANDRITA ST
LOS ANGELES, 90065
PHONE: (213) 258-5055
CONTINENTAL SCENERY
1022 N. LA BREA AVE
HOLLYWOOD, 90038
PHONE: (213) 461-4139
DESIGN SETTER CORPORATION
139 NORTH VICTORY BLVD
BURBANK,
PHONE: (818) 245-3008
LEXINGTON SCENERY
13005 SATICOY
NORTH HOLLYWOOD, 91605
PHONE: (818) 765-0443
PYRAMID SCENIC
5555 MELROSE
HOLLYWOOD, 90038
PHONE: (213) 468-5178
R.L. GROSH AND SONS SCENIC
4114 W. SUNSET BLVD
LOS ANGELES, 90029
PHONE: (213) 662-1134
SCENIC EXPRESS
3025 FLETCHER DR
LOS ANGELES, 90035
PHONE: (213) 254-4351
SCENIC SERVICES
695 SOUTH GLENWOOD PL
BURBANK, 91506
PHONE: (818) 842-7410
SUN-SETS INC
7600 WHEATLAND AVE
SUN VALLEY, 91352
PHONE: (818) 767-9099
VISTA SCENIC
3019 ANDRITA ST
LOS ANGELES, 90065
PHONE: (213) 258-8864/3494

SIREN LIGHTS
ELLIS MERCANTILE COMPANY
169 N. LA BREA AVE
LOS ANGLES 90036
PHONE: (213) 933-7334
FAX: (213) 930-1268
HAND PROP ROOM
5700 VENICE BLVD
LOS ANGLES 90019
PHONE: (213) 931-1534
THEATRE VISION, INC
5426 FAIR AVE
NORTH HOLLYWOOD 91601
PHONE: (818) 769-0928

SLOT MACHINES
HAND PROP ROOM
5700 VENICE BLVD
LOS ANGELES 90019
PHONE: (213) 931-1534
HISTORY FOR HIRE
7103 FAIR AVE
NORTH HOLLYWOOD 91605
PHONE: (818) 765-7767
FAX: (818) 765-7871
LENNIE MARVIN ENTERPRISES
1105 N. HOLLYWOOD WAY
BURBANK 91505
PHONE: (818) 841-5882

LUCKY ENTERTAINMENT
10271 ALMAYO AVE. #101
LOS ANGELES 90064
PHONE: (213) 277-9666
T.R. KING
400 WEST PICO BLVD
LOS ANGELES 90015
PHONE: (213) 749-7636

SNOW PLASTIC
AA SPECIAL EFFECTS
7017 HAYVENHURST BLVD
VAN NUYS 91406
PHONE: (818) 782-6558
OLESEN
1535 IVAR AVE
HOLLYWOOD 90028
PHONE: (213) 461-4631
FAX: (213) 464-0444
ROGER GEORGE RENTALS, INC
145 1/2 BESSEMER ST
VAN NUYS 91411
PHONE: (818) 994-3049
FAX: (818) 994-9432
SPECIAL EFFECTS UNLIMITED
752 N. CAHUENGA BLVD
HOLLYWOOD 90038
PHONE: (213) 466-3361
STUDIO SPECIALTIES LTD
3013 GILROY ST
LOS ANGELES 90039
PHONE: (213) 662-3031
TRI-ESS SCIENCES, INC
1020 W. CHESTNUT ST
BURBANK 91506
PHONE: (213) 245-7685/ (800) 274-6910
FAX: (818) 848-3521
WALTER ALLEN PLANT RENTALS
5500 MELROSE AVE
HOLLYWOOD 90029
PHONE: (213) 469-3621

SPACE SHIP COMPUTER PANEL
OMEGA/CINEMA PROPS
5857 SANTA MONICA BLVD
HOLLYWOOD 90038
PHONE: (213) 466-8201
FAX: (213) 461-3643

SPECIAL EFFECTS
AA SPECIAL EFFECTS
7017 HAYVENHURST BLVD
VAN NUYS, 91406
PHONE: (818) 782-6558
AUTOMATED STUDIO LIGHT-ING
545 RODIER DR
GLENDALE, 91201
PHONE: (818) 500-1646
BISCHOFFS
449 S. SAN FERNANDO BLVD
BURBANK, 91501
PHONE: (818) 843-7561
CINEMA SECRETS INC
4400 RIVERSIDE DR
BURBANK, 91505
PHONE: (818) 846-0579
CUSTOM NEON
2210 SOUTH LA BREA AVE
LOS ANGELES, 90016
PHONE: (213) 937-6366
FIRST TAKE PRODUCTIONS INC
8070 SAN FERNANDO RD
SUN VALLEY, 91352
PHONE: (818) 767-8261
NIGHTS OF NEON
7442 VARNA AVE
NORTH HOLLYWOOD, 91605
PHONE: (818) 982-3592
PLAYERS SPECIAL EFFECTS
1128 PASO ROBLES AVE
GRANADA HILLS, 91344
PHONE: (818) 360-4558

SCENIC EXPRESS
3025 FLETCHER DR
LOS ANGELES, 90035
PHONE: (213) 254-4351
FAX: (213) 254-4411
SPECIAL EFFECTS UNLIMITED
752 N. CAHUENGA BLVD
HOLLYWOOD, 90038
PHONE: (818) 466-3361
THE CINESPHERE CORPORATION
2443 LILLYVALE
LOS ANGELES, 90032
PHONE: (213) 221-6043
THE GREAT AMERICAN MARKET
826 N. COLE AVE
HOLLYWOOD, 90038
PHONE: (213) 461-0200
THEATRE VISION, INC
5426 FAIR AVE
NORTH HOLLYWOOD, 91601
PHONE: (818) 769-0928
UNITED AUTOMATION INC
1525 NORTH HOBART BLVD
LOS ANGELES, 90027
PHONE: (213) 469-5211
VISTA ELECTRONICS, INC
3019 ANDRITA ST
LOS ANGELES, 90065
PHONE: (213) 258-8864
WIZARDS SPECIAL EFFECTS INC.
18333 LAHEY ST
NORTHRIDGE, 91326
PHONE: (818) 368-5084
FAX: (818) 368-5084

SPECIAL EFFECTS LIGHTING SERVICES
PRODUCTION LIGHTING SYSTEMS INC.
ONE WEST ALAMEDA
BURBANK 91502
PHONE: (818) 845-1200

SPECIAL PROPS ANIMATION
BOB BAKER MARIONETTE THEATER
1345 WEST 1ST ST
LOS ANGELES 90026
PHONE: (213) 250-9955
NIGHTS OF NEON
7442 VARNA AVE
NORTH HOLLYWOOD '91605
PHONE: (818) 982-3592
FAX: (818) 503-1090
PAGEANTRY PRODUCTIONS
11122 WRIGHT RD
LYNWOOD 90262
PHONE: (213) 632-5600
FAX: (213) 632-3304
RENE & HIS ARTISTS PRODUCTIONS
707-A MAIN ST
BURBANK 91506
PHONE: (818) 849-4115
FAX: (818) 848-4861
ROSCHU
6514 SANTA MONICA BLVD
HOLLYWOOD 90038
PHONE: (213) 469-2749
SCENIC EXPRESS
3025 FLETCHER DR.
LOS ANGELES 90035
PHONE: (213) 254-4351
FAX: (213) 254-4411
SPECIAL EFFECTS SYSTEMS
24802 APPLE ST
NEWHALL 91321
PHONE: (805) 254-1985

STAGE LIGHTING EQUIPMENT
ACEY DECY EQUIPMENT CO
5420 VINELAND AVE
NORTH HOLLYWOOD 91601
PHONE: (818) 766-9445
FAX: (818) 766-4758
AMERICAN LIGHTING
8327 MELROSE AVE
LOS ANGELES 90069
PHONE: (213) 653-8555
AUTOMATED STUDIO LIGHTING
545 RODIER DR
GLENDALE 91201
PHONE: (818) 500-1646
FOUR STAR LIGHTING
3935 NORTH MISSION RD
LOS ANGELES 90031
PHONE: (213) 221-5114
GREAT AMERICAN MARKET
826 N. COLE ST
HOLLYWOOD 90038
PHONE: (213) 461-0200
HISTORY FOR HIRE
7103 FAIR AVE
NORTH HOLLYWOOD 91605
PHONE: (818) 765-7767
FAX: (818) 765-7871
KEYLITE P.S.I.
11200 SHERMAN WAY
SUN VALLEY 91352
PHONE: (818) 503-0900
FAX: (818) 503-9736
MOLE-RICHARDSON
937 NORTH SYCAMORE AVE
NORTH HOLLYWOOD 90038
PHONE: (213) 851-0111
NIGHTS OF NEON
7442 VARNA AVE
NORTH HOLLYWOOD 91605
PHONE: (818) 982-3592
OLESEN
1535 IVAR AVE
HOLLYWOOD 90028
PHONE: (213) 461-4631
FAX: (213) 464-0444
PRODUCTION LIGHTING SYSTEMS, INC
ONE WEST ALAMEDA AVE
BURBANK 91502
PHONE: (818) 845-1200
ROSCO
1135 N. HIGHLAND AVE
LOS ANGELES 90038
PHONE: (213) 462-2233
SCENIC EXPRESS
3025 FLETCHER DR
LOS ANGELES 90035
PHONE: (213) 254-4351
FAX: (213) 254-4411
THEATRE VISION, INC
5426 FAIR AVE
NORTH HOLLYWOOD 91601
PHONE: (818) 769-0928

STAGE LIGHTING SUPPLIER/ MANUFACTURER
BARDWELL & MCALISTER, INC
2601 EMPIRE AVE
BURBANK 91504
PHONE: (818) 843-6821
BERKEY-COLORTRAN
1015 CHESTNUT ST
BURBANK 91506
PHONE: (818) 843-1200
FLOURESCENT TUBE SERVICE
13107 SOUTH BROADWAY
LOS ANGELES 90061
PHONE: (213) 321-6900
STRAND LIGHTING
18111 SOUTH SANTA FE AVE.
RANCHO DOMINGUES 90221
PHONE: (213) 637-7500

STROBE LIGHTS

ACEY DECY EQUIPMENT CO
5420 VINELAND AVE
NORTH HOLLYWOOD 91601
PHONE: (818) 766-9445
FAX: (818) 766-4758

AUTOMATED STUDIO LIGHTING
545 RODIER DR
GLENDALE 91201
PHONE: (818) 500-1646

PRODUCTION LIGHTING SYSTEMS, INC
ONE WEST ALAMEDA AVE
BURBANK 91502
PHONE: (818) 845-1200

THEATRE VISION, INC
5426 FAIR AVE
NORTH HOLLYWOOD 91601

STYROFOAM

MOSKATELS
733 SOUTH SAN JULIAN
LOS ANGELES, 90014
PHONE: (213) 689-4830

SNOW FOAM
9917 GIDLY ST
EL MONTE, 91731
PHONE: (213) 283-0526

WALTER ALLEN PLANT RENTALS
5500 MELROSE AVE
HOLLYWOOD, 90029
PHONE: (213) 469-3621

SUIT OF ARMOR

ELLIS MERCANTILE COMPANY
169 N. LA BREA AVE
LOS ANGELES, 90036
PHONE: (213) 933-7334
FAX: (213) 930-1268

HAND PROP ROOM
5700 VENICE BLVD
LOS ANGELES, 90019
PHONE: (213) 931-1534

OMEGA/CINEMA PROPS
58857 SANTA MONICA BLVD
HOLLYWOOD, 90038
PHONE: (213) 466-8201
FAX: (213) 461-3643

PROP SERVICES WEST
915 N. CITRUS AVE
HOLLYWOOD, 90038
PHONE: (213) 461-3371

SWEEPING COMPOUND

IDEAL SAWDUST COMPANY
1516 GRANDE VISTA AVE.
EAST LOS ANGELES, 90023
PHONE: (213) 269-2195

SUPERIOR CHEMICAL PRODUCTS
2035 EAST VERNON BLVD
LOS ANGELES, 90058
PHONE: (213) 232-3521

TAXIDERMY

BISCHOFFS
449 SOUTH SAN FERNANDO BLVD
BURBANK, 91501
PHONE: (818) 843-7561

C.P.TWO
5755 SANTA MONICA BLVD
LOS ANGELES, 90038
PHONE: (213) 466-8201
FAX: (213) 461-3643

ELLIS MERCANTILE COMPANY
169 N. LA BREA AVE
LOS ANGELES, 90036
PHONE: (213) 933-7334
FAX: (213) 930-1268

HAND PROP ROOM
5700 VENICE BLVD
LOS ANGELES, 90019
PHONE: (213) 931-1534

HISTORY FOR HIRE
7103 FAIR AVE
NORTH HOLLYWOOD, 91605
PHONE: (818) 765-7767
FAX: (818) 765-7871

WOODLANDS SPORTSMAN TAXIDERMY
26 E. HUNTINGTON DR
ARCADIA, 91006
PHONE: (818) 574-9862

TELEPHONE BOOTHS

HISTORY FOR HIRE
7103 FAIR AVE
NORTH HOLLYWOOD 91605
PHONE: (818) 765-7767
FAX: (818) 765-7871

TELEPHONES

AT&T
4444 RIVERSIDE BLVD.
SUITE 110
BURBANK, 91505
PHONE: (818) 841-5801

CALIFORNIA ATTRACTIONS LIMITED
7023 CANOGA AVE. SUITE E
CANOGA PARK, 91303
PHONE: (818) 999-6255

HISTORY FOR HIRE
7103 FAIR AVE
NORTH HOLLYWOOD, 91605
PHONE: (818) 765-7767
FAX: (818) 765-7871

LENNIE MARVIN ENTERPRISES
1105 N. HOLLYWOOD WAY
BURBANK, 91505
PHONE: (213) 841-5881

OLDE TELEPHONE CO
PHONE: (818) 795-6145

PROP SERVICES WEST
915 N. CITRUS AVE
HOLLYWOOD, 90038
PHONE: (213) 461-3371

TELEPROMPTER

INTELLIPROMPT
746 N. CAHUENGA BLVD
HOLLYWOOD, 90038
PHONE: (213) 461-3113

Q-TV INC
7350 BEVERLY BLVD
LOS ANGELES, 90036
PHONE: (213) 936-6195

THEATRICAL PLATFORMS

THEATRE VISION, INC
5426 FAIR AVE
NORTH HOLLYWOOD 91601
PHONE: (818) 769-0928

THEATRICAL SPECIALTIES
4211 & 4209 VERDANT ST
LOS ANGELES 90039
PHONE: (213) 244-2232

TOOLS

ELLIS MERCANTILE COMPANY
169 N. LA BREA AVE
LOS ANGELES 90036
PHONE: (213) 933-7334
FAX: (213) 930-1268

FIRST STREET FURNITURE
1123 N. BRONSON AVE
HOLLYWOOD 90038
PHONE: (213) 462-6305

HISTORY FOR HIRE
7103 FAIR AVE
NORTH HOLLYWOOD 91605
PHONE: (818) 765-7767
FAX: (818) 765-7871

HOLLYWOOD CENTRAL PROPS
525 W. ELK AVE
GLENDALE 91204
PHONE: (818) 240-4504

TRUCKING

BALLIER TRANSFER INC
3021 5TH AVE
LOS ANGELES, 90018
PHONE: (213) 737-8192

COMBINED ENTERTAINMENT SERVICES INC
3019 ANDRITA ST
LOS ANGELES, 90065
PHONE: (213) 258-5055
FAX: (213) 258-6950

GILBERT PRODUCTION SERVICE
11156 SHERMAN WAY
SUN VALLEY, 91352
PHONE: (818) 764-5548

PINK CAT
15044 KESWICK AVE
VAN NUYS, 91405
PHONE: (818) 909-7465

SCENIC EXPRESS
3025 FLETCHER DR
LOS ANGELES, 90035
PHONE: (213) 254-4351
FAX: (213) 254-4411

WESTERN STUDIO SERVICE INC
ONE WEST ALAMEDA AVE
BURBANK, 91502
PHONE: (818) 842-9272

TURNTABLES

HISTORY FOR HIRE
7103 FAIR AVE
NORTH HOLLYWOOD, 91605
PHONE: (818) 765-7767
FAX: (818) 765-7871

ROGER GEORGE RENTALS, INC
145 1/2 BESSEMER ST
VAN NUYS, 91411
PHONE: (818) 994-3049/762-6478
FAX: (818) 994-9432

SPECIAL EFFECTS UNLIMITED
752 N. CAHUENGA BLVD
HOLLYWOOD, 90038
PHONE: (213) 466-3361

STAGE MECHANIX/LA.INC.
1634 E. 23RD ST
LOS ANGELES,
PHONE: (213) 232-8177

TRI-ESS SCIENCES, INC
1020 W. CHESTNUT ST
BURBANK, 91506
PHONE: (213) 245-7685/
(800)274-6910
FAX: (818) 848-3521

TURNTABLE RENTAL & SALE
752 NORTH CAHUENGA BLVD
HOLLYWOOD, 90038
PHONE: (213) 466-72178

VACU-FORMING

AMERICAN SCENERY
18555 EDDY ST
NORTHRIDGE, 91324
PHONE: (818) 886-1585

CINNABAR
1040 NORTH LAS PALMAS
HOLLYWOOD, 90038
PHONE: (213) 462-3737

SCENIC EXPRESS
3025 FLETCHER DR
LOS ANGELES, 90035
PHONE: (213) 245-4351
FAX: (213) 254-4411

SPECIAL EFFECTS SYSTEMS
23501 LLOYD HOUGHTON PLACE
NEWHALL, 91321
PHONE: (805) 254-1985

VIDEO GRAPHICS

VISTA ELECTRONICS, INC
3019 ANDRITA ST
LOS ANGELES, 90065
PHONE: (213) 258-8864

WELDING & METAL WORK

FIRST TAKE PRODUCTIONS, INC
8070 SAN FERNANDO RD
SUN VALLEY 91352
PHONE: (818) 767-8261

VISTA SCENERY, INC
3019 ANDRITA ST
LOS ANGELES 90065
PHONE: (213) 258-8864

WHEELCHAIRS

HISTORY FOR HIRE
7103 FAIR AVE
NORTH HOLLYWOOD, 91605
PHONE: (818) 765-7767
FAX: (818) 765-7871

WIGS

CINEMA SECRETS INC
44000 RIVERSIDE DR.
BURBANK, 91505
OR GIBEL (818) 846-0579

STAGES - PORTABLE

BUSS CARSON BLEACHERS
7905 LLOYD AVE
NORTH HOLLYWOOD, 91605
PHONE: (818) 780-1735/782-5758

PRODUCTION SERVICES HOLLYWOOD
8033 SUNSET BLVD., SUITE 5035
LOS ANGELES, 90046
PHONE: (213) 858-7610

S.I.R. RISER & STAGING
6048 SUNSET BLVD
HOLLYWOOD, 90028
PHONE: (213) 466-1314/3417
FAX: (213) 650-6866

SSI - STAGE SYSTEMS INTERNATIONAL
325 E. SOUTH ST.
LONG BEACH, 90805
PHONE: (213) 428-7428

THEATRE VISION, INC
5426 FAIR AVE
NORTH HOLLYWOOD, 91601
PHONE: (818) 769-0928/ (800) 431-2884
FAX: (818) 769-0627

THEATRICAL SPECIALTIES
4209 VERDANT ST
LOS ANGELES, 90039
PHONE (818) 244-4525/2232
FAX: (818) 244-6065

STUDIOS, MAJOR

COLUMBIA PICTURES
10202 W. WASHINGTON BLVD
CULVER CITY, 90232
PHONE: (213) 280-8000

MCA/UNIVERSAL
100 UNIVERSAL CITY PLAZA
UNIVERSAL CITY, 91608
PHONE: (818) 777-1000

PARAMOUNT PICTURES CORPORATION
5555 MELROSE AVE
HOLLYWOOD, 90038
PHONE: (213) 956-5000

TWENTIETH CENTURY FOX FILM CORPORATION
10201 W. PICO BLVD
LOS ANGELES, 90035
PHONE: (213) 277-2211

WALT DISNEY PICTURES & TELEVISION
500 S. BUENA VISTA ST
BURBANK, 91521
PHONE: (818) 560-5151

WARNER BROS. INC
4000 WARNER BVLD
BURBANK, 91522
PHONE: (818) 954-6000

STUDIOS/SOUND STAGES

A & M CHAPLIN STAGE
1416 NORTH LA BREA AVE
HOLLYWOOD, 90028
PHONE: (23) 856-2682

ABC TELEVISION CENTER
4151 PROSPECT AVE
LOS ANGELES, 90027
PHONE: (213) 557-7777

JAY AHREND PHOTOGRAPHY, INC
1046 N ORANGE DR
HOLLYWOOD, 90038
PHONE: (213) 462-5256

ALPHA STUDIOS VIDEO, INC
4720 WEST MAGNOLIA BLVD
BURBANK, 91505
PHONE: (818) 506-7443

AMETHYST STUDIOS
7000 SANTA MONICA BLVD
HOLLYWOOD, 90038
PHONE: (213) 859-5

APPLE CORPS, THE
6033 WEST CENTURY BLVD
LOS ANGELES, 90045
PHONE: (213) 337-7800

APRICOT ENTERTAINMENT INC
940 NORTH ORANGE DR
LOS ANGELES, 90038
PHONE: (213) 469-4000

JAMES ATIEE STUDIO
922 N FORMOSA AVE
LOS ANGELES, 90046
PHONE: (213) 850-6112

BADLANDS
11174 FLEETWOOD ST
SUN VALLEY, 91352
PHONE: (818) 504-2404

BEN KITAY PRODUCTIONS, STAGES 10 & 15
1015 NORTH CAHUENGA BLVD
HOLLYWOOD, 90038
PHONE: (213) 466-9015

BOWEN STAGE
760 NORTH LAKE ST
BURBANK, 91502
PHONE: (818) 841-0295
FAX: (818) 841-1048

BUENA VISTA STUDIOS
500 SOUTH BUENA VISTA ST
BURBANK, 91506
PHONE: (818) 8560-450
FAX: (818) 841-8328

BURBANK PRODUCTION PLAZA
801 SOUTH MAIN ST
BURBANK, 91506
PHONE: (818) 846-7677
FAX: (818) 841-1572

BURBANK STUDIOS
4000 WARNER BLVD
BURBANK, 91522
PHONE: (213) 954-6000

CBS/MTM
4024 REDFORD AVE
STUDIO CITY, 91604
PHONE: (818) 760-5000
FAX: (818) 760-5400

C.F.I. STAGE
959 N SEWARD ST
HOLLYWOOD, 90038
PHONE: (213) 462-3161
TELEX: 06-74257

CAPITOL RECORDING STUDIOS
1750 NORTH VINE ST
HOLLYWOOD, 90028
PHONE: (213) 462-6252

CARTHAY STUDIOS
5907 WEST PICO BLVD
LOS ANGELES, 90035
PHONE: (213) 938-2101

CHANDLER STUDIOS
11405 CHANDLER BLVD
NORTH HOLLYWOOD, 91601
PHONE: (818) 763-3650

CHAPLIN STAGE
146 NORTH LA BREA AVE
HOLLYWOOD, 90028
PHONE: (213) 469-2411

CHASE PRODUCTIONS
7080 HOWLLYWOOD BLVD
LOS ANGELES, 90028
PHONE: (213) 466-3946

CINE RENT WEST
991 TENNESSEE ST
SAN FRANCISCO, 94107
PHONE: (415) 864-4644
FAX: (415) 826-4522

CINE VIDEO MOTION PICTURE EQUIPMENT CORP
948 NORTH CAHUENGA BLVD
LOS ANGELES, 90038
PHONE: (213) 464-6200

CINEMA DIGITAL SOUND
1300 OPTICAL DR
AZUSA, 91702
PHONE: (818) 969-3344

CINETYP, INC
843 SEWARD ST
HOLLYWOOD, 90038
PHONE: (213) 463-8569

CITY STUDIOS
7700 BALBOA AVE
VAN NUYS, 91406
PHONE: (818) 909-7001

THE COMPLEX
2323 CORINTH ST
WEST LOS ANGELES 90064
PHONE: (23) 477-1938
FAX: (213) 473-2485

CONSOLIDATED FILM INDUSTRIES (CFI)
959 SEWARD ST
HOLLYWOOD, 90038
PHONE: (213) 960-7444
FAX: (213) 460-4885

COURT THEATRE CO
722 NORTH LA CIENEGA BLVD
LOS ANGELES, 90069
PHONE: (213) 652-4035

COX CABLE
5159 FEDERAL BLVD
SAN DIEGO, 92105
PHONE: (619) 263-9251

CRYSTAL SOUND
1014 NORTH VINE
HOLLYWOOD, 90038
PHONE: (213) 466-6452

CULVER STUDIOS
9336 WEST WASHINGTON BLVD
CULVER CITY, 90230
PHONE: (213) 202-3396
FAX: (213) 202-3272

DEL ORO RECORDS & FILMWORKS
PHONE: (213) 202-3396

DESIGN-ARTS STUDIOS
1128 NORTH LAS PALMAS
HOLLYWOOD, 90038
PHONE: (213) 851-1090

DEVONSHIRE SOUND STUDIOS
10729 MAGNOLIA BLVD
NORTH HOLLYWOOD,91502
PHONE: (818) 985-1945

EFX SYSTEMS
919 NORTH VICTORY BLVD
BURBANK, 91502
PHONE: (818) 843-4762

ERECTER SET STUDIOS
1150 SOUTH LA BREA AVE
LOS ANGELES, 90019
PHONE: (213) 938-4762
FAX: (213) 931-9565

EVERGREEN RECORDING STUDIOS
4403 WEST MAGNOLIA BLVD
BURBANK, 91505
PHONE: (818) 841-6800

FILMTRIX, INC
11054 CHANDLER BLVD
NORTH HOLLYWOOD, 91601
PHONE: (818) 980-3700
FAX: (818) 980-3703

FOUR SQUARE PRODUCTIONS
PHONE: (213) 474-5566

FOX TAPE
5746 SUNSET BLVD
HOLLYWOOD, 90028
PHONE: (213) 856-1253
FAX: (213) 463-6239

GABOO MUSIC
6709 LA TIJERA BLVD.,
SUITE 418
LOS ANGELES, 90045
PHONE: (213) 337-0876

GLENDALE STUDIOS
1239 SOUTH GLENDALE AVE
GLENDALE, 91205
PHONE: (818) 502-5300
FAX: (818) 502-5311

GMT STUDIOS
5751 BUCKINGHAM PARKWAY
CULVER CITY, 90230
PHONE: (213) 649-3733
FAX: (23) 216-0056

GROUP IV RECORDING
1541 NORTH WILCOX AVE
LOS ANGELES, 90028
PHONE: (213) 466-6444

HARBOR STAR STAGE
399 NAVY WAY, BLDG 575
TERMINAL ISLAND, 90731
PHONE: (213) 833-7712

HARRIER, KEITH PRODUCTION SERVICE
7070 HAYVENHURST AVE
VAN NUYS, 91406
PHONE: (213) 930-2720

HAYVENHURST STUDIOS
7017 HAYVENHURST AVE
VAN NUYS, 91406
PHONE: (818) 782-6560
FAX: (818) 782-0635

HELMS PRODUCTIONS
8741 WEST WASHINGTON BLVD
CULVER CITY, 90230
PHONE: (213) 838-1344

HENDERSON LIGHTING AND GRIP
PHONE: (619) 270-7660

HLC
6528 SUNSET BLVD
HOLLYWOOD, 90028
PHONE: (213) 464-6333

HOLLYWOOD CENTER STUDIOS
1040 NORTH LAS PALMAS AVE
HOLLYWOOD, 90038
PHONE: (213) 469-5000
FAX: (213) 871-8105

HOLLYWOOD NATIONAL STUDIOS
6605 ELEANOR AVE
HOLLYWOOD, 90038
PHONE: (213) 467-6272

THE HOLLYWOOD STAGE
6650 SANTA MONICA BLVD
HOLLYWOOD, 90038
PHONE: (213) 466-4393

INTER VIDEO/TRITRONICS, INC
733 N. VICTORY BLVD
BURBANK, 91502
PHONE: (818) 569-4000/
(800) 843-3626

INTERNATIONAL VIDEO PRODUCTIONS
12401 W. OLYMPIC BLVD
LOS ANGELES, 90064
PHONE: (213) 478-1818
FAX: (213) 479-8118

J.M. TELEVISION PRODUCTIONS
PHONE: (619) 434-3363

KCET
4401 SUNSET BLVD
LOS ANGELES, 90027
PHONE: (23) 666-6500/
667-9258

KFMB - CHANNEL 8 (CBS)
7677 ENGINEER RD
SAN DIEGO, 9211
PHONE: (619) 571-8888

KGTV - CHANNEL 10 (ABC)
4600 AIR WAY
SAN DIEGO, 92102
PHONE: (619) 237-1010

KNSD - CHANNEL 39/CABLE 7 (NBC)
8330 ENGINEER RD
SAN DIEGO, 92111
PHONE: (619) 279-3939

KPBS - CHANNEL 15 (PBS)
SAN DIEGO STATE UNIVERSITY
SAN DIEGO, 92182
PHONE: (619) 594-1515

KTLA VIDEOTAPE DIVISION
5800 SUNSET BLVD
LOS ANGELES, 90028
PHONE: (213) 460-5500

KTTY - CHANNEL 69
1696 FRONTAGE RD
CHULA VISTA, 92011
PHONE: (619) 575-6969

KUSI - CHANNEL 51
7377 CONVOY COURT
SAN DIEGO, 92111
PHONE: (619) 571-5151

BEN KITAY PRODUCTIONS - STAGES 5,10 & 15
1015 N CAHUENGA BLVD
HOLLYWOOD, 90038
PHONE: (213) 466-9015
FAX: (213) 466-4421

KITCHEN CONSULTANTS
261 E. IMPERIAL HWY.,
SUITE 530
FULLERTON, 92635
PHONE: (714) 871-9944

L.A. STAGE
8451 MELROSE AVE.
LOS ANGELES, 90069
PHONE: (213) 651-5184

L.A. STUDIOS, THE
3453 CAHUENGA BLVD WEST
HOLLYWOOD, 9068
PHONE: (213) 851-6351

LA BREA STUDIOS
1028 N LA BREA AVE
HOLLYWOOD, 90038
PHONE: (213) 462-7210/6

LANDMARK STUDIOS
1455 N. GORDON ST
HOLLYWOOD, 90028
PHONE: (213) 962-4702

LIGHTING CORPORATION
PHONE: (619) 565-6494

LINDSEY STUDIOS
26030 AVENUE HALL
VALENCIA, 91335
PHONE: (805) 257-9292

LIONS GATE STUDIOS
1861 SOUTH BUNDY DR
LOS ANGELES, 90025
PHONE: (213) 820-7751

LORIMER STUDIOS
10202 WEST WASHINGTON BLD
CULVER CITY, 90232
PHONE: (213) 280-5500

MACK SENNETT STAGE/ TRIANGLE
1215 BATES AVE
LOS ANGELES, 90029
PHONE: (213) 660-8466

MAGA LINK INC.
1968 W ADAMS BLVD
LOS ANGELES, 90018
PHONE: (213) 732-06005

MANSFIELD STUDIOS
1041 N MANSFIELD AVE
HOLLYWOOD, 90038
PHONE: (23) 461-3393

MASON STUDIO SERVICES
430 COLOMA ST
SAUSALITO, 94965
PHONE: (415) 332-4230

MEGA PRODUCTIONS
1714 N WILTON PL
HOLLYWOOD, 90028
PHONE: (213) 462-6342
FAX: (213)462-7572

METROPOLIS STUDIOS
AT HOLLYWOOD & VINE
HOLLYWOOD, 90028
PHONE: (213) 461-1771

MINCEY PRODUCTIONS, INC
8050 RONSON RD
SAN DIEGO, 92111
PHONE: (619) 292-0337

MORO LANDIS DUPUY STUDIOS
10960 VENTURA BLVD
STUDIO CITY, 91604
PHONE: (818) 761-9510

MOVIE TECH STUDIOS
832 N SEWARD ST
HOLLYWOOD, 90038
PHONE: (213) 467-8491
FAX: (213) 467-8471

MTC PRODUCTION CENTER
4150 GLENCOE AVE
MARINA DEL REY, 90292
PHONE: (213) 823-8000
FAX: (213) 823-6991

MULTIMEDIA STUDIOS
10401 W. JEFFERSON BLVD
CULVER CITY, 90232
PHONE: (213) 202-0135
FAX: (213) 202-8219

MUSIC CONSULTANT GROUP
4209 WEST BURBANK BLVD
BURBANK, 91505
PHONE: (818) 841-9100

MUSIC FOR FILMS & TV
241 WEST ALAMEDA AVE.,
SUITE 3
BURBANK, 91502
PHONE: (818) 846-6042

NBC STUDIOS
3000 WEST ALAMEDA AVE
BURBANK, 91523
PHONE: (818) 840-3243

NEWHALL RANCH
23823 VALENCIA BLVD
VALENCIA, 91355
PHONE: (805) 255-4004/ (818)
362-1515

NORWOOD-STAGE
9023 WASHINGTON BLVD
CULVER CITY, 90266
PHONE: (213) 204-3323

OCCIDENTAL STUDIOS, INC
201 N. OCCIDENTAL BLVD
LOS ANGELES, 90026
PHONE: (213) 384-3331
FAX: (213) 464-3861

PAN PACIFIC WAREHOUSE
120 S. HEWITT ST
LOS ANGELES, 90012
PHONE: (213) 626-6000
FAX: (213) 617-1372

PARADISE STUDIOS
604 MOULTON AVE
LOS ANGELES, 90031
PHONE: (213) 837-2572/
224-8191
FAX: (213) 224-8153

PARAMOUNT PICTURES CORPORATION
5555 MELROSE AVE
LOS ANGELES, 90036
PHONE: (213) 956-5284
FAX: (213) 468-5555

PASADENA PRODUCTION STUDIOS
39 E. WALNUT ST
PASADENA, 91103
PHONE: (818) 584-4090
FAX: (818) 584-4099

PKE STUDIOS
8621 HAYDEN PLACE
CULVER CITY, 90232
PHONE: (213) 838-7000
FAX: (213) 838-8430

PICO BRONSON STUDIO
1272 SOUTH BRONSON AVE
LOS ANGELES, 90019
PHONE: PHONE: (213) 732-0605

POST LOGIC
1800 NORTH VINE, SUITE 1
LOS ANGELES,90028
PHONE: (213) 461-7887

PRAXIS STAGE WORKS
6920 TUJUNGA AVE
NORTH HOLLYWOOD, 91605
PHONE: (818) 508-0402
FAX: (818) 508-0988

PRODUCERS RECORDING STUDIOS
6035 HOLLYWOOD BLVD
LOS ANGELES, 90028
PHONE: (213) 466-7766

PRODUCTION GROUP, THE
1330 NORTH VINE
LOS ANGELES, 90028
PHONE: (213) 469-8111
FAX: (213) 962-2182

RALEIGH STUDIOS
5300 MELROSE AVE
HOLLYWOOD, 90038
PHONE: (213) 466-3111
FAX: (213) 871-4428

REAL TO REEL LOCATIONS
6922 HOLLYWOOD BLVD,
SUITE 612
HOLLYWOOD, 90028
PHONE: (213) 461-0038

REN-MAR STUDIOS
846 N CAHUENGA BLVD
HOLLYWOOD, 90038
PHONE: (213) 455-8173

DEBBIE REYNOLDS STUDIOS
6514 LANKERSHIM BLVD
NORTH HOLLYWOOD, 91606
PHONE: (818) 985-3193

ROSEBUD STUDIOS
7336 HINDS AVE
NORTH HOLLYWOOD, 91605
PHONE: (818) 503-8808
FAX: (818) 982-8565

S & A STUDIOS INC
201 NORTH OCCIDENTAL BLVD
LOS ANGELES, 90026
PHONE: (213) 384-3331

S.I.R. STAGES
6048 SUNSET BLVD
HOLLYWOOD, 90028
PHONE: (213) 466-3417/ 467-
0034
FAX: (213) 650-6366

S.I.R. STAGES/REHEARSAL
6235 SANTA MONICA BLVD
LOS ANGELES, 90038
PHONE: (213) 462-3186/
466-3417

SAMY'S CAMERA
7122 BEVERLY BLVD
LOS ANGELES, 90036
PHONE: (213) 938-2420

SAMY'S CAMERA - CULVER CITY
8659 HAYDEN PL
CULVER CITY, 90230
PHONE: (213) 841-0240
FAX: (213) 841-0245

SAMY'S CAMERA - DOWNTOWN
610 MOULTON AVE
LOS ANGELES, 90031
PHONE: (213) 222-9530
FAX: (213) 222-0128

SAN FRANCISCO STUDIOS INC
375 SEVENTH ST
SAN FRANCISCO, 94103
PHONE: (415) 621-6900

SANTA CLARITA STUDIOS
25135 ANZA DR
SANTA CLARITA, 91355
PHONE: (805) 294-2000

SANTA FE COMMUNICATIONS
2525 N. NAOMI ST
BURBANK, 91504
PHONE: (818) 848-5800
FAX: (818) 848-6454

SANTA FE STAGE
5121 SANTA FE ST
SAN DIEGO, 92109
PHONE: (619) 270-7660
FAX: (619) 270-8722

SCHULMAN VIDEO CENTER
861 NORTH SEWARD ST
HOLLYWOOD, 90038
PHONE: (213) 465-8110
FAX: (213) 465-1874

SCREENMUSIC STUDIOS
11700 VENTURA BLVD
STUDIO CITY, 91406
PHONE: (818) 985-0900/ (213)
877-0300

SHRINE AUDITORIUM
649 W JEFFERSON BLVD
LOS ANGELES 90007
PHONE: (213) 784-5116

SIDEWINDER STUDIOS LTD
3334 LA CIENEGA PL
LOS ANGELES, 90016
PHONE: (213) 559-2721

SOUTH BAY STUDIOS
20434 SOUTH SANTA FE AVE
CARSON, 90810
PHONE: (213) 762-1360
FAX: (213) 639-2055

SPRINGBOARD STUDIOS
1229 MONTAGUE ST
ARLETA, 91331
PHONE: (818) 896-4321
FAX: (818) 890-2092

STAGE BY DESIGN
28343 AVENUE CROCKER
VALENCIA, 91355
PHONE: (805) 254-3164
FAX: (805) 257-1002

STETZ, WILLIAM, DESIGN
1108 SEWARD ST
LOS ANGELES, 90038
PHONE: (213) 461-4267

STUDIO 46
8646 W. PICO BLVD
LOS ANGELES, 90035
PHONE: (213) 855-7060
FAX: (213) 855-0645

STUDIO 905
905 N. COLE AVE
HOLLYWOOD, 90038
PHONE: (213) 463-1134
FAX: (213) 462-7116

STUDIO ONE
PHONE: (619) 660-1981

STUDIO INSTRUMENT RENTALS
6048 SUNSET BLVD
HOLLYWOOD, 90028
PHONE: (213) 466-1314

SUNSET-GOWER STUDIOS
1438 NORTH GOWER ST
LOS ANGELES, 90028
PHONE: (213) 467-1001
FAX: (213) 467-2717

SUPERSTAGE
1119 N HUDSON AVE
HOLLYWOOD, 90038
PHONE: (213) 464-0296

TELE-CUE
1306 ARIZONA AVE., SUITE F
SANTA MONICA, 90404
PHONE: (213) 395-1576

TRANS-AMERICAN VIDEO (TAV)
541 N VINE ST
HOLLYWOOD, 90028
PHONE: (213) 466-2141

VALENCIA STUDIOS
28343 AVENUE CROCKER
VALENCIA, 91355
PHONE: (805) 257-1202/
(300) 782-4248
FAX: (805) 257-1002

VALLEY PRODUCTION CENTER
6633 VAN NUYS BLVD
VAN NUYS, 91405
PHONE: (818) 988-6601
FAX: (818) 988-7120

VENICE INSERT STAGE
8705 W. WASHINGTON BLVD
CULVER CITY, 90230
PHONE: (213) 559-1711

VENICE STUDIOS
2017 PACIFIC AVE
VENICE, 90291
PHONE: (213) 822-2400

VIDEO IMAGE STAGES
4121 REDWOOD AVE
LOS ANGELES, 90066
PHONE: (213) 322-8872

VIDEO GENERAL
20432 SOUTH SANTA FE, UNIT K
LONG BEACH, 90810
PHONE: (213) 763-7781

VINE STREET VIDEO CENTER
8471 UNIVERSAL PLAZA
UNIVERSAL CITY, 91608
PHONE: (213) 462-1099

VPS STUDIOS
800 N SEWARD ST
HOLLYWOOD, 90038
PHONE: (213) 469-7244

WARNER BROS./STUDIO FACILITIES
4000 WARNER BLVD
BURBANK, 91522
PHONE: (818) 954-1131/2577
FAX: (818) 954-4467

THE WARNER WAREHOUSE
8461 WARNER DR
CULVER CITY, 90230
PHONE: (213) 274-5618/
652-9635

WATERHOUSE & CLAYTON
855 W VICTORIA ST., SUITE G
DOMINGUEZ HILLS, 90220
PHONE: (213) 635-0911

XETV - CHANNEL 6
8253 RONSON RD
SAN DIEGO, 92111
PHONE: (619) 279-6666

ZAENTZ, SAUL CO. FILM CENTER
10TH & PARKER
BERKELEY, 94710
PHONE: (415) 549-2500

UNIONS AND ASSOCIATIONS

ACADEMY OF MOTION PICTURE ARTS & SCIENCES(AMPAS)
8949 WILSHIRE BL
BEVERLY HILLS 90211
PHONE: (213) 278-8990

ACADEMY OF SCIENCE FICTION
FANTASY & HORROR FILMS
334 W 54TH ST
LOS ANGELES 90037
PHONE: (213) 752-5811

ACADEMY OF TEEN FILMMAKING
P.O. BOX 895
HERMOSA BEACH 90254
PHONE: (213) 719-1884

ACADEMY OF TELEVISION ARTS & SCIENCES(ATAS)
3500 W OLIVE #700
BURBANK 91505
PHONE: (818) 953-7575
FAX: (818) 953-1182

ACTORS EQUITY ASSOCIATION
6430 SUNSET BL #1002
LOS ANGELES 90028
PHONE: (213) 462-2334

AFL-CIO
2102 ALMADEN RD #305C
SAN JOSE 95125
PHONE: (408) 264-6007

AFTRA
6922 HOLLYWOOD BL 8TH FLOOR
HOLLYWOOD 90028
PHONE: (213) 461-8111

AFTRA
7827 CONVOY CT #400
SAN DIEGO 92111
PHONE: (619) 278-2918

AFFILIATED PROPERTY CRAFTSPERSON (IA LOCAL 44)
11500 BURBANK BLVD
NORTH HOLLYWOOD 91601
PHONE: (818) 769-2500
FAX: (818) 769-1739

ALLIANCE OF MP & TV PRODUCERS (AMPTP)
14144 VENTURA BL
SHERMAN OAKS, 91423
PHONE: (818) 995-3600
FAX: (818) 789-7431

AMERICAN CINEMA EDITORS
1041 N FORMOSA AVE
LOS ANGELES 90046
PHONE: (213) 850-2900

AMERICAN FEDERATION OF TV & RADIO ARTISTS (AFTRA)
6922 HOLLYWOOD BLVD., 8TH FL
HOLLYWOOD 90028
PHONE: (213) 461-8111
FAX: (213) 461-1377

THE AMERICAN FILM INSTITUTE (AFI)
2021 N. WESTERN AVE
HOLLYWOOD 90027
PHONE: (213) 856-7600

AMERICAN GUILD OF VARIETY ARTISTS
4741 LAUREL CYN BL #208
NORTH HOLLYWOOD 91607
PHONE: (818) 508-9984

AMERICAN SOCIETY OF CINEMATOGRAPHERS (ASC)
P.O. BOX 2230
HOLLYWOOD 90078
PHONE: (213) 876-5080
FAX: (213) 876-4973

ASIFA/HOLLYWOOD
P.O. BOX 787
BURBANK 91503
PHONE: (818) 988-6505

ASSISTANT DIRECTORS TRAINING PROG
14144 VENTURA BL
SHERMAN OAKS 91423
PHONE: (818) 995-3600

ASSOCIATION OF FILM CRAFTSMEN
LOCAL 531, NABET, AFL-CIO, CLC
1800 N ARGYLE AVE #501
LOS ANGELES 90028
PHONE: (213) 462-7484

ASSOCIATION OF INDEPENDENT COMMERCIAL PRODUCERS (AICP)
P.O. BOX 6188
BURBANK 91510
PHONE: (818) 763-2427
FAX: (818) 763-3258

BEHIND THE LENS (ASSOC. OF PROFESSIONAL CAMERAWOMEN)
P.O. BOX 868
SANTA MONICA 90406

COMMUNICATION WORKERS OF AMERICA
LOCAL 9502
3598 BEVERLY BL
LOS ANGELES 90004
PHONE: (213) 931-9000

COSTUME DESIGNERS' GUILD
LOCAL 892
13949 VENTURA BL #309
SHERMAN OAKS 91423
PHONE: (818) 905-1557
FAX: (818) 905-1560

DIRECTORS GUILD OF AMERICA, INC (DGA)
7920 SUNSET BLVD
LOS ANGELES 90046
PHONE: (213) 289-2000

DIRECTORS GUILD OF AMERICA
PRODUCERS PENSION, HEALTH & WELFARE PLAN
8436 W. THIRD ST., SUITE 900
LOS ANGELES 90048
PHONE: (213) 653-2991

FILM EXCHANGE EMPLOYEES
LOCAL 861, IATSE
13949 VENTURA BL #304
PHONE: (818) 906-7977

FILM INDUSTRY WORKSHOP INC
4047 RADFORD AVE
STUDIO CITY 91604
PHONE: (818) 769-4146

FILM/VIDEO TECHNICIANS
LOCAL 683, IATSE, AFL
P.O. BOX 7429
BURBANK 91510
PHONE: (213) 935-1123

GREEK AMERICANS IN THE ARTS & ENTERTAINMENT
1551 MIDVALE AVE
LOS ANGELES 90024
PHONE: (213) 477-7188

HOLLYWOOD ARTS COUNCIL
P.O. BOX 931056
HOLLYWOOD 90093
PHONE: (213) 462-2355

HOLLYWOOD CHAMBER OF COMMERCE
6255 SUNSET BL #911
HOLLYWOOD 90028
PHONE: (213) 469-8311

HOLLYWOOD PRESS & ENTERTAINMENT INDUSTRY CLUB
P.O. BOX 3381
HOLLYWOOD 90028
PHONE: (213) 466-1212

I.A.T.S.E. & M.P.M.O. (AFL-CIO)
14724 VENTURA BLVD., PENTHOUSE
SHERMAN OAKS 91403
PHONE: (818) 905-8999
FAX: (818) 905-6297

I.A.T.S.E. LOCAL 122
3760 FAIRMOUNT AVE
SAN DIEGO 92105
PHONE: (619) 283-6407
FAX: (619) 283-0309

IATSE LOCAL 134
P.O. BOX 28585
SAN JOSE 95159
PHONE: (408) 225-3293

IATSE LOCAL 297
3760 FAIRMOUNT DR
SAN DIEGO 92105
PHONE: (619) 283-3488

IBEW LOCAL 40, AFL-CIO
5643 VINELAND AVE
NORTH HOLLYWOOD 91601
PHONE: (818) 762-4239

INTERNATIONAL PHOTOGRAPHERS GUILD
LOCAL 659, IATSE, MPMO
7715 SUNSET BL #300
LOS ANGELES 90046
PHONE: (213) 876-0160

INTERNATIONAL SOUND TECHNICIANS (IA LOCAL 695)
11331 VENTURA BLVD., SUITE 201
STUDIO CITY 91604
PHONE: (818) 985-9204/ (213) 877-1052

INTERNATIONAL STUNT ASSOC
3518 CAHUENGA BL W #300
HOLLYWOOD 90068
PHONE: (213) 874-3174

LOS ANGELES ADVERTISING WOMEN (LAW)
3900 W. ALAMEDA AVE., SUITE 700
BURBANK 91505
PHONE: (818) 972-1771
FAX: (818) 972-9021

MAKE-UP ARTISTS & HAIR STYLISTS (IA LOCAL 706)
11519 CHANDLER BLVD
NORTH HOLLYWOOD 91601
PHONE: (818) 984-1700/ (213) 877-2776

MASQUERS
11110 VICTORY BL
NORTH HOLLYWOOD 91608
PHONE: (213) 856-4554

MOTION PICTURE & TELEVISION FUND
23388 MULHOLLAND DR
WOODLAND HILLS 91364
PHONE: (818) 347-1591

MOTION PICTURE ASSOCIATION OF AMERICA (MPAA)
14144 VENTURA BLVD., SUITE 210
SHERMAN OAKS 91423
PHONE: (818) 995-3600

MOTION PICTURE COSTUMERS
LOCAL 705, IATSE, MPMO, AFL
1427 N LA BREA AVE
LOS ANGELES 90028
PHONE: (213) 851-0220

MOTION PICTURE CRAFTS SERVICE
LOCAL 727, IATSE
14629 NORDHOFF ST
PANORAMA CITY 91402
PHONE: (818) 891-0717

MOTION PICTURE EDITORS GUILD (IA LOCAL 776)
7715 SUNSET BLVD., SUITE 220
HOLLYWOOD 90046
PHONE: (213) 876-4770
FAX: (213) 876-0861

MOTION PICTURE FIRST AID EMPLOYEES (IA LOCAL 767)
8303 GUSTAV LANE
CANOGA PARK 91304
PHONE: (818) 884-8894/ 760-5341

MOTION PICTURE SET PAINTERS (IA LOCAL 729)
11365 VENTURA BLVD., SUITE 202
STUDIO CITY 91604
PHONE: (818) 984-3000

MOTION PICTURE SOUND EDITORS
P.O. BOX 8306
UNIVERSAL CITY 91608
PHONE: (818) 762-2816

MOTION PICTURE STUDIO GRIPS
LOCAL 80, IATSE, AFL
6926 MELROSE AVE
LOS ANGELES 90038
PHONE: (213) 931-1419

MOTION PICTURE STUDIO PROJECTIONISTS
LOCAL 165, IATSE-MPMO-US-CANADA
17424 VENTURA BL PH5
SHERMAN OAKS 91403
PHONE: (818) 905-5221

MP ILLUSTRATORS & MATTE ARTISTS
LOCAL 790, IATSE
13949 VENTURA BL #301
SHERMAN OAKS 91423
PHONE: (818) 784-6555

MP & VIDEOTAPE EDITORS GUILD
LOCAL 776, IATSE
7715 SUNSET BL #220
LOS ANGELES 90046
PHONE: (213) 876-4770

N.A.B.E.T. - A.F.C. (LOCAL 531)
2501 W. BURBANK BLVD., SUITE 301
BURBANK 91505
PHONE: (818) 563-3772
FAX: (818) 563-4712

ORNAMENTAL PLASTERERS, SCULPTORS & MODEL MAKERS (LOCAL 755)
13949 VENTURA BLVD., SUITE 305
SHERMAN OAKS 91423
PHONE: (818) 379-9711

PRODUCERS GUILD OF AMERICA
400 S BEVERLY DR #211
BEVERLY HILLS 90212
PHONE: (213) 557-0807

PRODUCERS REPRESENTATIVE INC
9911 W PICO BL
LOS ANGELES 90035
PHONE: (213) 553-0084

PRODUCERS-WRITERS GUILD PENSION PLAN
1015 N HOLLYWOOD WAY
BURBANK 91505
PHONE: (818) 846-1015

PRODUCTION OFFICE COORDINATORS
LOCAL 717 IATSE
13949 VENTURA BL #306
PHONE: (818) 906-9986

SCENIC & TITLE ARTISTS
LOCAL 816, IBT, MPMO
6180 LAUREL CYN BL #275
NORTH HOLLYWOOD 91606
PHONE: (818) 769-0816

SCREEN ACTORS GUILD (SAG)
7065 HOLLYWOOD BLVD
HOLLYWOOD 90028
PHONE: (213) 465-4600
FAX: (213) 856-6603

SCREEN ACTORS GUILD
7827 CONVOY CT #400
SAN DIEGO 92111
PHONE: (619) 278-2918

SCREEN EXTRAS GUILD (SEG)
3629 CAHUENGA BL W
LOS ANGELES 90068
PHONE: (213) 851-4301
FAX: (213) 851-0262

SCREEN EXTRAS GUILD
3045 ROSECRANS ST #308
SAN DIEGO 92110
PHONE: (619) 222-1161

SCRIPT SUPERVISORS
LOCAL 871, IATSE
7061-B HAYVENHURST
VAN NUYS 91406
PHONE: (818) 782-7063

SET DESIGNERS & MODEL MAKERS
LOCAL 847, IATSE
13949 VENTURA BL #301
SHERMAN OAKS 91423
PHONE: (818) 784-6555

SOCIETY OF MP & TV ART DIRECTORS
LOCAL 876, IATSE
11365 VENTURA BL #315
STUDIO CITY 91604
PHONE: (818) 762-9995

SOUND CONST. INST. & MAINT TECH
LOCAL 40, IBEW, AFL-CIO
5643 VINELAND AVE
NORTH HOLLYWOOD 91601
PHONE: (818) 762-4239

SOUTHERN CALIF. MOTION PICTURE COUNCIL
1922 N WESTERN AVE
LOS ANGELES 90027
PHONE: (213) 467-7332

STUDIO ELECTRICAL LIGHTING TECH
LOCAL 728 IATSE, MPMO, AFL-CIO
14629 NORDHOFF ST
PANORAMA CITY 91402
PHONE: (818) 891-0728

THE STUDIO TEACHERS (LOCAL 884)
3601 W. OLIVE AVE., 7TH FL
BURBANK 91505
PHONE: (818) 953-8899
FAX: (818) 842-5321

STUDIO TRANSPORTATION DRIVERS (TEAMSTER'S LOCAL 399)
4747 VINELAND AVE., SUITE E
NORTH HOLLYWOOD 91602
PHONE: (818) 985-7374/ (213) 877-3277

STUDIO UTILITY EMPLOYEES
LOCAL 724
6700 MELROSE AVE
HOLLYWOOD 90038
PHONE: (213) 938-6277

STUNTMEN'S ASSOCIATION OF MOTION PICTURES
4810 WHITSETT AVE
NORTH HOLLYWOOD 91607
PHONE: (818) 766-4334

STUNTWOMEN'S ASSOC. OF MP, INC
202 VANCE RD
PACIFIC PALISADES 90272
PHONE: (213) 462-1605

TEAMSTERS LOCAL 287
1452 N. 4TH ST
SAN JOSE 95112
PHONE: (408) 453-0287

TEAMSTERS LOCAL 542
4602 MERCURY ST
SAN DIEGO 92111
PHONE: (619) 278-1920

THEATER AUTHORITY INC
6464 SUNSET BL #640
LOS ANGELES 90028
PHONE: (213) 462-5761

UNITED SCENIC ARTISTS
LOCAL 829
5410 WILSHIRE BL #407
LOS ANGELES 90036
PHONE: (213) 965-0957

VISUAL ARTISTS ASSOCIATION
2550 BEVERLY BLVD
LOS ANGELES 90067
PHONE: (213) 388-0477

WOMEN IN FILM (WIF)
6464 SUNSET BL #900
HOLLYWOOD 90028
PHONE: (213) 463-6040

WOMEN IN SHOW BUSINESS
P.O. BOX 2535
NORTH HOLLYWOOD 91602
PHONE: (213) 271-3415

WRITERS GUILD OF AMERICA WEST (WGA)
8955 BEVERLY BL
WEST HOLLYWOOD 90048
PHONE: (213) 550-1000

COLORADO

SPECIAL EFFECTS

CONQUISTADORES SKI AREA
P.O. BOX 347
WESTCLIFFE, 81252
PHONE: (303) 783-9206

CREATIVE MATRIX
2435 TOPAZ DR
BOULDER, 80302
PHONE: (303) 443-8215

FASTLANE PRODUCTIONS
820 16TH STREET, SUITE 335
DENVER, 80223
PHONE: (303) 778-0045

GROUND ZERO
610 SOLANO DR.
COLORADO SPRINGS, 80906
PHONE: (303)473-0059

IMAGE-CORP
1863 SOUTH PEARL STREET
DENVER, 80210
PHONE: (303)698-1866

LANDIS, FRANK
P.O. BOX 1585
BASALT, 81621
PHONE: (303)927-4057/923-436

ON CAMERA, INC
2435 TOPAZ DR
BOULDER, 80302
PHONE: (303)443-8215

STOTT, ROBERT
WINTER STORMS, INC
859 B E. GRAND
FRUITA, 81521
PHONE: (303) 858-3403

SULLAIR ROCKY MOUNTAIN, INC
15680 WEST 6TH AVE
GOLDEN, 80401
PHONE: (303) 278-7450

TOOLE, TERRENCE
2421 BRYANT
DENVER, 80211
PHONE: (303)433-9092

EXPENDABLES

FILM/VIDEO EQUIPMENT SERVICE CO.
800 SOUTH JASON STREET
DENVER, 80210
PHONE: (303)778-8616
FAX: (303)778-8657

TTI TRUE-TEMP, INC
505 WEST 40TH AVE
DENVER, 80216
PHONE: (303)480-0531

MAKE-UP

MAKING FACES INC
3132 EAST SECOND
DENVER, 80206
PHONE: (303)377-7799

ROCKY MOUNTAIN CINE SUPPORT
255 WASHINGTON
DENVER, 80203
PHONE: (303)795-9713

MODEL MAKERS

COLORADO RAILROAD MUSEUM
P.O. BOX 10
GOLDEN, 80401
PHONE: (303)279-4591

FOUNTAIN, ROGER PROPS AND MODELS
490 SOUTH LOGAN
DENVER, 80209
PHONE: (303)722-3641

PERRY MODEL BUILDERS
130 WEST 70TH
DENVER, 80221
PHONE: (303) 427-7627

STAMES, CHRISTOPHER
1217 SOUTH JOSEPHINE
DENVER, 80210
PHONE: (303) 733-3327/ 393-1293

PROPS

ADMI FURNITURE RENTAL
1197 WEST ALAMEDA AVE
DENVER, 80223
PHONE: (303) 778-6888

BLUMENBERG, PAUL G.
8399 WEST 75TH WAY
AVRADA, 80005
PHONE: (303) 431-5105

BOULDER VALLEY RANCH
3700 LONGHORN ROAD
BOULDER, 80302
PHONE: (303)442-6219

BUDGET FURNITURE RENTAL
5795 EAST EVANS AVENUE
DENVER, 80222
PHONE: (303) 753-1166

BUICK CLUB OF AMERICA
4735 SOUTH GALAPAGO
ENGLEWOOD, 80110
PHONE: (303)781-0169

CLASSIC CAR CLUB OF AMERICA, THE
1400 RIDGE ROAD
LITTLETON, 80120
PHONE: (303)794-3702

CREATIVE MATRIX
2435 TOPAZ DR
BOULDER, 80302
PHONE: (303) 443-8215

DRAPER, GEORGE
P.O. BOX 27
WETMORE, 81253
PHONE: (303)784-3162

FANTASY FARMS
11089 EAST STALLION DR
PARKER, 80134
PHONE: (303) 841-2077

FORNER TRANSPORTATION MUSEUM
1416 PLATTE
DENVER, 80202
PHONE: (303)433-3643

FOUNTAIN, ROGER PROPS AND MODELS
490 SOUTH LOGAN
DENVER, 80209
PHONE: (303)722-3641

GRANTREE
2495 SOUTH HAVANA STREET
AURORA, 80014
PHONE: (303) 696-8840/ 297-8831

GRANTREE FURNITURE RENTAL CORP
5700 NORTH BROADWAY
DENVER, 80216
PHONE: (303)297-8831/0303

M & L BUSINESS MACHINES COMPANY
350 HAVANA STREET
AURORA, 80010
PHONE: (303)344-4104

MURPHY, JIM
BOX 490
EVERGREEN, 80439
PHONE: (303)674-3554

NURSERY SUPPLY COMPANY
4255 YARROW STREET
WHEAT RIDGE, 80033
PHONE: (303)424-3968

ON CAMERA, INC
2435 TOPAZ DR
BOULDER, 80302
PHONE: (303)443-3215

ORDNANCE PARK CORPORATION
657 20 1/2 ROAD
GRAND JUNCTION, 81503
PHONE: (303)242-3135

PRESIDENTS TRANSPORTATION
P.O. BOX 427
GOLDEN, 80402
PHONE: (303) 278-7779

ROBOT FACTORY, THE
P.O. BOX 11
CASCADE, 80809
PHONE: (303)687-6203/687-6244

ROCKY MOUNTAIN THUNDERBIRD CLUB
15600 WEST FIRST DR.
GOLDEN, 80401
PHONE: (303) 278-9034/ 295-2297

SOMBRERO RANCH
BOULDER AIRPORT, 3100
AIRPORT ROAD
BOULDER, 80301
PHONE: (303) 442-0258/ 586-4577

SUPPLY LINE
1245 RIVERSIDE AVE
BOULDER, 80302
PHONE: (303)449-6277

T-LAZY-7 RANCH
BOX 858
ASPEN, 81612
PHONE: (303)925-7040

VETERAN MOTOR CAR CLUB OF AMERICA
FIKES PEAK CHAPTER
PIONEER'S MUSEUM, 215
SOUTH TEJON ST
COLORADO SPRINGS, 80903
PHONE: (303)578-6650

WAGONS UNLIMITED/MAYDAY LIVERY
4317 COUNTY ROAD 124
HESPERUS, 81326
PHONE: (303) 385-4585

WESTERN WHIP
401 WATSON LANE
LITTLETON, 80123
PHONE: (303) 795-6535

STUDIOS/SOUND STAGES

ALHAMBRA PHOTO PRODUCTION STUDIOS
751 SANTA FE DRIVE
DENVER, 80204
PHONE: (303)571-1700

ANOTHER PRODUCTION COMPANY
1420 BLAKE ST
DENVER, 80202
PHONE: (303) 623-6615

BENSON PRODUCTIONS, INC
410 SEVENTEENTH STREET, SUITE 1120
DENVER,
PHONE: (303) 325-3355

COMMUNI CREATIONS, INC
2130 SOUTH BELLAIRE ST
DENVER, 80222
PHONE: (303) 759-1155

DENVER CENTER FOR THE PERFORMING ARTS
1245 CHAMPA ST
DENVER, 80204
PHONE: (303)853-4000

KCNC - TV/CHANNEL 4 (NBC)
1044 LINCOLN ST
DENVER, 80217
PHONE: (303)861-4444

KMGH - TV/CHANNEL 7 (CBS)
123 SPEER BLVD
DENVER, 80217
PHONE: (303)832-7777

KWGN - TV/CHANNEL 2 (IND)
6160 SOUTH WABASH WAY
ENGELWOOD, 80111
PHONE: (303)740-2222/2804

LA PRODUCTIONS
201 SOUTH CHEROKEE ST
DENVER, 80223
PHONE: (303)733-3456

MILE HI CABLEVISION
2505 WEST 16TH AVE
DENVER, 80204
PHONE: (303) 770-4500

PIKES PEAK CENTER
P.O. BOX 2007
COLORADO SPRINGS, 80901
PHONE: (303)520-7453

TELEMATION PRODUCTIONS
7700 EAST ILIFF, SUITE M
DENVER, 80231
PHONE: (303)751-6000

WICKERWORKS VIDEO PRODS
7342 SOUTH ALTON WAY,
SUITE 1
ENGLEWOOD, 80112
PHONE: (303) 741-3400

WICKERWORKS VIDEO
PRODUCTIONS, INC
6900 SOUTH YOSEMITE,
SUITE 100
ENGLEWOOD, 80112
PHONE: (303)741-3400

WINDSTAR STUDIOS, INC
525 COMMUNICATION CIRCLE
COLORADO SPRINGS, 80905
PHONE: (303)635-2400

UNIONS AND ASSOCIATIONS

ASSOCIATION OF
INDEPENDENT COMMERCIAL
PRODUCERS (AICP)
5251 DTC PARKWAY,
SUITE 1100
ENGLEWOOD, 80111
PHONE: (303) 796-9380

COLORADO FILM & VIDEO
ASSOCIATION
P.O. BOX 9846, BRUCE
HENRICKSON,
DENVER, 80209
PHONE: (303) 573-1999

DENVER THEATRICAL STAGE
IATSE 7
910 FIFTEENTH ST #751
DENVER, 80202
PHONE: (303) 534-2423

IATSE LOCAL 47
P.O. BOX 1427
PUEBLO, 81002
PHONE: (719) 543-1041

IATSE LOCAL 62
P.O. BOX 147
COLORADO SPRINGS, 80901
PHONE: (719) 663-6572

IATSE LOCAL 689
P.O. BOX 484
GRAND JUNCTION, 81501
PHONE: (303) 242-5286

MACHINE OPERATORS IATSE
230
43 W FOURTH AVE
DENVER, 80223
PHONE: (303) 722-5151

SAG/AFTRA
950 S CHERRY ST #502
DENVER, 80222
PHONE: (303) 757-6226

STAGE HANDS IATSE LOCAL
229
P.O. BOX 677
FORT COLLINS, 80522
PHONE: (303) 482-4503

TEAMSTERS LOCAL 17
3245 ELLIOT ST
DENVER, 80211
PHONE: (303) 433-6497

CONNECTICUT

SPECIAL EFFECTS

ALTERED EGOS LTD
18 MARSHALL ST.,
SECOND FLOOR
S. NORWALK, 06854
PHONE: (203) 866-1166

ANDIN TUCKER DESIGN/
COMMUNICATIONS
56 ARBOR ST
HARTFORD, 06106
PHONE: (203) 232-4885

BRODAX FILM GROUP
45 CEDAR HILLS
WESTON, 06883
PHONE: (203) 227-8875

FANTASTIC FRIGHTS
149 CENTER ST
WEST HAVEN, 06516
PHONE: (203) 933-3823

ILLICIUM FERMANAGH
RFD 3, BOX 6, NORTH MAIN ST
STONINGTON, 06378
PHONE: (203) 535-1286

NEW ENGLAND STAR WORKS
150 THOMPSON ST
HAMDEN, 06518
PHONE: (203) 288-7695

RIMIFI TECHINCAL
PRODUCTION SERVICES
48 MADISON DR, P.O. BOX 70
KENSINGTON, 06037
PHONE: (203) 828-4834

STUNT INCORPORATED
116 UNION ST
NORWICH, 06360
PHONE: (203) 887-6505/739-2266

THIN AIR
250 HARBOR PLAZA DR., P.O.
BOX 10210
STAMFORD, 06904

TROC UNLIMITED
BATTLE ST, RFD 6
BRISTOL, 06010
PHONE: (203) 583-8306

WOWHAUS SPECIAL EFFECTS
1440 WHALLEY AVE,
P.O. BOX 63
NEW HAVEN, 06515

MAKE-UP SUPPLIERS

ALTERED EGOS LTD.
18 MARSHALL ST., SECOND
FLOOR
S. NORWALK, 06854
PHONE: (203)866-1166

WIZARD OF WESTPORT
236 POST RD. E
WESTPORT, 06880
PHONE: (203) 227-7753

PROPS

ANDIN TUCKER DESIGN/
COMMUNICATIONS
56 ARBOR ST
HARTFORD, 06106
PHONE: (203) 232-4885

ATLAS SCENIC STUDIOS, LTD
46 BROOKFIELD AVE
BRIDGEPORT, 06610
PHONE: (203) 334-2130

BAKER, J.L. ASSOCIATES
478 N. MAIN STREET
WALLINGFORD, 06492
PHONE: (203) 265-9231

BRODAX FILM GROUP
45 CEDAR HILLS
WESTON, 06883
PHONE: (203) 227-8875

CLOCKWORK REPERTORY
THEATRE
133 MAIN ST
OAKVILLE, 06679
PHONE: (203) 274-7247

CONNECTICUT SEA GRANT
MARINE ADVISORY SERVICE
UCONN AT AVERY POINT
GROTON, 06340
PHONE: (203) 445-8664/3458

COST PLUS FURNITURE
RENTALS
417 SHIPPAN AVE
STAMFORD, 06902
PHONE: (203) 353-0400
FAX: (203) 353-1249

CREATIVE TALENT LIMITED
P.O. BOX 143
EAST GLASTONBURY, 06025
PHONE: (203) 633-636

DAVID JOHNSON DESIGNS
1324 PALISADO AVE
WINDSOR, 06095-1439
PHONE: (203) 688-7701

ENTERTAINING DESIGN
2701 SUMMER ST
STAMFORD, 06905
PHONE: (203) 348-8111

FANTASTIC FRIGHTS
149 CENTER ST.
WEST HAVEN, 06516
PHONE: (203) 933-3823

HEPBURN GLASS
P.O. BOX 156
OLD SAYBROOK, 06498
PHONE: (203) 663-1169

ILLICIUM FERMANAGH
RFD 3, BOX 6, NORTH MAIN ST
STONINGTON, 06378
PHONE: (203) 535-1286

KELLINGTON FARM
BOX 145, KING'S HILL RD
SHARON, 06069
PHONE: (203) 364-5466

LOUIS NICHOLE INC
1 NICHOLE COURT,
54 NEW HAVEN RD
PROSPECT, 06712
PHONE: (203) 758-3160

MARIS WAY
92 SILVERMINE AVE
NORWALK, 06850
PHONE: (203) 847-9488

NEW ENGLAND STAR WORKS
150 THOMPSON ST
HAMDEN, 06518
PHONE: (203) 288-7695

KAREN E. PEARLSTEIN
144 GOODHILL RD
WESTON, 06883
PHONE: (203) 226-3192

PLANTATIONS, INC
102 OLD POQUONOVK RD
BLOOMFIELD, 06002
PHONE: (203) 242-2554

ANN POWDERLY
7 SOUTH BARN HILL RD
BLOOMFIELD, 06002
PHONE: (203) 242-6120

PUPPET PROJECTS
35 POWDER HORN HILL RD
WILTON, 06897
PHONE: (203) 834-0800

R.W. COMMERFORDS & SONS
ROUTE 4
GOSHEN, 06750
PHONE: (203) 491-3421

RIMIFI TECHINCAL
PRODUCTION SERVICES
48 MADISON DR, P.O. BOX 70
KENSINGTON, 06037
PHONE: (203) 828-4834

STUNT INCORPORATED
116 UNION ST
NORWICH, 06360
PHONE: (203) 887-6505/739-2266

THIN AIR
250 HARBOR PLAZA DR.,
P.O. BOX 10210
STAMFORD, 06904

TROC ULIMITED
BATTLE ST., RFD 3
BRISTOL, 06010
PHONE: (203) 583-8306

UNITED HOUSE WRECKING
535 HOPE ST
STAMFORD, 06906
PHONE: (203) 348-5371

WOWHAUS SPECIAL EFFECTS
1440 WHALLEY AVE,
P.O. BOX 63
NEW HAVEN, 06515

STUDIOS/SOUND STAGES

AMERICAN FILM STUDIOS, INC
297 DUNBAR HILL RD.
HAMDEN, 06514
PHONE: (203) 288-1753

AHMPHION ENTERPRISES
43 W. MAIN ST.
ROCKVILLE, 06066
PHONE: (203) 871-1786

ANGELSEA PRODUCTIONS, INC
55 RUSS ST
HARTFORD, 06106
PHONE: (203) 241-8111

AUDIO PRODUCTIONS
152 WINFIELD DR
STRATFORD, 06497
PHONE: (203) 378-1855

BJEJ PRODUCTIONS, INC
532 MAIN ST
CROMWELL, 06416
PHONE: (203) 635-4780
FAX: (203) 635-6979

CANDLEWOOD PLAYHOUSE
RTE. 37 & RTE. 39,
P.O. BOX 8209
NEW FAIRFIELD,06812
PHONE: (203) 746-6557
FAX: (203) 746-6550

CIN COM CORPORATION
140 BRADLEY ST
NEW HAVEN, 06511
PHONE: (203) 624-6324

CINEMED
129 MAIN ST. N.
WOODBURN, 06798
PHONE: (203) 263-0006
FAX: (203) 263-483

CLOCKWORK REPERTORY
THEATRE
133 MAIN ST
OAKVILLE, 06779
PHONE: (203) 274-7247

COMMERCIALWORKS
10 MIDDLE ST
BRIDGEPORT, 06604
PHONE: (203) 384-9443
FAX: (203) 367-9346

CONNECTICUT RECORDING
STUDIOS, INC
1122 MAIN ST
BRIDGEPORT, 06604
PHONE: (203) 366-9168

CONTINENTAL CABLEVISION
5 NIBLICK RD
ENFIELD, 06082
PHONE: (203) 741-3541

CORPORATE VIDEO CENTER
240 NEW BRITAIN AVE
HARTFORD, 06106
PHONE: (203) 249-2424
FAX: (203) 278-2157

CORPORATE VIDEO CENTER
250 HARBOR PLAZA DR,
P.O. BOX 10210
STAMFORD, 06904
PHONE: (203) 965-6666/6507
FAX: (203) 353-9086

CREATIVE EFFORT
PUTNAM GREEN
GREENWICH, 06830
PHONE: (203) 531-8685

D & K SOUND SERVICES, INC
842 SILAS DEANE HWY
WETHERSFIELD, 06109
PHONE: (203) 529-8353

DARIEN DINNER THEATRE
65 TOKENEEKE RD
DARIEN, 06820
PHONE: (203) 655-6812

DEBEDEDETTO RECORDING CO.
105 WOOD AVE
BRIDGEPORT, 06605
PHONE: (203) 384-0076

DON ELLIOT PRODUCTIONS
15 BRIDGE ST
WESTON, 06883
PHONE: (203) 226-4200

EAGLEVISION, INC
880 CANAL ST
STAMFORD, 06902
PHONE: (203) 359-8777

ENCORE TELEPRODUCTIONS CORPORATION
600 MAIN ST, RTE. 25
MONROE, 06468
PHONE: (203) 268-7487/3574

ESPN
ESPN PLAZA
BRISTOL, 06010
PHONE: (203) 585-2000
FAX: (203) 585-2217

FRED WEINBERG PRODUC- TIONS/WORLDWIDE AUDIO VIDEO ENTERPRISES
16 DUNDEE RD.
STAMFORD, 06903
PHONE: (203) 322-5778
FAX: (203) 329-7838

THE GALLERY
87 CHURCH ST
EAST HARTFORD, 06108
PHONE: (203) 528-9009

GUYMARK STUDIOS
3019 DIXWELL AVE.,
P.O. BOX 5037
HAMDEN, 06518
PHONE: (203) 248-9323
FAX: (203) 249-9325

HARRIS PRODUCTIONS INTERNATIONAL
2228 SHEPARD AVE
HAMDEN, 06518
PHONE: (203) 288-6622

HARTFORD STAGE COMPANY
50 CHURCH
S. NORWALK, 06854
PHONE: (203) 866-1166

DELAWARE

MODEL MAKERS

ALLSTATES DESIGN & DEV. CO. INC.
201 RUTHAR DR.
NEWARK, 19711
PHONE: (302) 366-1752

FIRST CONTACT DESIGN, INC
RT.2
HOCKESSIN, 19707
PHONE: (302) 239-5008

SUPERIOR MODELS INC
2600 PHILADEPHIA PK., P.O. BOX 99
CLAYMONT, 19703
PHONE: (302) 798-0291

STUDIOS/SOUND STAGES

AV3 DIGITAL
53 MCCULLOUCH DR.
NEW CASTLE, 19720
PHONE: (302) 324-5300/ (800) EDIT-601

STEPHEN PALA PRODUCTIONS
1806 LOVERING AVE
WILMINGTON, 19806
PHONE: (302) 652-1105

TELEPRODUCTION ASSOCIATES INC
305 A ST.
WILMINGTON, 19801
PHONE: (302) 429-0303

DISTRICT OF COLUMBIA

SPECIAL EFFECTS

INTERFACE VIDEO SYSTEMS, INC
1233 20TH ST, N.W.
WASHINGTON, D.C., 20036
PHONE: (202) 861-0500
FAX: (202) 296-4492

JOHNSON, ANNA, AV PRODUCTION SPECIALIST
1498 DOUGLAS STREET NORTH EAST
WASHINGTON, D.C., 20018
PHONE: (202) 529-9295

WEST END FILM
1825 Q ST. NW
WASHINGTON, D.C., 20009
PHONE: (202) 232-7753

FLORIDA

SPECIAL EFFECTS

ACCORD PRODUCTIONS
2000 SOUTH DIXIE HIGHWAY,
SUITE 112
MIAMI, 33133
PHONE: (305) 856-1245
FAX: (305) 856-9101

ARCHITECTE MINIATURA
P.O. BOX 1783
SARASOTA, 34230-1783
PHONE: (813) 955-5555

ARTIFAKES
P.O. BOX 10112
LARGO, 34643
PHONE: (813) 577-9595

BOLAND HARDWARE
1088 HAVENDALE BLVD
WINTER HAVEN, 33881
PHONE: (813) 293-7355

CENTAUR ENTERTAINMENT PRODUCTIONS, INC
4559 34TH ST
ORLANDO, 32811
PHONE: (407) 649-8810
FAX: (407) 649-8812

CINEMAGIC MAKE-UP & PROFESSIONAL BEAUTY SUPPLIES
7492 REPUBLIC DR
ORLANDO, 32819
PHONE: (407) 351-3330
FAX: (407) 363-4409

CINNABAR
4560 L. B. MCLEAOD RD.,
SUITE B
ORLANDO, 32811
PHONE: (407) 649-7633
FAX: (407) 872-3616

PETER D. CIPORKIN, D.D.S.
7280 W. PALMETTO PARK RD.,
SUITE 206-N
BOCA RATON, 33433
PHONE: (407) 393-1770/ (800) 780-SMILE

CRAWFORD PRODUCTIONS
P.O. BOX 1192
NEW SMYRNA BEACH, 32170
PHONE: (904) 427-6626/ (800) 745-0363

DUBOIS PRODUCTIONS, INC
511 N.E. 3RD AVE
FT. LAUDERDALE, 33301
PHONE: (305) 463-5950
FAX: (305) 463-5952

EXHIBIT BUILDERS, INC
150 WILDWOOD RD., DEPT. F
DELAND, 32720
PHONE: (904) 734-3196
FAX: (904) 734-9391

EXHIBIT RESOURCES U.S.A., INC
803 PRICE ST
JACKSONVILLE, 32204
PHONE: (904) 353-ERUS
FAX: (904) 353-4489

FREELANCE STAGING
8681 N.W. 66TH ST
MIAMI, 33166
PHONE: (305) 477-3138
FAX: (305) 594-9959

GEARHART RACING PRODUCTS
6001 JOHNS RD
TAMPA, 33634
PHONE: (813) 886-8223

THE GIBSON GROUP
P.O. BOX 2254
HAINES CITY, 33845
PHONE: (813) 676-6856

HERTZ EQUIPMENT RENTAL
3838 NAVY BLVD
PENSACOLA, 32507
PHONE: (800) 537-4501
FAX: (904) 455-3165

ILLUSION MASTERS INC
1621 BOWMAN ST
CLERMONT, 34711
PHONE: (904) 394-7436

SALLY INDUSTRIES, INC
803 PRICE ST
JACKSONVILLE, 32204
PHONE: (904) 353-5051
FAX: (904) 353-4489

J.B. JONES, INC
13815 N.W. 19TH AVE
MIAMI, 33054
PHONE: (305) 681-7627
FAX: (305) 687-4737

LAST STAGE OUT OF TOWN
P.O. BOX 626
LAKE ALFRED, 33850
PHONE: (800) 942-3448/ (813) 956-3448
FAX: (813) 956-5302

LESTER KALMANSON AGENCY, INC
P.O. BOX 940008
MAITLAND, 32794-0008
PHONE: (407) 645-5000
FAX: (407) 645-2810

KEN KARBOSWKI
1840 N.W. 42ND TERRACE
LAUDERHILL, 33313
PHONE: (305) 485-4935

LASER PRODUCTIONS
P.O. BOX 141411
ORLANDO, 32814
PHONE: (407) 321-5673

LIGHTING SYSTEMS DESIGN, INC
4625 OLD WINTER GARDEN RD.,
B-6
ORLANDO, 32811-1777
PHONE: (407) 299-9504
FAX: (407) 299-3965

LIGHTING EFEX CORPORATION
8242 SANDBERRY BLVD
ORLANDO, 32819
PHONE: (407) 351-6512

LUMONICS
3017 N.W. 60TH ST
FT. LAUDERDALE, 33319
PHONE: (305) 979-3161

BRUCE MERLIN
5810 WAYT CT
ORLANDO, 32810
PHONE: (407) 293-4315

THE MODEL FACTORY, INC
2011 S.W. 70TH AVE ,
SUITE A-22
DAVIE, 33317
PHONE: (305) 475-1905
FAX: (305) 475-1010

ANDREW NICHOLLS
3820 LAGUNA DR
ORLANDO, 32805
PHONE: (407) 648-1857

ONCAMERA
P.O. BOX 413
WINDERMERE, 34786
PHONE: (407) 256-5916

ONLINE BY DESIGN, INC
10300 SUNSET DR
MIAMI, 33173
PHONE: (305) 595-7307
FAX: (305) 595-7457

ORLANDO SPECIAL EFFECTS
3820 LAGUNA DR
ORLANDO, 32805
PHONE: (407) 648-1867

PACIFIC TITLE & ART STUDIO
608 S. MAIN AVE #12
CLERMONT, 34711
PHONE: (904) 334-8487/ (800) 343-3547
FAX: (904) 394-8489

ROBERT PELLEGRINI & ASSOCIATES
5780 S.W. 62ND ST
MIAMI, 33143
PHONE: (305) 666-3117

POLYEFFECTS, INC
2000 UNIVERSAL STUDIOS PLAZA, SUITE 730
ORLANDO, 32819
PHONE: (407) 363-0666
FAX: (407) 363-0666

PROMOTIONAL TECHNOLOGIES
3643 N.E. 25TH ST
OCALA, 32670
PHONE: (800) 755-7626
FAX: (904) 629-4563

PROPMASTERS MIAMI
9940 N.W. 79TH AVE
MIAMI, 33016
PHONE: (305) 826-1900
FAX: (305) 826-1850

THE PYRO CREW CO
2085 KINGSWOOD AVE
DELTONA, 32725
PHONE: (800) 722-7976

REDD FROGE-PURVEYOR OF CHEAP THRILLS
P.O. BOX 10402
LARGO, 34643
PHONE: (813) 393-6610

SANTORE & SONS FIREWORKS
P.O. BOX 1127
BUNNELL, 32010-1127
PHONE: (904) 437-2242

SCALE REPRESENTATIONS
2645 12TH ST. N.
ST. PETERSBURG, 33704
PHONE: (813) 821-8541

STAGE EQUIPMENT AND LIGHTING, INC
12231 N.E. 13TH CT
MIAMI, 33161
PHONE: (305) 891-0291
FAX: (305) 893-2828

STAGE EQUIPMENT & LIGHTING INC
4602 S.W. 25TH ST
ORLANDO, 32811
PHONE: (407) 425-2010
FAX: (407) 648-2604

MICHAEL E. STEWART
P.O. BOX 4426
CLEARWATER, 34618
PHONE: (813) 796-1982

STUNT ACTION COORDINATORS, INC
P.O. BOX 127
GROVELAND, 34736
PHONE: (904) 394-8893
FAX: (904) 429-4029

UNIQUE PRODUCERS SERVICE INC
13815 N.W. 19TH AVE
MIAMI, 33054
PHONE: (305) 681-7627
FAX: (305) 687-4737

VIDEO BROKERS, INC
5516 COMMERCE DR
ORLANDO, 32809
PHONE: (407) 851-4595
FAX: (407) 851-7497

ZELLER INTERNATIONAL
623 ELLEN DR
WINTER PARK, 32789
PHONE: (407) 629-2905

EXPENDABLES

A - 1 MARINE, INC
P.O. BOX 4338 (010
5601 W HWY 98
PANAMA CITY, 32405
PHONE: (904) 785-2567/7440

ARIZONA CHEMICAL COMPANY
1001 EAST BUSINESS 98
PANAMA CITY, 32402
PHONE: (904) 785-6700
FAX: (904) 785-4599

ART ZONE/DAUGHDRILL INC
933 CHERRY
P.O. BOX 12064 (02)
PANAMA CITY,32401
PHONE: (904) 769-7384

BAY TANK & FABRICATING
1810 INDUSTRIAL DR
P.O. BOX 2418 (02)
PANAMA CITY, 32405
PHONE: (904) 763-7696

BERG STEEL PIPE CORP
1415 C AVENUE
P.O. BOX 2029 (02)
PANAMA CITY, 32401
PHONE: (904) 769-2273

C.B. SAILS - NEIL PRYDE SAILS
5308 E. HWY 98
PANAMA CITY, 32404
PHONE: (904) 871-1921

C.E. CAMP ROOF & SHEET METAL
1406 MINNESOTA AVE
LYNN HAVEN, 32444
PHONE: (904) 265-0003

CALLAWAY FOUNDRY/ MACHINE WORKS
5412 HWY 22
P.O. BOX 6803
PANAMA CITY, 32404
PHONE: (904) 763-5141

CARGILL STEEL & WIRE
1800 INDUSTRIAL DR
P.O. BOX 16548
PANAMA CITY, 32405
PHONE: (904) 784-6721
FAX: (904) 784-9472

CAROLINA CHAIN & CABLE
1901 EAST AVE
P.O. BOX 15879 (02)
PANAMA CITY, 32401
PHONE: (904) 769-5201

CHEVRON USA
500 W. 5TH ST
PANAMA CITY, 32401
PHONE: (904) 785-7426
FAX: (904) 769-7140

CUSTOM TRAILER
3716 E. 3RD ST
PANAMA CITY, 32401
PHONE: (904) 763-5880

EASTERN SHIPYARDS, INC
13332 ALLANTON RD
P.O. BOX 1171
ALLANTON,
PHONE: (904) 871-4800
FAX: (904) 871-1889

ETHERIDGE CABINET SHOP, INC
902 E BALDWIN RD
PANAMA CITY, 32405
PHONE: (904) 769-0201

FLAGALA CORPORATION
9700 FRONT BEACH
PANAMA CITY, 32407
PHONE: (904) 234-2141
FAX: (904) 235-4000

FLORIDA MINING & MATERIAL
1819 N. COVE BLVD
P.O. BOX 1819 (02)
PANAMA CITY, 32401
PHONE: (904) 763-5301

FREEMAN ELECTRIC COMPANY
534 OAK AVE
P.O. BOX 2267 (02)
PANAMA CITY, 32401

GENERAL MARINE INDUSTRIES
6725 BAY LINE DR.
BAY INDUSTRIAL PARK
P.O. BOX 15488
PANAMA CITY, 32406
PHONE: (904) 769-0311
FAX: (904) 769-0731

GULF POWER COMPANY
1230 W. 15TH ST
P.O. BOX 2448
PANAMA CITY, 32401
PHONE: (904) 872-3200
FAX: (904) 872-3292

INDUSTRIAL FIBERGLASS CORP
4511 E. 11TH ST
PANAMA CITY, 32401
PHONE: (904) 785-1212

J J'S MARINE SERVICE & CUSTOM ENGINEERING
2505 THOMAS DR
PANAMA CITY, 32407
PHONE: (904) 234-8048
FAX: (904) 234-8048

K & K PRECISION MFG
2307 INDUSTRIAL DR
PANAMA CITY, 32405
PHONE: (904) 769-9080
FAX: (904) 769-2479

KURT SCHMIDT ENTERPRISES INC
730 HIGHLAND DR
PANAMA CITY, 32404
PHONE: (904) 769-5192

LOUISIANA-PACIFIC CORP
HWY 79 & STEELFIELD RD
P.O. BOX 11160 W. BAY STATION
PANAMA CITY, 32413
PHONE: (904) 234-6692
FAX: (904) 235-1769

LOUISIANA PLASTIC
P.O. BOX 15908
PANAMA CITY, 32406
PHONE: (904) 769-3285
FAX: (904) 763-8098

LYNN HAVEN STEEL & FABRICATION
3604 HWY 390
PANAMA CITY, 32405
PHONE: (904) 769-6347

MERRICK CORP.
P.O. BOX S
LYNN HAVEN, 32444
PHONE: (904) 265-3611
FAX: (904) 265-9768

MIDWEST PIPE COATING
1412 C AVENUE
P.O. BOX 15367
PANAMA CITY, 32401
PHONE: (904) 763-0244
FAX: (904) 769-4344

PANAMA FIRE APPARATUS
4643 HWY 231
PANAMA CITY, 32405
PHONE: (904) 763-0741
FAX: (904) 872-8722

PROFESSIONAL WINDOWS
2808 N EAST AVE
PANAMA CITY, 32405
PHONE: (904) 872-8702

QUEEN CRAFT, INC
3615 CALHOUN AVE
PANAMA CITY, 32405
PHONE: (904) 769-2391
FAX: (904) 769-9290

RUBBER & SPECIALTIES INC
2427 INDUSTRIAL DR
PANAMA CITY, 32405
PHONE: (904) 769-3450
FAX: (904) 769-3459

SPURLIN INDUSTRIES INC
700 JACKSON WAY
PANAMA CITY, 32402
PHONE: (904) 785-1535
FAX: (904) 785-9781

STEEL CITY, INC
749 E. 15TH ST
P.O. BOX 15666
PANAMA CITY, 32401
PHONE: (904) 785-9596

SUN INDUSTRIES
1816 ALLISON AVE
PANAMA CITY, 32407
PHONE: (904) 234-3292
FAX: (904) 234-3609

SUNSHINE PIPING INC
6513 BAY LINE DR
BAY INDUSTRIAL PARK
PANAMA CITY, 32404
PHONE: (904) 763-4834

TARPON DOCK METAL CRAFT
1721 E. 11TH
P.O. DRAWER 1730
PANAMA CITY,32401
PHONE: (904) 785-9568

WELLSTRAN INC
1700 C AVE
PANAMA CITY,32401
PHONE: (904) 769-9471
FAX: (904) 769-9065

MAKE-UP SUPPLIERS

ALTERNATIVES HAIR & MORE INC
2838 S. TAMIAMI RD
SARASOTA, 34239
PHONE: (813) 366-0160

BALLONATICS
170 CANAVERAL PLAZA BLVD
COCOA BEACH, 32923
PHONE: (407) 783-2225

CINEMAGIC MAKE-UP & PROFESSIONAL BEAUTY SUPPLIES
7492 REPUBLIC DR
ORLANDO, 32819
PHONE: (407) 351-3330
FAX: (407) 363-4409

CINEMA MAKE-UP MADNESS
464 LAKESIDE BLVD
BOCA RATON, 33434
PHONE: (407) 483-6404

DANCIN DUDS
1450 N. COURTENAY PARK-WAY, SUITE 13B
MERRITT ISLAND, 32952
PHONE: (407) 452-1899<

DESIGN LINE
5518 COMMERCE PARK BLVD
TAMPA, 33610
PHONE: (813) 626-5991
FAX: (813) 626-5475

GOODY PRODUCTS
1405 FOSTER AVE
PANAMA CITY, 32401
PHONE: (904) 769-2214

LE TROPIQUE INC
2679 TIGERTAIL AVE
COCONUT GROVE, 33133
PHONE: (305) 856-4066
FAX: (305) 858-2407

THE MAKE-UP STUDIO
6394 HAMPTON DR. N
ST. PETERSBURG, 33710
PHONE: (813) 347-3883

REDD FROGE-PURVEYOR OF CHEAP THRILLS
P.O. BOX 10402
LARGO, 34643
PHONE: (813) 393-6610

STUDIO MAGIC INC
1417-2 DEL PRADO BLVD.
SUITE 480
CAPE CORAL, 33990-3749
PHONE: (813) 283-5000/
(800) 749-5002
FAX: (813) 772-1313

NOREEN YOUNG, INC
2313 UNIVERSITY BLVD W
JACKSONVILLE, 32217
PHONE: (904) 739-2560/
(800) 950-4YOU
FAX: (904) 384-8175

MODELS/MINIATURES

ARCHITECTE MINIATURA
P.O. BOX 1783
SARASOTA, 34230-1783
PHONE: (813) 955-5555

CINNABAR
4560 L. B. MCLEAOD RD.,
SUITE B
ORLANDO, 32811
PHONE: (407) 649-7633
FAX: (407) 872-3616

DOLAN & COMPANY PRODUCTIONS
1101-17 S. ROGERS CIRCLE
BOCA RATON, 33487
PHONE: (407) 994-6331
FAX: (407) 994-6332

EXHIBIT BUILDERS, INC
150 WILDWOOD RD., DEPT. F
DELAND, 32720
PHONE: (904) 734-3196
FAX: (904) 734-9391

FANTASTIC PROPS, INC
4235 N.E. 6TH AVE
OAKLAND PARK, 33334
PHONE: (305) 537-9068
FAX: (305) 772-6476

ILLUSION MASTERS, INC
1621 BOWMAN ST
CLERMONT, 34711
PHONE: (904) 394-7436

KEN KARBOSWKI
1840 N.W. 42ND TERRACE
LAUDERHILL, 33313
PHONE: (305) 485-4935

MAKO PRODUCTIONS
545 S. ATLANTIC BLVD.,
SUITE 402
FT. LAUDERDALE, 33316
PHONE: (305) 523-7714

THE MODEL FACTORY, INC
2011 S.W. 70TH AVE., SUITE A-22
DAVIE, 33317
PHONE: (305) 475-1905
FAX: (305) 475-1010

ONCAMERA
P.O. BOX 413
WINDERMERE, 34786
PHONE: (407) 256-5916

ORLANDO SPECIAL EFFECTS
3820 LAGUNA DR
ORLANDO, 32805
PHONE: (407) 648-1867

POLYEFFECTS, INC
2000 UNIVERSAL STUDIOS
PLAZA, SUITE 700
ORLANDO, 32819
PHONE: (407) 363-0666
FAX: (407) 363-0666

THE PROPTOLOGISTS
P.O. BOX 593921
ORLANDO, 32859
PHONE: (407) 240-7330

SCALE REPRESENTATIONS
2645 12TH ST. N.
ST. PETERSBURG, 33704
PHONE: (813) 821-8541

PHILIP E. THIBODEAU
3236 PARR ST.
JACKSONVILLE, 32205
PHONE: (904) 388-3011

PROPS

A CERTAIN AMBIANCE
1094 N.W. FEDERAL HWY
STUART, 34994
PHONE: (407) 692-3054
FAX: (407) 287-4591

ARCHITECTE MINIATURA
P.O. BOX 1783
SARASOTA, 34230-1783
PHONE: (813) 955-5555

ARTIFAKES
P.O. BOX 10112
LARGO, 34643
PHONE: (813) 577-9595

ASOLO SCENIC STUDIO
5555 N. TAMIAMI TRAIL
SARASOTA, 34243
PHONE: (813) 351-9010
FAX: (813) 351-5796

AT & T
4451 BAYOU BLVD
PENSACOLA, 32509
PHONE: (904) 477-6648

**BEST WEST FLORIDA
PRODUCTIONS**
1435 STATE ST
SARASOTA, 34236
PHONE: (813) 954-1122

BISCAYNE HELICOPTERS, INC
12760 S.W. 137TH AVE., S-2
MIAMI, 33186
PHONE: (305) 252-3883
FAX: (305) 378-4407

BOBBY AMOR
2329 CLUBHOUSE DR
W. PALM BEACH, 33409
PHONE: (407) 687-3792
FAX: (407) 697-8254

BOLAND HARDWARE
1088 HAVENDALE BLVD
WINTER HAVEN, 33881
PHONE: (813) 293-7355

SAM BULLARA
4203 CARTNAL AVE
TAMPA, 33624
PHONE: (813) 962-6688/
871-4645
FAX: (813) 871-4769

**CATTLEMANS ASSOCIATION
(LIVESTOCK)**
4115 SOUTH FISKE BLVD
ROCKLEDGE, 32955
PHONE: (407) 636-2390/6840

**CHAPMAN REPTILES FOR FILM
& VIDEO (EXOTIC ANIMALS)**
P.O. BOX 351
ROCKLEDGE, 32955
PHONE: (407) 632-7125

CINNABAR
4560 L. B. MCLEAOD RD.,
SUITE B
ORLANDO, 32811
PHONE: (407) 649-7633
FAX: (407) 872-3616

CHRYSLER STUDIO PROGRAM
2000 UNIVERSAL STUDIO
PLAZA, SUITE 628
ORLANDO, 32819
PHONE: (407) 363-2026

CLASSIC JUKE BOXES
P.O. BOX 547035
ORLANDO, 32804
PHONE: (407) 422-8801

CLASSIC PROPS
128 N. LINE DRIVE
APOPKA, 32703
PHONE: (407) 889-4994

**CLUB NAUTICO NATIONAL
POWERBOAT RENTALS**
46 LOCATIONS IN FLORIDA
PHONE: (800) BOAT-RENT
FAX: (305) 739-9892

JENNIE CURLAND INTERIORS
3718 S.E. OCEAN BLVD
STUART, 34996
PHONE: (407) 287-1696
FAX: (407) 237-6254

DESIGN LINE
5518 COMMERCE PARK BLVD
TAMPA, 33610
PHONE: (813) 626-5991
FAX: (813) 626-5475

DESIGNSDESIGNS
4613/C N. CLARK AVE
TAMPA, 33614
PHONE: (813) 876-2167
FAX: (813) 876-2167

RICHARD DIMMLER
690 PURDY DR
SANIBEL, 33957
PHONE: (813) 472-1410

EXHIBIT BUILDERS, INC
150 WILDWOOD RD., DEPT. F
DELAND, 32720
PHONE: (904) 734-3196
FAX: (904) 734-9391

EXHIBIT RESOURCES U.S.A., INC
803 PRICE ST
JACKSONVILLE, 32204
PHONE: (904) 353-ERUS
FAX: (904) 353-4489

FANTASTIC PROPS, INC
4235 N.E. 6TH AVE
OAKLAND PARK, 33334
PHONE: (305) 537-9068
FAX: (305) 772-6476

**FIRE SAFE OF CENTRAL
FLORIDS INC**
2644 C MICHIGAN AVE
KISSIMMEE, 34733
PHONE: (407) 870-1940
FAX: (407) 870-8052

FLORIDA DEEP SEA FISHING, INC
P.O. BOX 46421
ST. PETERSBURG BEACH, 33741
PHONE: (813) 360-2082

FLOWERS BY SQUIRES
2502 N.E. JACKSONVILLE RD.,
SUITE 102
OCALA, 32670-3762
PHONE: (904) 629-1288

FREELANCE STAGING
8681 N.W. 66TH ST
MIAMI, 33166
PHONE: (305) 477-3138
FAX: (305) 594-9959

RICHARD FREEMAN
5900 N. 9TH AVE
PENSACOLA, 32503
PHONE: (904) 476-8866

FUN MAKERS INC
12328 WOODLEIGH AVE
TAMPA, 33612
PHONE: (813) 933-8272

**GREAT AMERICAN PROP
COMPANY**
720 N. ORANGE AVE
ORLANDO, 32801
PHONE: (407) 648-8971
FAX: (407) 649-4137

**GREAT AMERICAN SALVAGE
COMPANY**
1722 HENDRICKS AVE
JACKSONVILLE, 32207
PHONE: (904) 396-8081

**GREAT AMERICAN SIGNS &
PROPS**
7033 NORTON AVE. #6
NORTON COMMERCIAL
CENTER
LAKE WORTH, 33405
PHONE: (407) 433-0784
FAX: (407) 585-5644

GREAT SOUTHERN STUDIOS
15221 N.E. 21ST AVE
N. MIAMI BEACH, 33162
PHONE: (305) 944-2464
FAX: (305) 944-9920

DANNY HAZZINGTON
220 N.W. 140TH ST
MIAMI, 33168
PHONE: (305) 769-0460/
353-2664

HIBEX CENTRAL INC
6949 VENTURE CIRCLE
ORLANDO, 32807
PHONE: (407) 679-4001
FAX: (407) 679-9637

**HOLLYWOOD EAST MOVIE
CARS**
4651 36TH ST., SUITE 600
ORLANDO, 32811
PHONE: (407) 423-4640

RALPH HORN
5649 STONERIDGE CIRCLE
ORLANDO, 32809
PHONE: (407) 855-3439

**INDEPENDENT STUDIO
SERVICES**
4651 36TH ST., S SUITE 600
ORLANDO, 32811
PHONE: (407) 423-4321

SALLY INDUSTRIES, INC
803 PRICE ST
JACKSONVILLE, 32204
PHONE: (904) 353-5051
FAX: (904) 353-4489

**I.S.S. - INDEPENDENT STUDIO
SERVICES, INC**
4651 36TH ST
ORLANDO, 32811
PHONE: (407) 423-4321
FAX: (407) 872-0891

KEN KARBOSWKI
1840 N.W. 42ND TERRACE
LAUDERHILL, 33313
PHONE: (305) 485-4935

JORDAN KLEIN PRODUCTIONS
10197 S.E. 144TH PL
SUMMERFIELD, 32691
PHONE: (904) 288-6060

**LOCAMOTION SUPPORT
SERVICES, INC**
4232 S.W. 75TH AVE
MIAMI, 33155
PHONE: (305) 261-5242
FAX: (305) 261-9809

LOOKING GLASS COTTAGE
1558 SOUTH WICKHAM RD
MELBOURNE,
PHONE: (407) 724-5011

**DERALD MARTIN
CORPORATION**
4613-C N. CLARK AVE
TAMPA, 33614
PHONE: (800) 633-1586
FAX: (813) 876-2167

THE MODEL FACTORY, INC
2011 S.W. 70TH AVE.,
SUITE A-22
DAVIE, 33317
PHONE: (305) 475-1905
FAX: (305) 475-1010

MONAHANS AUTO CENTERS
506 IRENE ST
ORLANDO, 32808
PHONE: (407) 295-2535/2635

**NEW ATMOSPHERE
PRODUCTIONS INC**
P.O. BOX 2263
SARASOTA, 34230
PHONE: (813) 954-1264

ORLANDO PROP GALLERY
7500 EXCHANGE DR
ORLANDO, 32809
PHONE: (407) 438-0734
FAX: (407) 438-0736

POLYEFFECTS, INC
2000 UNIVERSAL STUDIOS
PLAZA, SUITE 700
ORLANDO, 32819
PHONE: (407) 363-0666
FAX: (407) 363-0666

**PRODUCTION RESOURCES INC.
OF SOUTH FLORIDA**
5710 S.W. 89TH AVE
MIAMI, 33173
PHONE: (305) 332-4803

PRODUCTION TECHNIQUES INC
7517 CAROLTON CIRCLE
TAMPA, 33619
PHONE: (813) 626-5990

**PROMISE ART &
ENTERTAINMENT**
713 FLAG WAY
KISSIMMEE, 34758
PHONE: (407) 841-5679
FAX: (407) 872-1258

**PROMOTIONAL
TECHNOLOGIES**
3643 N.E. 25TH ST
OCALA, 32670
PHONE: (800) 765-7626
FAX: (904) 629-4563

THE PROPTOLOGISTS
P.O. BOX 593921
ORLANDO, 32859
PHONE: (407) 240-7330

PUTTIN ON THE RITZ
4865 NORTH HARBOR CITY
BLVD
MELBOURNE, 32935
PHONE: (407) 259-0083

**THE RED BEARON (ANTIQUE
AIRCRAFT)**
1901 E. NEW HAVEN AVE
MELBOURNE, 32901
PHONE: (407) 727-8290

**REDD FROGE-PURVEYOR OF
CHEAP THRILLS**
P.O. BOX 10402
LARGO, 34643
PHONE: (813) 393-6510

RENTAL MART
1040 AURORA RD
MELBOURNE, 32935
PHONE: (407) 254-5386

**RINGS STUDIO GUN RENTAL
INC**
4651 36TH ST., SUITE 600
ORLANDO, 32811
PHONE: (407) 423-4079
FAX: (407) 872-0891

**ROYAL CHARTER AND
TRADING COMPANY (LUXURY
MOTOR YACHT)**
P.O. BOX 510851
MELBOURNE BEACH, 32951-0851
PHONE: (407) 984-3347

**ROSE MARIE, INC (COMMER-
CIAL FISHING FLEETS)**
2152 DUMAS ST
MERRITT ISLAND, 32952
PHONE: (407) 452-9271

RYDER TRUCK RENTAL
1180 S.W. 36TH AVE
POMPANO BEACH, 33069
PHONE: (305) 978-0033
FAX: (305) 978-0944

SHOW CAR-PIX CARS
160 E. 25TH ST
HIALEAH, 33013
PHONE: (305) 885-7909
FAX: (305) 821-0609

SOUTHERN SCENIC & THEATRICAL SUPPLY, INC
3320-C VINELAND RD
ORLANDO, 32811
PHONE: (407) 425-5787
FAX: (407) 872-1258

SUPPLY ROOM (MILITARY PROPS)
2110 SOUTH WASHINGTON AVE
TITUSVILLE, 32780
PHONE: (407) 267-4070

TANDOVA, INC
6010 N. ARMENIA AVE
TAMPA, 33604-5704
PHONE: (813) 877-6204

TANIT CORPORATION (RESEARCH VESSEL)
205 ORLANDO BLVD
INDIALANTIC, 32903
PHONE: (407) 951-7606

TOMMY THOMAS CHEVROLET
P.O.BOX 490
PANAMA CITY, 32402
PHONE: (904) 785-5521
FAX: (904) 763-0353

TYNDALL AIR FORCE BASE
USAF/ADWC/CC
TYNDALL AIR FORCE BASE, 32403
PHONE: (904) 283-4271

UNIQUELY FLORIDA INC
7530 CURRENCY DR
ORLANDO, 32809
PHONE: (407) 240-3333
FAX: (407) 240-7720

VALIANT AIR COMMAND (VINTAGE AIRCRAFT)
6600 TICO RD
TITUSVILLE, 32780
PHONE: (407) 268-1941

VILLAGE TRADING POST
313 DELANNEY AVE
COCOA, 32922
PHONE: (407) 783-5860
FAX: (407) 783-3847

ZELLER INTERNATIONAL LTD
623 ELLEN DR
WINTER PARK, 32789
PHONE: (407) 629-2905

LEWIS ZUCKER
3911 S.W. 60TH CT
MIAMI, 33155
PHONE: (305) 666-8305

PYROTECHNICS

AVP, INC
12155 METRO PARKWAY, S.E. SUITE 1
FT. MYERS, 33912-1332
PHONE: (813) 768-0500
FAX: (813) 768-0503

BOLAND HARDWARE
1088 HAVENDALE BLVD
WINTER HAVEN, 33881
PHONE: (813) 293-7355

JOIE CHITWOOD STUNT SHOW
4410 W. ALVA ST
TAMPA, 33614
PHONE: (813) 876-8946

FIRE SAFE OF CENTRAL FLORIDS INC
2644 C MICHIGAN AVE
KISSIMMEE, 34733
PHONE: (407) 870-1940
FAX: (407) 870-8052

THE GIBSON GROUP
P.O. BOX 2254
HAINES CITY, 33845
PHONE: (813) 676-6856

J.B. JONES, INC
13815 N.W. 19TH AVE
MIAMI, 33054
PHONE: (305) 681-7627
FAX: (305) 687-4737

SHAWN T. LAWLESS
231 TALLWOOD DR
CASSELBERRY, 32707
PHONE: (407) 354-6071
FAX: (407) 875-1115

LEE SLY EXPLOSIVES CONSULTANT
P.O. BOX 667
ONECO, 34264
PHONE: (813) 748-6635

LESTER KALMANSON AGENCY, INC
P.O. BOX 940008
MAITLAND, 32794-0008
PHONE: (407) 645-5000
FAX: (407) 645-2810

LIGHTS, CABLES & HEAVY STUFF
1300 N.W. 31ST AVE
FT. LAUDERDALE, 33311
PHONE: (305) 584-8108
FAX: (305) 584-1256

BRUCE E. MERLIN
5810 WAYT CT
ORLANDO, 32810
PHONE: (407) 293-4315

ORLANDO SPECIAL EFFECTS
3820 LAGUNA DR
ORLANDO, 32805
PHONE: (407) 648-1867

RINGS STUDIO GUN RENTAL INC
4651 36TH ST., SUITE 600
ORLANDO, 32811
PHONE: (407) 423-4079
FAX: (407) 872-0891

SANTORE & SONS FIREWORKS
P.O. BOX 1127
BUNNELL, 32010-1127
PHONE: (904) 437-2242

STAGE EQUIPMENT & LIGHTING INC
4602 S.W. 25TH ST
ORLANDO, 32811
PHONE: (407) 425-2010
FAX: (407) 648-2604

STUNT ACTION COORDINATORS INC
P.O. BOX 127
GROVELAND, 34736
PHONE: (904) 394-8893
FAX: (904) 429-4029

THE STUNT COMPANY INC
4651 36TH ST #600
ORLANDO, 32811
PHONE: (407) 422-3176/ (800) 782-5171
FAX: (407) 872-0891

VALIANT AIR COMMAND
6600 TICO RD
TITUSVILLE, 32780-8009
PHONE: (407) 268-1941
FAX: (407) 267-0327

ZELLER INTERNATIONAL
623 ELLEN DR
WINTER PARK, 32789
PHONE: (407) 629-2905

STUDIOS/SOUND STAGES

AMERICAN TELEVISION PROD.
1400 GULF SHORE BLVD. N. SUITE 208
NAPLES, 33940
PHONE: (813) 263-3383/ (800) 258-1298

CREATIVE MEDIA RECORDING
10051 5TH ST N SUITE 108
ST. PETERSBURG, 33702
PHONE: (813) 578-1926

CYPRESS PRODUCTIONS
5301 W. CYPRESS ST. #109
TAMPA, 33607
PHONE: (813) 875-8145

DISNEY-MGM STUDIOS
P.O. BOX 10,200
LAKE BUENA VISTA, 32830
PHONE: (407) 560-7299
FAX: (407) 827-5168

DOLAN & COMPANY PRODUCTIONS
1101-17 S. ROGERS CIRCLE
BOCA RATON, 33487
PHONE: (407) 994-6331
FAX: (407) 994-6332

EAGLE PRODUCTIONS CO.
6854 N.W. 77TH COURT
MIAMI,
PHONE: (305) 594-5674
FAX: (305) 477-5219

F & F PRODUCTIONS
10393 GANDY BLVD
ST. PETERSBERG, 33702
PHONE: (813) 576-7676

FT. LAUDERDALE PRODUC-TION CENTRAL
1300 N.W. 31ST AVE
FT. LAUDERDALE, 33311
PHONE: (305) 792-8108
FAX: (305) 584-1256

GREAT SOUTHERN STUDIOS
15221 N.E. 21ST AVE
N. MIAMI BEACH, 33162
PHONE: (305) 944-2464
FAX: (305) 944-9920

GREENWICH STUDIO CITY
12100 N.W. 16TH AVE
NORTH MIAMI, 33161
PHONE: (305) 899-9467
FAX: (305) 895-0560

IMAGE FILM & TAPE
5456 W. CRENSHAW ST
TAMPA, 33634
PHONE: (813) 885-7793
FAX: (813) 886-3034

INSIGHT & SOUND, INC
105 N.W. 2ND AVE
FT. LAUDERDALE, 33311
PHONE: (305) 767-0117/8
FAX: (305) 767-0119

LAGUNA STUDIOS
6854 N.W. 77TH COURT
MIAMI, 33166
PHONE: (305) 594-5674
FAX: (305) 477-5219

LAST STAGE OUT OF TOWN
2152 STATE RD., 557 N.
POLK CITY, 33868
PHONE: (800) 942-3448
FAX: (81) 956-5302

LIMELITE STUDIOS, INC
7355 N.W. 41ST ST
MIAMI, 33166
PHONE: (305) 593-6969
FAX: (305) 593-9785

MAKO PRODUCTIONS
545 S. ATLANTIC BLVD., SUITE 402
FT. LAUDERDALE, 33316
PHONE: (305) 523-7714

MANATEE 43 SMILE TV
925 26 AVE. E
BRADENTON, 34208
PHONE: (813) 747-1264

MIKE'S PRO-TRONICS
45 IRWIN ST
SAFETY HARBOR, 33572
PHONE: (813) 956-3448

MOON MANAGEMENT, INC
P.O. BOX 1717
TALLAHASSEE, 32302
PHONE: (904) 878-6900

NORTH FLORIDA SOUNDSTAGE
100 FESTIVAL PARK AVE
JACKSONVILLE, 32202
PHONE: (904) 353-7770
FAX: (904) 358-6381

PARALLAX PRODUCTIONS
4264 WESTROADS DR
W. PALM BEACH, 33407
PHONE: (407) 842-7788
FAX: (407) 842-4566

PATTERSON STUDIO
600 OVERLOOK DR
WINTER HAVEN, 33884
PHONE: (813) 324-3696

THE PRODUCER'S STUDIO
1323 63RD AVE. E
BRADENTON, 34203
PHONE: (813) 753-7277

PROPMASTERS MIAMI
9940 N.W. 79TH AVE
MIAMI, 33016
PHONE: (305) 826-1900
FAX: (305) 826-1850

REEL TO REEL RECORDING
970 E. LAKE DR
BARTOW, 33830
PHONE: (813) 533-4650

RITZ THEATERS, INC
217 HYDE PARK AVE
TAMPA, 33606
PHONE: (813) 251-6496

RON ROSE PRODUCTIONS
3409 W. LEMON ST., SUITE 6
TAMPA, 33609
PHONE: (813) 873-7700

PHIL SEIFLINE
P.O. BOX 187
CLEARWATER, 94617
PHONE: (813) 797-3663

T. SKORMAN PRODUCTIONS, INC
4700 L.B. MCLEOD RD
ORLANDO, 32811
PHONE: (407) 843-4300
FAX: (407) 843-7161

UNIVERSAL STUDIOS FLORIDA
1000 UNIVERSAL STUDIOS PLAZA
ORLANDO, 32819
PHONE :(407) 363-8400
FAX: (407) 363-8490

VERTICAL HOLD, INC
P.O. BOX 518
TAMPA, 33601
PHONE: (813) 222-1046
FAX: (813) 222-1057

WOFL PRODUCTIONS
35 SKYLINE DR
LAKE MARY, 32476
PHONE: (407) 644-3535
FAX: (407) 333-3535

WXEL-TV 42
P.O. BOX 6607
W. PALM BEACH, 33405
PHONE: (407) 737-8000

CHARLES ZELLEY
5110 FORMBY DR
ORLANDO, 32812

UNIONS AND ASSOCIATIONS

AFTRA
20401 N.W. 2ND AVE., SUITE 102
MIAMI, 33169
PHONE: (305) 652-4824

AICP, FLORIDA CHAPTER
830 LINCOLN RD
MIAMI BEACH, 33139
PHONE: (305) 672-4288

D.G.A.
136 MARGATE MEWS
ORLANDO, 32779
PHONE: (407) 682-6482

DIRECTORS GUILD OF AMERICA
2791 BIRD AVE
MIAMI, 33133
PHONE: (305) 448-6716

**FFRI - FLORIDA FILM &
RECORDING INSTITUTE**
1031 NORTH MIAMI BEACH BLVD
NORTH MIAMI, 33162
PHONE: (305) 947-0800
FILM SOCIETY OF MIAMI, INC
444 BRICKELL AVE., #229
MIAMI, 33131
PHONE: (305) 377-FILM
FAX: (305) 577-9768
**FLORIDA COUNCIL OF
CARPENTERS**
333 E. OAKRIDGE RD
ORLANDO, 32809
PHONE: (407) 859-4231
**FLORIDA FILM INDUSTRY
ALLIANCE**
P.O. BOX 2847
WEST PALM BEACH, 33402-2847
PHONE: (407) 650-1217
FMPTA - STATEWIDE
335 BEARD ST
TALLAHASSEE, 32302
PHONE: (904) 222-6000
FAX: (904) 681-2890
**FMPTA - GREATER
FT. LAUDERDALE CHAPTER**
7100 RADICE COURT, #106
LAUDERHILL, 33319
PHONE: (305) 486-6215
FMPTA - GAINSVILLE CHAPTER
2104 WEIMER HALL,
UNIVERSITY OF FLORIDA
GAINSVILLE, 32611
PHONE: (904) 392-5200
**FMPTA - MIAMI/SOUTH
CHAPTER**
9315 PARK DR., B-1
MIAMI, 33156
PHONE: (305) 757-8978
**FMPTA - NORTHEAST
JACKSONVILLE CHAPTER**
352 STOWE AVE
ORANGE PARK, 32073
PHONE: (904) 264-5610
FAX: (904) 264-3224
**FMPTA - NORTHEAST/
PENSACOLA CHAPTER**
616 SHEPPARD DR.,
P.O. BOX 17093
PENSACOLA, 32522-7093
PHONE: (904) 438-4444
FAX: (904) 438-4306
FMPTA
P.O. BOX 231
ORLANDO, 32802-0231
PHONE: (407) 843-7860
FMPTA - ORLANDO CHAPTER
1040 HUNTINGTON CT
LONGWOOD, 32750-2944
PHONE: (407) 834-3230
**FMPTA - PALM BEACH AREA
CHAPTER**
143 LIGHTHOUSE DR
NORTH PALM BEACH, 33408
PHONE: (407) 848-3100
FMPTA - POLK COUNTY CHAPTER
610 RALPH ST
AUBURNDALE, 33823
PHONE: (813) 967-8232
FMPTA - SARASOTA CHAPTER
1518 PINE BAY DR
SARASOTA, 34231
PHONE: (813) 921-1983
**FMPTA - SOUTHWEST/FT.
MYERS CHAPTER**
1417-2 DEL PRADO BLVD.,
SUITE 480
CAPE CORAL, 33990-3749
PHONE: (813) 283-5000
**FMPTA - SPACE COAST/
BREVARD CHAPTER**
C/O PHILLIPS JUNIOR COLLEGE
2401 N. HARBOR CITY BLVD
MELBOURNE, 32935
PHONE: (407) 254-6459

**FMPTA - ST. PETERSBURG/
CLEARWATER CHAPTER**
966 BRUCE AVE
CLEARWATER BEACH, 34630
PHONE: (813) 447-3080
**FMPTA - TALLAHASSEE
CHAPTER**
3710 RANDALL ST
TALLAHASSEE, 32308
PHONE: (904) 668-3373
FMPTA - TAMPA CHAPTER
P.O. BOX 2106
LUTZ, 33549
PHONE: (813) 949-7363
IATSE LOCAL 60 (MIXED)
105 GEORGIA DR
PENSACOLA, 32505
PHONE: (904) 455-5369
IATSE LOCAL 115 (STAGE)
3610 RIVER HALL DR
JACKSONVILLE, 32217
PHONE: (904) 731-7163
IATSE LOCAL 360
RT. 2, BOX 631M
BELL, 32619
PHONE: (904) 935-3699
IATSE LOCAL 361
1216 E. COLONIAL DR. #9
ORLANDO, 32803
PHONE: (407) 422-2747
IATSE LOCAL 412 (MIXED)
P.O. DRAWER Q
SARASOTA, 34230
PHONE: (813) 351-2145
IATSE STATEWIDE LOCAL 477
1035 N.E. 125TH ST., SUITE 205
NORTH MIAMI, 33161
PHONE: (305) 893-8585
IATSE STATEWIDE LOCAL 477
3610 RIVER HALL DR
JACKSONVILLE, 32217
PHONE: (904) 731-7163
IATSE LOCAL 545
1190 N.E. 125TH ST #25
NORTH MIAMI, 33161
PHONE: (305) 895-0026
IATSE LOCAL 631 (MIXED)
1206 33RD ST
ORLANDO, 32803
PHONE: (407) 422-2747
FAX: (407) 843-9170
IATSE LOCAL 646 (MIXED)
15 S.W. 7TH ST
FORT LAUDERDALE, 33301
PHONE: (305) 463-6175
IATSE LOCAL 798
3113 S. OCEAN DR
HALLANDALE, 33009
PHONE: (305) 453-7852
**IATSE LOCAL 853 (WARD-
ROBE)**
2851 N.E. 183RD ST., #1206
MIAMI, 33160
PHONE: (305) 931-6664
IBEW LOCAL 108
RT. 2 BOX 309 A
TAMPA, 33610
PHONE: (813) 621-2418
IBEW LOCAL 349
1657 N.W. 17TH AVE
MIAMI, 33125
PHONE: (305) 325-1330
IBEW LOCAL 606
820 VIRGINIA DR
ORLANDO, 32803
PHONE: (407) 896-7271
ITVA
1 SOUTHEAST FINANCIAL
CENTER MS 1042
MIAMI, 33131
PHONE: (305) 375-7265
**ITVA - JACKSONVILLE
CHAPTER**
1719 EMERSON ST
JACKSONVILLE, 32207
PHONE: (904) 396-7744

ITVA - MIAMI CHAPTER
136 UNIVERSITY PARK
MIAMI, 33199
PHONE: (305) 348-3669
ITVA - ORLANDO CHAPTER
4364 35TH ST
ORLANDO, 32811-6502
PHONE: (407) 649-0008
ITVA - PALM BEACH CHAPTER
P.O. BOX 24680
WEST PALM BEACH, 33416
PHONE: (407) 687-6069
ITVA - TAMPA BAY CHAPTER
5000 PARK ST. NORTH
ST. PETERSBURG, 33709
PHONE: (813) 541-7571
**INTERNATIONAL
PHOTOGRAPHERS LOCAL 666**
7061 GRAND NATIONAL DR.,
SUITE 105
ORLANDO, 32819
PHONE: (407) 352-1161
FAX: (407) 352-8902
MIAMI FILM SOCIETY
444 BRICKELL AVE., SUITE 229
MIAMI, 33131
PHONE: (305) 377-3456
NABET 15
P.O. BOX 40-2543
MIAMI BEACH, 33140-0543
PHONE: (305) 868-6658
**PAAF - PROFESSIONAL
ACTORS ASSOCIATION OF
FLORIDA**
P.O. BOX 610366
NORTH MIAMI, 33261-0266
PHONE: (800) 780-1427/ (305)
532-7735
**SAG - SCREEN ACTORS GUILD,
INC**
2299 DOUGLAS DR., SUITE 200
MIAMI, 33145
PHONE: (305) 444-7677
SAG
1190 NE 125 ST., SUITE 30
N. MIAMI, 33020
PHONE: (305) 891-9714
**SARASOTA FRENCH FILM
FESTIVAL**
555 N. TAMIAMI TRAIL
SARASOTA, 34243
PHONE: (813) 351-3446
**SOUTH FLORIDA SCREEN-
WRITERS GUILD**
800 WEST AVE., SUITE 745
MIAMI BEACH, 33139
PHONE: (305) 532-5600
TEAMSTERS
2940 N.W. 7TH ST
MIAMI, 33125
PHONE: (305) 642-6255
TEAMSTERS LOCAL 79
5818 E. BUFFALO AVE
TAMPA, 33619
PHONE: (813) 621-1391
FAX: (813) 626-7915
TEAMSTER LOCAL 385
122 N. KIRKMAN RD
ORLANDO, 32811
PHONE: (407) 298-7037
TEAMSTER LOCAL 512
10478 ATLANTIC BLVD
JACKSONVILLE, 32216
PHONE: (904) 642-1594
UNITED SCENIC ARTISTS
455 S.W. 185 TERRACE
MIAMI, 33157
PHONE: (305) 232-6801
**WOMPI - WOMEN IN THE
MOTION PICTURE INDUSTRY**
2870 DOUGLAS RD
MIAMI, 33133
PHONE: (305) 443-9213

GEORGIA

SPECIAL EFFECTS

ATLANTA FILM EFFECTS
2207 FAULKNER RD NE
ATLANTA, 30324
PHONE: (404) 634-5021
FAX: (404) 634-5021
**BLACK EAGLE SPECIAL
EFFECTS CO. INC.**
4005 WILLOW RIDGE RD
DOUGLASVILLE, 30135
PHONE: (404) 489-9945
**BOB SHELLEYS SPECIAL
EFFECTS INTERNATIONAL INC**
2000 LAKEWOOD WAY, SE
ATLANTA, 30315
PHONE: (404) 622-9245
FAX: (404) 622-9246
BRUCE LARSEN
1038 MCLENDON DR
DECATUR, 30033
PHONE: (404) 297-0336
**DICK CROSS SPECIAL EFFECTS
INC**
P.O. BOX 1198
CONYERS, 30207
PHONE: (404) 760-0435/0253
DOLPH F/X AND STUFF
2850 DELK RD, BLDG 7 STE. F
MARIETTA, 30067
PHONE: (404) 988-0378
DWIGHT BENJAMIN-CREEL
334 MARK AVE
MARIETTA, 30066
PHONE: (404) 425-4354
JOHN D. GREEN
409 1ST ST
POOLER, 31322
PHONE: (912) 748-4547/
925-3410
**MACK CHAPMAN SPECIAL
EFFECTS**
2000 LAKEWOOD AVE, BLDG 8
ATLANTA, 30315
PHONE: (404) 627-8679
ONE BALLOON PLACE INC
3908 SHIRLEY DR
ATLANTA, 30336
PHONE: (404) 696-9548/
(800) 642-5666
FAX: (404) 696-2226
RIVERWOOD STUDIO
600 CHESTLE HURST RD
SENOIA, 30276
PHONE: (404) 599-4000
FAX: (404) 599-4099
RON GOLDSMITH
2547 SCALPEM COURT
DULUTH, 30136
PHONE: (404) 449-7420
**SPECTACULAR EFFECTS
INTERNATIONAL
INCORPORATED**
2000 LAKEWOOD WAY
ATLANTA, 30315
PHONE: (404) 622-4015
TIM BARRETT DESIGNS
165 MANGUM ST
ATLANTA, 30313
PHONE: (404) 522-5508
FAX: (404) 222-0756

EQUIPMENT RENTALS

**BOB SHELLEYS SPECIAL
EFFECTS INTERNATIONAL INC.**
2000 LAKEWOOD WAY
BUILDING 2
S.E. ATLANTA, 30329
PHONE: (404) 622-9245

HAWAII

STUDIOS/SOUND STAGES

ALOHA STUDIOS
345 QUEEN ST., SUITE 400
HONOLULU, 96813
PHONE: (808) 523-6145
HAWAII PRODUCTION CENTER
1524 KAPIOLANI BLVD
HONOLULU, 96814
PHONE: (808) 944-5200

IDAHO

SPECIAL EFFECTS

MONGE, MARION C.
BOX 1391
KETCHUM, 83340
PHONE: (208) 788-5171/(805) 378-0077
TAYLOR, TONY
BOX 129
HAILEY, 83333
PHONE: (208) 788-2751
SWIGERT, KEVIN
BOX 411
KETCHUM, 83340
PHONE: (208) 774-3369
TURNER, JOE
103 S. 4TH ST
COEUR D'ALENE, 83814
PHONE: (208) 664-2466/ 667-6312

ILLINOIS

SPECIAL EFFECTS

FILMFAIR INC
22 WEST HUBBARD
CHICAGO, 60610
PHONE: (312) 822-9200
PRODUCTION DEPARTMENT INC
600 W. ROOSEVELT
WEST CHICAGO, 60185
PHONE: (312) 231-5077
STANDARD PHOTO SUPPLY
520 W. EIRIE
CHICAGO, 60610
PHONE: (312) 440-4920
VANNI, ROBERT
12 OLD COACH ROAD
SOUTH BARRINGTON, 60016
PHONE: (708) 382-3364

STUDIOS/SOUND STAGES

ACADEMY OF MOVEMENT & MUSIC
605 LAKE STREET
OAK PARK, 60302
PHONE: (708) 848-2329
AIR FAX PRODUCTIONS INC.
727 NORTH HUDSON
CHICAGO, 60610
PHONE: (312) 944-5577
ALTSCHULL, GILBERT
930 PITNER STREET
EVANSTON, 60202
PHONE: (798) 328-6700
BLOODWORTH, LAMAR
220 WEST LOCUST
CHICAGO, 60610
PHONE: (312) 751-7500

CAMPBELL STREET STUDIO
1934 N. CAMPBELL STREET
CHICAGO, 60647
PHONE: (312) 644-2371
CATTELL
1057 W. COLUMBIA
CHICAGO,
PHONE: (312) 222-0836
CENTER CITY STUDIOS
32 WEST RANDOLPH, SUITE 1400
CHICAGO, 60610
PHONE: (312) 984-3470
CENTRE EAST
7701 NORTH LINCOLN AVE
SKOKIE, 60077
PHONE: (708) 673-6305
CHICAGO STUDIO CITY
6656 W. TAYLOR
CHICAGO, 60644
PHONE: (312) 261-3400
CINEMA VIDEO CENTER
211 EAST GRAND AVE
CHICAGO, 60611
PHONE: (312) 644-1650
CUMBERLAND IMAGE WORKS
1317 W. GRAND
CHICAGO, 60622
PHONE: (312) 226-1616
DUPLICATION PLUS LTD
214 W. OHIO
CHICAGO, 60610
PHONE: (312) 337-4900
EDITEL GROUP
301 EAST EIRIE
CHICAGO, 60611
PHONE: (312) 440-2360
ESSANAY STUDIO
1345 WEST ARGYLE ST
CHICAGO, 60640
PHONE: (312) 989-8808
FILM CHICAGO
909 WEST DIVERSEY PARKWAY
CHICAGO, 60614
PHONE: (312) 528-8200
FILMFAIR INC.
22 WEST HUBBARD
CHICAGO, 60610
PHONE: (312) 822-9200
FREESE & FRIENDS, INCORPORATED
1429 NORTH WELLS
CHICAGO, 60610
PHONE: (312) 642-4475
GOLDEN PICTURES, LTD
125 SOUTH RACINE
CHICAGO, 60607
PHONE: (312) 226-8240
HARPO STUDIOS, INC
1058 W. WASHINGTON
CHICAGO, 60607
PHONE: (312) 738-3456
HOLLYWOOD STUDIOS
1932 S. HALSTED, 4TH FLOOR
CHICAGO, 60608
PHONE: (312) 226-2680
HORWICH STUDIOS INC
1801 N. HALSTED
CHICAGO, 60614
PHONE: (312) 944-9402
METROPOLITAN CHICAGO CORP
2500 W. ROOSEVELT ROAD
CHICAGO, 60608
PHONE: (312) 226-3434
MOTIVATION MEDIA, INC
1245 MILWAUKEE
GLENVIEW, 60025
PHONE: (708) 297-4740
NORTH AVENUE STAGE
455 NORTH AVENUE
CHICAGO, 60610
PHONE: (312) 642-3232
OPTIMUS
161 EAST GRAND AVE
CHICAGO, 60610
PHONE: (312) 321-0880

OSWEGO STREET STAGE
1648 W. KINZIE
CHICAGO,
PHONE: (312) 266-2292
POLYCOM CORPORATION OF AMERICA
142 E. ONTERIO
CHICAGO, 60611
PHONE: (312) 337-6000
POST EFFECTS
400 W. EIRIE STREET
CHICAGO, 60610
PHONE: (312) 944-1690
PRODUCTION DEPARTMENT INC
600 W. ROOSEVELT
WEST CHICAGO, 60185
PHONE: (312) 231-5077
R.A.H. PRODUCERS CENTER
700 SOUTH DES PLAINES
CHICAGO, 60607
PHONE: (312) 461-0445
SWELL PICTURES
233 EAST WACKER DR
CHICAGO, 60601
PHONE: (312) 649-9000
TELEMATION PRODUCTIONS
100 S. SANGAMON
CHICAGO, 60607
PHONE: (312) 421-4111
VIDEO DUB
214 WEST OHIO
CHICAGO,
PHONE: (312) 337-4900
WEBSTER PRODUCTIONS
220 WEST LOCUST
CHICAGO, 60610
PHONE: (312) 951-7500

UNIONS AND ASSOCIATIONS

AMERICAN FED OF TV & RADIO ARTISTS (AFTRA)
307 N MICHIGAN AVE #320
CHICAGO, 60601
PHONE: (312) 372-8081
CAMERAMEN IATSE LOCAL 666
327 S LASALLE ST #1122
CHICAGO, 60603
PHONE: (312) 341-0966
CHICAGO STUNTMEN'S ASSOCIATION
8800 S HARLEM AVE #1208
BRIDGEVIEW, 60455
PHONE: (312) 430-2815
MP LAB & FILM EDITORS IATSE
LOCAL 780
327 S LASALLE
CHICAGO, 60604
PHONE: (312) 922-7105
SCREEN ACTORS GUILD CHICAGO
307 N MICHIGAN AVE #200
CHICAGO, 60601
PHONE: (312) 372-8081
STUDIO MECHANICS IATSE LOCAL 476
327 S LASALLE
CHICAGO, 60604
PHONE: (312) 922-5215
TEAMSTERS (LOCAL 714)
6815 W. ROOSEVELT
BERWYN, 60402
PHONE: (312) 242-3215
THEATRICAL WARDROBE IATSE #769
3500 N LAKE SHORE DR
CHICAGO, 60657
PHONE: (312) 346-6111
UNITED SCENIC ARTISTS
343 S DEARBORN RM. 1114
CHICAGO, 60604
PHONE: (312) 431-0790

INDIANA

PROPS

BLUE MOON STUDIOS
2019 ALLISON ROAD
CAMBY, 46113
PHONE: (317) 831-2585
INDIANA REPERTORY THEATRE
140 W. WASHINGTON STREET
INDIANAPOLIS, 46204
PHONE: (317) 635-5277
SET-UP AND COMPANY INC., THE
1049 E. MICHIGAN
INDIANAPOLIS, 46202
PHONE: (317) 635-2323

STUDIOS/SOUND STAGES

AUDIO-VISUAL ASSOCS.
4760 E. 65TH ST
INDIANAPOLIS, 46220
PHONE: (317) 255-6457
BIG FISH PRODUCTIONS
111 E MAIN ST
GREENFIELD, 46140
PHONE: (317) 462-2644
CASSELL PRODUCTIONS
2950 E. 55TH ST
INDIANAPOLIS, 46220
PHONE: (317) 251-1201
DEAN CROW PRODUCTIONS
3445 N. WASHINGTON BLVD
INDIANAPOLIS, 46205
PHONE: (317) 926-3335
GENERAL PRODUCTION STUDIOS
10 N. TACOMA AVE
INDIANAPOLIS, 46201
PHONE: (317) 637-8771
GOLDEN DOME PRODUCTIONS
P.O. BOX 1616
SOUTH BEND, 46634
PHONE: (219) 239-1616
INDIANA REPERTORY THEATRE
140 W. WASHINGTON STREET
INDIANAPOLIS, 46204
PHONE: (317) 635-5277
QUANTUM GROUP, THE
101 W. 10TH STREET
INDIANAPOLIS, 46208
PHONE: (317) 639-6001
SANDERS & COMPANY
3610 N. MERIDIAN
INDIANAPOLIS, 46208
PHONE: (317) 926-2841F
USA TELPRODUCTIONS
1440 N. MERIDIAN ST
INDIANAPOLIS, 46202
PHONE: (317) 632-5900

IOWA

SPECIAL EFFECTS

ASSOCIATE PRODUCERS, INC
6545 BLOOMFIELD RD
DES MOINES, 50320
PHONE: (515) 285-1209
IOWA FIREWORKS PRODUCTIONS
RTE. 2, BOX 203F
SOLON, 52333
PHONE: (319) 848-4075
J.E.S. VIDEO
112 W. 4TH ST, P.O. BOX 254
WILTON, 52778
PHONE: (319) 732-3340

OGLESBY ENTPS/SCORPIO PROD.
3430 DODGE ST.
DUBUQUE, 52002
PHONE: (319) 588-4595/
(800) 827-1672
PYRO-TECHNIQUS
R.R. 1, BOX 144
WOODWARD, 50276
PHONE: (515) 438-2625

MAKE-UP

AMES COMMUNITY THEATER
427 DOUGLAS
AMES, 50010
PHONE: (515) 292-2636
DANE ART STUDIOS
3138 SW 9TH ST
DES MOINES, 50315
PHONE: (515) 280-9927
DES MOINES THEATRICAL SHOP
145 5TH ST
WEST DES MOINES, 50265
PHONE: (515) 274-3661
GLOBE THEATRICAL SUPPLY
813 PEARL ST
SIOUX CITY, 51101
PHONE: (712) 255-0972
IMAGE ARTISTS
213 4TH ST, SUITE 100
DES MOINES, 50309
MANGELSENS
3457 SOUTH 84TH
OMAHA, NE, 68124
PHONE: (402) 391-6225
MID COAST COMMUNICA-TIONS CENTRE
616 WEST 35TH ST
DAVENPORT, 52806
PHONE: (319) 388-9131
ROSES THEATRICAL SUPPLY
RR #10
WEST DES MOINES, 50265
PHONE: (515) 987-1105
TALENT/IOWA
6545 SE BLOOMFIELD RD
DES MOINES, 50320
PHONE: (515) 285-1209
THEATRE DEPARTMENT
UNIVERSITY OF NORTHERN IOWA
CEDAR FALLS, 50614
PHONE: (319) 273-6386
UNI
CAC 257
CEDAR FALLS, 50614-0357
PHONE: (319) 273-273-2217
UNIVERSITY OF IOWA
N. RIVERSIDE DR. - THEATRE BLDG.
DEPT. OF THEATRE ARTS
IOWA CITY, 52242
PHONE: (319) 335-2700

PROPS

A-1 RENTAL COMPANY
5320 23RD AVE
MOLINE, 61265
PHONE: (309) 762-7573
CIRCA 21 PRODUCTIONS
P.O. BOX 3784
ROCK ISLAND, 61204-3784
PHONE: (309) 786-2667
CYS RENTAL
312 E LOCUST
DAVENPORT, 52803
PHONE: (319) 324-1307
DOWRY ANTIQUES & COSTUMES
2001 LEECH AVE
SIOUX CITY, 51107
PHONE: (712) 255-8007

GIERKE-ROBINSON COMPANY
3929 W. RIVER DR
DAVENPORT, 52802
PHONE: (319) 322-1725
GLOBE THEATRICAL SUPPLY
813 PEARL ST
SIOUX CITY, 51101
PHONE: (712) 255-0972
GREGORY ANDERSON
507 E. 1ST #101
HUXLEY, 50124
PHONE: (515) 597-3488/
243-0766
HICKS, JENNIFER
1828 NW PINE ROAD
ANKENY, 50021
PHONE: (515) 964-8514
MARION ANTIQUE DEALERS ASSOCIATION
786 8TH AVE
MARION, 52302
PHONE: (319) 322-1437
OMAHA SCENIC STUDIOS
4641 NORTH 82ND
OMAHA, 68134
PHONE: (402) 572-8014
RAINBOW RECORDING STUDIOS, INC
2322 SOUTH 64TH AVE
RAINBOW PRODUCTIONS
OMAHA, 68106
PHONE: (402) 554-0123
RENT AMERICA
708 N BRADY ST
DAVENPORT, 52803
PHONE: (319) 324-0811
RIVER BEND ANTIQUES
425 BRADY ST
DAVENPORT, 52801
PHONE: (319) 323-8622/
326-0843
SIOUX CITY COMMUNITY THEATRE
P.O.BOX 512
SIOUX CITY, 51102
PHONE: (712) 233-2719
THE OLD CREAMERY THEATRE CO
P.O. BOX 160
GARRISON, 52229-0160
PHONE: (319) 477-3925
THE RENTAL CENTER
2411 W. CENTRAL PARK
DAVENPORT, 52804
PHONE: (319) 386-0177/
(309) 788-5615
THEATRE CR
102 3RD ST SE
CEDAR RAPIDS, 52401
PHONE: (319) 366-8592
UNIVERSITY OF IOWA
N. RIVERSIDE DR. - THEATRE BLDG. - DEPT OF THEATRE ARTS
IOWA CITY, 52242
PHONE: (319) 335-2700
VEST, RICHARD
P.O. BOX 542
IOWA FALLS, 50126
PHONE: (515) 648-3576/9220

PYROTECHNICS

ASSOCIATE PRODUCERS, INC
6545 BLOOMFIELD RD
DES MOINES, 50320
PHONE: (515) 285-1209
DANE ART STUDIOS
3138 SW 9TH ST
DES MOINES, 50315
PHONE: (515) 280-9927
IOWA FIREWORKS PRODUCTIONS
RTE. 2, BOX 203F
SOLON, 52333
PHONE: (319) 848-4075

PYRO-TECHNIQUS
R.R. 1, BOX 144
WOODWARD, 50276
PHONE: (515) 438-2625

UNIONS AND ASSOCIATIONS

CARPENTERS LOCAL 678
1638 CENTRAL
DUBUQUE, 52001
PHONE: (319) 582-8521
COMMUNICATION OF WORKERS OF AMERICA
4685 MERLE HAY RD., SUITE 203
DES MOINES, 50322
PHONE: (515) 278-5551
COMMUNICATION OF WORKERS OF AMERICA - LOCAL 7103
610 13TH ST
SIOUX CITY, 51105
PHONE: (712) 252-1308
COMMUNICATION OF WORKERS OF AMERICA - LOCAL 7108
1695 BURTON AVE
WATERLOO, 50703
PHONE: (319) 291-6601
COMMUNICATION OF WORKERS OF AMERICA - LOCAL 7117
1416 WEST 16TH ST
DAVENPORT, 52804
PHONE: (319) 324-0479
IBEW #109
1630 5TH AVE
MOLINE, 61265
PHONE: (309) 764-1163
IBEW - LOCAL UNION #145
2835 7TH AVE
ROCK ISLAND, 61201
PHONE: (309) 788-9591
IBEW - LOCAL #231
308 BADGEROW BLDG
SIOUX CITY, 51101
PHONE: (712) 255-8138
IBEW - LOCAL #704
1610 GARFIELD
DUBUQUE, 52001
PHONE: (319) 582-5947
INTERNATIONAL BROTHER-HOOD OF ELECTRICIANS #22
8946 L STREET
OMAHA, 68127
PHONE: (402) 331-8147
LABOR LOCAL 659
1638 CENTRAL
DUBUQUE, 52001
PHONE: (319) 583-0686
LOCAL UNION #238
5000 J STREET SW
CHAUFFERS, TEAMSTERS & HELPERS
CEDAR RAPIDS, 52404
PHONE: (319) 365-1461
MACHINISTS UNION LOCAL 1238
1610 GARFIELD
DUBUQUE, 52001
PHONE: (319) 583-0122
TEAMSTERS LOCAL 90
2425 DELAWARE
DES MOINES, 50317
PHONE: (515) 262-3194
TEAMSTERS LOCAL 383
700 COURT ST
SIOUX CITY, 51105
PHONE: (712) 252-2751
TEAMSTERS LOCAL 421
P.O. BOX 714
DUBUQUE, 52001
PHONE: (319) 583-9149

TEAMSTERS LOCAL 70/LOCAL 554
4359 SOUTH 90TH
OMAHA, 68127
PHONE: (402) 331-0550
TEAMSTERS LOCAL 828
404 15TH ST N.W.
MASON CITY, 50401
PHONE: (515) 424-3771
TB>UNION LABOR TEMPLE
1610 GARFIELD
DUBUQUE, 52001
PHONE: (319) 556-9410
WOODBURY COUNTY LABOR COUNCIL
1209 PIERCE, AFL-C O
SIOUX CITY, 51105
PHONE: (712) 258-7773
UNION LABOR TEMPLE
1610 GARFIELD
DUBUQUE, 52001
PHONE: (319) 556-9410
WOODBURY COUNTY LABOR COUNCIL
1209 PIERCE, AFL-C O
SIOUX CITY, 51105
PHONE: (712) 258-7773

KANSAS

SPECIAL EFFECTS

KAKE PRODUCTIONS (SC)
1500 NORTH WEST STREET
WICHITA, 67201
PHONE: (316) 943-4221

STUDIOS/SOUND STAGES

KWCH - TV (SC)
2815 EAST 37TH STREET
NORTH WICHITA 67201
PHONE: (316) 943-1212

KENTUCKY

MAKE-UP SUPPLIERS

TELEVISION THEATRICAL, STAGE LIGHTING SERVICES
67 BAXTER AVE
LOUISVILLE, 40204
PHONE: (502) 589-9675

PROPS

DREW, JAMES
P.O. BOX 4122
LOUISVILLE, 40204
PHONE: (502) 584-1725
HENSON, EBEN
PIONEER PLAYHOUSE, WILDERNESS ROAD
DANVILLE, 40422
PHONE: (606) 236-2747

STUDIOS/SOUND STAGES

ALLEN-MARTIN VIDEO PRODUCTIONS INC.
9701 TAYLORSVILLE ROAD
LOUISVILLE, 40299
PHONE: (502) 267-9653
CHANNEL 27 PRODUCTIONS
WKYT TELEVISION, 2851 WINCHESTER ROAD
LEXINGTON, 40505
PHONE: (606) 299-0411

IMAGE 32
WLKY TELEVISION, 1918
MELLWOOD AVE
LOUISVILLE, 40206
PHONE: (502) 893-3671
**KENTUCKY PRODUCTION
CENTER**
P.O. BOX 5590
LEXINGTON, 40555
PHONE: (606) 293-0539
LOUISVILLE PRODUCTIONS
520 WEST CHESTNUT STREET
LOUISVILLE, 40202
PHONE: (502)582-7744
**MERIDIAN TELECOMMUNICA-
TION SERVICES**
1401 N. MERIDIAN ST
INDIANAPOLIS, 46202
PHONE: (317) 636-2020
VIDEO EDITING SERVICES
134 NORTH LIMESTONE ST.,
SUITE 505
LEXINGTON, 40507
PHONE: (606) 255-9049
WDRB TELEVISION
624 WEST MOHAMMED ALI
BLVD
LOUISVILLE, 40203
PHONE: (502)584-6441

LOUISIANA

SPECIAL EFFECTS

ADAMS, PHILLIP
1013 RIDGEFIELD RD
THIBODAUX,70301
PHONE: (504) 447-5674
COOPER, ROBERT
1922 BULL RUN ROAD
SCHRIEVER, 70395
PHONE: (504) 447-5577
FIELDER, HENRY
96 WEEKS DR
YOUNGSVILLE, 70592
PHONE: (318) 856-9969
MCINTYRE, EDWARD
1554 MAGAZINE, APT. 2
NEW ORLEANS, 70130
PHONE: (504) 529-7261
SCHLESINGER, DAVID
1126 ALINE, APT.A
NEW ORLEANS, 70115
PHONE: (504) 891-0832

SPECIAL EFFECTS -
MATTE/OPTICAL &
ANIMATION

MORRISON PRODUCTIONS
8140 FORSHEY STREET
NEW ORLEANS, 70118
PHONE: (504) 486-8150

PROPS

A-1 SIGNS
3950 METROPOLITAN ST
NEW ORLEANS, 70126
PHONE; (504) 947-8381
FAX: (504) 947-8790
**ADRIAN'S PHOTOGRAPHY &
GUNS**
316 EAU CLAIR DR
THIBODAUX, 70301
**ATCHAFALAYA BASIN BACK-
WATER TOURS**
P. O. BOX128
GIBSON,70356
CELLUSETS
6320 W. 70TH ST
SHREVEPORT, 711239
PHONE; (318) 687-5773

DIBOLL SCALE MODELS
2624 ST. PHILIP ST
NEW ORLEANS, 70119
FIRESIDE ANTIQUES
14007 PERKINS RD
BATON ROUGE, 70810
PHONE: (504)292-9565/291-
7603
**HARTS LANDSCAPING
COMPANY**
12655 HARRELL'S FERRY RD
BATON ROUGE, 70816
PHONE: (504) 293-4922
SILENT STARS
206 ROOKS DR
SLIDELL, 70458
PHONE: (504) 649-6069
SMALL WONDERS
17202 CULPS BLUFF AVE
BATON ROUGE, 70817
PHONE: (504) 295-0660
**SPECTRUM ENGINEERING
CORPORATION**
9181 INTERLINE AVE
BATON ROUGE, 70809
PHONE: (504) 231-4600
THE POOL AND PATIO CENTER
3740 N. CAUSEWAY BLVD
METAIRIE 70002
PHONE: (504) 831-2022

UNIONS AND ASSOCIATIONS

AFTRA
2475 CANAL, STE. 108
NEW ORLEANS, 70119
PHONE: (504) 822-6568
IATSE #260
RT. 2, BOX 206D
LAKE CHARLES, 70605
PHONE: (318) 598-3455
IATSE #293
P. O. BOX 50293
NEW ORLEANS, 70150
PHONE: (504) 835-4329/831-
4514
IATSE #298
P. O. BOX 8767
SHREVEPORT, 71108
PHONE: (318) 227-2914
IATSE #39
P. O. BOX 19289
NEW ORLEANS, 70179
PHONE: (504) 486-2192
IATSE #400
P. O. BOX 4535
ALEXANDRIA, 71301
PHONE: (318) 442-6738
IATSE #668
200 CARLETON AVE
MONROE, 71202
PHONE: (318) 323-6253
IATSE #798
419 STONEYCREEK
BATON ROUGE, 70808
PHONE: (504) 767-2626/945-
5708
IATSE #840
4540 FOLSE DR
METAIRIE, 70002
PHONE: (504) 455-6003
**INT'L BROTHERHOOD OF
TEAMSTERS #270**
2207 ROYAL ST
NEW ORLEANS, 70117
PHONE: (504) 945-3152
**INT'L BROTHERHOOD OF
TEAMSTERS #969**
2004 HODGES ST
LAKE CHARLES, 70601
PHONE: (318) 433-5297
**INT'L BROTHERHOOD OF
TEAMSTERS #568**
P. O. BOX 7805
SHREVEPORT, 71137
PHONE: (318) 222-4681

**INT'L BROTHERHOOD OF
TEAMSTERS #5**
P. O. BOX 526, 1772 DALLS DR
BATON ROUGE, 70821
PHONE: (504) 924-3886/1757
SCREENACTORS GUILD (SAG)
2299 DOUGLAS RD., SUITE 200
MIAMI, FL 33145
PHONE: (305) 444-7677

MAINE

PROPS

ARCHITECTURAL ANTIQUITIES
HARBORSIDE, 04642
PHONE: (207) 326-4938
BATTLEMASTER
1 ALLEN LANE
BATH, 04530
PHONE: (207) 443-1035
CATS PAW PRODUCTIONS
P.O.BOX 328
EAST MACHIAS, 04630
PHONE: (207) 255-8006
COLLECTORS CAROUSEL
84 WARREN AVE
WESTBROOK, 04092
PHONE: (207) 854-0343
HANDS ON
P.O.BOX 430
WARREN, 04864
PHONE: (207) 273-2126
MIKAN TRACY THEATRICALS
150 HIGH ST
PORTLAND, 04101
PHONE: (207) 772-8860
MINOR ROOTS
U. SOUTHERN MAIN THEATER
GORHAM, 04038
PHONE; (207) 780-5148/
775-2941
NATIVE SON PRODUCTIONS
40 BEVERLY TERRACE
CAPE ELIZABETH, 04107
**OWLS HEAD TRANSPORTA-
TION MUSEUM**
P.O.BOX 277
OWLSHEAD 04854
PHONE: (207) 594-4418
SUPERS JUNKIN CO.
P.O.BOX 2425, RT. 102
BAR HARBOR, 04609
PHONE: (208) 288-5740

MAKE-UP SUPPLIERS

KOSMETICKOS
P.O.BOX 2355
AUGUSTA, 03228
PHONE: (207) 622-1329

UNIONS AND ASSOCIATIONS

AFTRA/SAG
11 BEACON ST. #512
BOSTON, MA 02108
IATSE LOCAL 114
P.O.BOX 993 25 GRANITE ST
PORTLAND 04104
PHONE:(207) 761-5878
IATSE LOCAL 161
(SCRIPT SUPERVISORS,
PRODUCTION OFFICE,
PRODUCTION AUDITORS)
1697 BROADWAY, SUITE 902
NEW YORK, NY 10019
PHONE: (212) 956-5410
FAX: (212) 489-7325

**IATSE LOCAL 481 - NEW
ENGLAND STUDIO MECHANICS**
PHONE: (617) 482-7370/ (802)
893-7067
FAX: (802) 893-6517
**IATSE LOCAL 644 - INTERNA-
TIONAL PHOTOGRAPHERS OF
MP & TV**
505 8TH AVE., 16TH FLOOR
NEW YORK, NY 10018
PHONE: (212) 581-0771
FAX: (212) 581-0825
**IATSE LOCAL 798 - MAKE-UP
ARTISTS AND HAIRSTYLISTS**
31 WEST 21ST ST
NEW YORK, NY 10010
PHONE: (212) 627-0660
FAX: (212) 627-0664
IATSE LOCAL 921
815 WASHINGTON ST
NEWTONVILLE, MA 02160
PHONE: (617) 244-8179
FAX: (617) 244-5854
**TEAMSTERS UNION LOCAL
2340**
144 THADEUS ST.,
P.O. BOX 2290
SOUTH PORTLAND, 04106
PHONE: (207) 767-2106
FAX: (207) 767-7315

MARYLAND

EXPENDABLES

BALTIMORE STAGE LIGHTING INC
10-16 AZAR COURT
BALTIMORE 21227
PHONE: (301) 242-3322
FAX: (301) 242-5242

MODEL MAKERS

E.B. ASSOCIATES
6 WOODHUE COURT
BALTIMORE, 21207
PHONE: (301) 922-7442
FAX: (301) 922-1613
MODEL MAKERS
2114 NORTH CHARLES ST
BALTIMORE, 21218
PHONE: (301) 547-8989
FAX: (301) 783-1239
WAYNE NICOLETTE
5813 CLEARSPRING RD
BALTIMORE 21212
PHONE: (301) 435-1546
CAROLYN THOME
12628 ST. JAMES RD
ROCKVILLE, 20850
PHONE: (301) 340-0755

PROPS

**BALDY MORES HARDCORE &
LOOK-ALIKE CASTING**
P.O. BOX 241
SUNSHINE AT US #1
KINGSVILLE, 21087
PHONE: (301) 335-2270
E.B. ASSOCIATES
6 WOODHUE COURT
BALTIMORE, 21207
PHONE: (301) 922-7442
FAX: (301) 922-1613
TE.J. GRANT ANTIQUES
P.O. BOX 124
SAVAGE, 20763
PHONE: (301) 792-4538
FILM ARTS, INC
3500 PARKDALE AVE
BALTIMORE, 21211
PHONE: (301) 728-8570
FAX: (301) 728-8573

MODEL MAKERS
2114 NORTH CHARLES ST
BALTIMORE, 21218
PHONE: (301) 547-8989
FAX: (301) 783-1239
PROPS AND SETS
913 SOUTH WOLFE ST
BALTIMORE, 21231
PHONE: (301) 342-9482
FAX: (301) 675-9348

PYROTECHNICS

GDC/FX
P.O. BOX 9783
ARNOLD, 21012
PHONE: (301) 647-5722
THEATRE EFFECTS, INC
642 FREDERICK ST
HAGERSTOWN, 21740
PHONE: (301) 791-7646
FAX: (301) 791-7719

MASSACHUSETTES

SPECIAL EFFECTS

D4 FILM STUDIOS
749 CHARLES RIVER STREET
NEEDHAM, MA, 02192
PHONE: (617) 235-1119
FALLMAN PRODUCTIONS
55 GRACE STREET
MALDEN, MA, 02148
PHONE: (617) 322-4571
FRAME SHOP
25 LOS ANGELES STREET
NEWTON,02158
PHONE: (617) 964-0300
LOGO MOTION, INC.
1380 SOLDIERS FIELD ROAD
BOSTON, MA, 02135
PHONE: (617) 783-3535
PENPOINT ANIMATION
331 NEWBURY STREET
BOSTON,02115
PHONE: (617) 266-1331
VIDEOCRAFT PRODUCTIONS INC
29 NEWBURY ST
BOSTON,02116
PHONE: (617) 236-2200

STUDIOS/SOUND STAGES

BOYD ESTUS PRODUCTIONS
HELITROPE STUDIOS LTD., 21
ERIE STREET
CAMBRIDGE, 02139
PHONE: (617) 868-0171
TELEX: 499 7268
CHARLES RIVER STUDIOS
1380 SOLDIER'S FIELD ROAD
BOSTON, 02135
PHONE: (617) 783-3535
GBH PRODUCTIONS
125 WESTERN AVE
BOSTON, 02134
PHONE: (617) 492-9273
GILMORE, ROBERT ASSOCI-ATES
990 WASHINGTON ST
DEDHAM, 02026
PHONE: (617) 329-6633
SEPTEMBER PRODUCTIONS INC
171 NEWBURY STREET
BOSTON, 02116
PHONE: (617) 262-6090
VIDEOCOM, INC
502 SPRAGUE ST
DEDHAM, 02026
PHONE: (617) 329-4080

VIZWIZ, INC
115 DUMMER STREET
BROOKLINE, 02146
PHONE: (617) 739-6400
WBZ-TV
1170 SOLDIERS FIELD RD
ALLSTON, 02134
PHONE: (617) 787-7000
WGBY-TV
44 HAMPDEN ST
SPRINGFIELD, 01103
PHONE: (413) 781-2801
WGGB-TV
1300 LIBERTY SQUARE
SPRINGFIELD, 01104
PHONE: (413) 733-4040
WLVI-TV
75 MORRISSEY BLVD
DORCHESTER, 02125
PHONE: (617) 265-5656
WNEV-TV
7 BULFINCH PLACE
BOSTON, 02114
PHONE: (617) 725-0777
WSBK-TV
83 LEO BIRMINGHAM
PARKWAY
BOSTON, 02135
PHONE: (617) 783-3838
WAVE INC.
72 CAMBRIDGE ST
WORCESTER, 01603
PHONE: (508) 795-7100

UNIONS AND ASSOCIATIONS

AFTRA-SAG
11 BEACON ST
BOSTON. 02116
PHONE: (617) 752-2688
WOMEN IN FILM & VIDEO
NEW ENGLAND CHAPTER
71 CHERRY ST
CAMBRIDGE, 02139
PHONE: (617) 876-3821

MICHIGAN

STUDIOS/SOUND STAGES

AVP, INC.
2330 BRYD DR
KALAMAZOO, 49002
PHONE: (616) 332-5030
ASSOCIATES COMMUNICA-TIONS INC
2101 SOUTH TELEGRAPH
BLOOMFIELD HILLS, 48013
PHONE: (313) 332-8009
AUDIO GRAPHIC SERVICES
1516 FERRIS AVE
ROYAL OAK, 48067
PHONE: (313) 544-1793
BOULEVARD PHOTOGRAPHIC INC
151 VICTOR
HIGHLAND PARK, 48203
PHONE: (313) 868-2200
CAPTURED LIVE PRODUC-TIONS, INC.
4911 FERNLEE
ROYAL OAK, 48073
PHONE: (313) 288-4080
CITY ANIMATION CO
57 PARK STREET
TROY, 48083-2753
PHONE: (313) 589-0600
CITY LIGHTS STAGE, INC
679 E. MANDOLINE
MADISON HEIGHTS, 48071
PHONE: (313) 589-9000

COMBERMERE STAGE, THE
1350 COMBERMERE
TROY, 48083
PHONE: (313) 583-2800
CONTINENTAL CABLEVISION
27800 FRANKLIN ROAD
SOUTHFIELD, 48034
PHONE: (313)353-3908
FRANKLIN, DAVID STUDIOS, INC
23953 RESEARCH DR
FARMINGTON HILLS, 48024
PHONE: (313) 471-9000
FUTURE MEDIA CORPORATION
2853 W. JOLLY ROAD
OKEMOS, LANSING, 48864
PHONE: (517) 332-5560
FAX: (517) 332-5080
GENERAL TELEVISION NETWORK
13225 CAPITAL AVENUE
OAK PARK, 48237
PHONE: (313) 548-2500
FAX: (313) 548-8614
GRACE & WILD STUDIOS
23689 IND. PARK DR
FARMINGTON HILLS, 48024
PHONE: (313) 471-6010
GREENWELL, ANDY
37448 HILLS TECH DR
FARMINGTON HILLS, 48331
PHONE: (313)489-5777
GROUP W. CABLE OF WAYNE
35102, MICHICAN AVE
WAYNE, 48184
PHONE: (313) 729-1960
JONES-RASIKAS INC
800 BOND AVE NW
GRAND RAPIDS, 49503
PHONE: (616) 556-1855
KASPER STUDIOS
17903 KURON RIVER DR
NEW BOSTON, 48164
PHONE: (313) 753-9100
KEN MUSIC PHOTOGRAPHY
326 E. FOURTH ST
ROYAL OAK, 48067
PHONE: (313) 544-2441
LAROCHE, ANDREW PHOTOGRAPHY INC.
32588 DEQUINDRE
WARREN, 48092
PHONE: (313)978-8932
MVP COMMUNICATIONS INC
1075 RANKIN
TROY, 48083
PHONE: (313) 588-7600
FAX: (313)588-1899
MARITZ COMMUNICATIONS
600 W.LAFFAYETTE
DETROIT, 48226
PHONE: (313) 963-1200
MOTION PICTURE MAKERS INC
6660 28TH ST. SOUTH EAST
GRAND RAPIDS, 49506
PHONE: (616) 949-5744
PRODUCERS COLOR SERVICE
24242 NORTHWESTERN
HIGHWAY
SOUTHFIELD, 48075
PHONE: (313) 352-5353
PRODUCTION CENTER, THE
151 VICTOR AVENUE
HIGHLAND PARK, 48203
PHONE: (313)868-6600
PROJECTIONS INCORPORATED
2855 BOARDWALK
ANN ARBOR, 48104
PHONE: (313)665-8015
SGA PRODUCION STAGING
2222 SPIKES
LANSING, 48906
PHONE: (517) 372-5278
SPECIAL RECORDINGS INC
1600 NORTH WOODWARD
BIRMINGHAM. 48012
PHONE: (313) 644-1352

STAGE III PRODUCTIONS, INC
32588 DEQUINDRE
WATTEN, 48092
PHONE: (313) 978-7373
STUDIO CENTER
1 STUDIO CENER, 23801 IND.
PARK DR
FARMINGTON HILLS, 48024
PHONE: (313) 471-1110
T.G.A. RECORDING CO.
295 URBANDALE AVE
BENTON HARBOUR, 49022
PHONE: (616) 926-7581
VISION COMMUNICATIONS/ WTVS-56
7441 SECOND BLVD
DETROIT, 43202
PHONE: (313) 873-7200/876-8110
FAX: (313) 876-8118
WARD, LES PHOTOGRAPHY, INC
21477 BRIDGE ST , SUITE C
SOUTHFIELD, 48034
PHONE: (313) 350-8665

MINNESOTA

STUDIOS/SOUND STAGES

COMPUTER VIDEO PRODUC-TIONS/CMG
1317 CLOVER DRIVE SOUTH
BLOOMINGTON, 55420
PHONE: (612) 888-2388
FAX: (612) 888-679*
CY DECOSSE INCORPORATED
5900 GREEN OAK DR
MINNETONKA, 55343
PHONE: (612) 935-4788/00
FAX: (612) 933-1456
GANNETT PRODUCTION SERVICES
8811 OLSON MEMORIAL
HIGHWAY
MINNEAPOLIS, 55427
PHONE: (612) 541-8054
FAX: (612) 541-8019
GREER & ASSOCIATES INC
905 PARK AVE
MINNEAPOLIS, 55404
PHONE: (612) 333-6171
FAX: (612) 338-2522
ICE HOUSE STUDIO, INC
2540 NICOLLET AVE
MINNEAPOLIS, 55404
PHONE: (612) 870-0716
LIGHTHOUSE INC
1401 3RD AVE SOUTH
MINNEAPOLIS, 55404
PHONE: (612)872-4565
MAINSTREAM COMMUNICA-TIONS, INC
9555 JAMES AVENUE SOUTH,
SUITE 235
BLOOMINGTON, 55431
PHONE: (612) 888-9000
FAX: (612) 888-3494
METRO TELEPRODUCTIONS
4808 PARK GLENN RD
MINNEAPOLIS, 55416
PHONE: (612) 922-3454
NORTHWEST TELEPRODUCTIONS INC
4455 WEST 77TH ST
MINNEAPOLIS, 55435
PHONE: (612)835-4455
PAISLEY PARK STUDIOS
7801 AUDOBON RD
CHANHASSEN, 55317
PHONE: (612) 474-8555

MISSOURI

SPECIAL EFFECTS

CINE SERVICES INC
1326 HANLEY IND. COURT
ST. LOUIS, 63144
PHONE: (314) 968-2200

GOLDBLATT, STEVE
4904 CENTRAL STREET
KANSAS CITY, 64111
PHONE: (816) 531-4889

HORSEDRAWN PRODUCTIONS
ROUTE 2, BOX 358
GARDEN CITY, 64747
PHONE: (816) 862-6536

JENKINS, JOEL
3605 CENTRAL STREET
KANSAS CITY, 64111
PHONE: (816) 931-0580

LILLARD, STEVEN W.
324 SOUTH 25TH ST
LEXINGTON, 64067
PHONE: (816) 259-3476

OWENS ENTERPRIZES & ASSOCIATES
1800 CHERRY ST
KANSAS CITY, 64108
PHONE: (816) 842-1820
FAX: (816) 471-3478

PROFIT, SCOTT
5500 WARD PARKWAY
KANSAS CITY, 64113
PHONE: (816) 523-2699

STAMPS, ED
ROUTE 2, BOX 128-A
GARDEN CITY, 64747
PHONE: (314) 862-6536

STREETT, J.D.
2100 NORTH 2ND AVE
ST. LOUIS, 63015
PHONE: (314) 727-4554

WINCHESTER, BARRON
7766 ATTINGHAM
ST. LOUIS, 63110
PHONE: (314) 962-8008

MAKE-UP SUPPLIERS

A T C
307 WEST 80TH ST
KANSAS CITY, MO, 64114
PHONE: (816)523-1655

ALLIED THEATRE CRAFT
224 WEST 5TH ST
KANSAS CITY, 64105
PHONE: (816)421-3980

S E C T THEATRICAL SUPPLIES INC
406 EAST 18TH ST
KANSAS CITY, MO, 64108
PHONE: (816) 471-1239

MODEL MAKERS

AMERICAN MODEL BUILDERS, INC
1408-10 HANLEY INDUSTRIAL COURT
ST. LOUIS, 63144
PHONE: (314) 968-3076

HAUCK, WALTER
4022 WEST PINE
ST. LOUIS, 63108
PHONE: (314) 535-3978

ROSLEVICH DESIGNS
484 LAKE AVE
ST. LOUIS, 63108
PHONE: (314) 361-5174

TURNER & ASSOCIATES
5251 PATTISON
ST. LOUIS, 63110
PHONE: (314) 771-9132
FAX: (314) 773-5886

PROPS

BARTA, CAROL
800 WEST GREGORY BLVD
KANSAS CITY, 64114
PHONE: (816) 444-4040

EBY-PAIDRICK DESIGNS, INC
7202 VIRGINIA AVE
ST. LOUIS, 63116
PHONE: (314) 351-1705

EISTERHOLDLLEWELLYN
218 DELAWARE #110
KANSAS CITY, 64105
PHONE: (816) 283-3537

GARRISON WAGNER & COMPANY
2020 DELMAR
ST. LOUIS, 63013
PHONE: (314) 241-9010

HORSEDRAWN PRODUCTIONS
ROUTE 2, BOX 358
GARDEN CITY, 64747
PHONE: (816) 862-6536

IMAGINATION SHOP, THE
1326 HANELY INDUSTRIAL COURT
ST. LOUIS, 63119
PHONE: (314) 968-2200

LILLARD, STEVEN W.
324 SOUTH 25TH ST
LEXINGTON, 64067
PHONE: (816) 259-3476

MATERIAL POSSESSIONS
4111 MCGEE
KANSAS CITY, 64111
PHONE: (816) 756-1929

OLDE THEATRE ARCHITECTURE & SALVAGE COMPANY
2045 BROADWAY
KANSAS CITY, 64108
PHONE: (816) 283-3740

OLD WORLD ANTIQUES LTD
1715 SUMMIT
KANSAS CITY, 64108
PHONE: (816) 472-0815

OMNI MODELS, INC
106 SOUTHWEST BLVD
KANSAS CITY, 64108
PHONE: (816) 474-9747

RAN ADLER FLOWERS
504 CAMPBELL ST
KANSAS CITY, 64106-1212
PHONE: (816) 842-4255

SAM BEAUS LTD
4724 MCPHERSON AVE
ST. LOUIS, 63108
PHONE: (314) 361-4636
FAX: (314) 361-2216

S E C T THEATRICAL SUPPLIES INC
406 EAST 18TH ST
KANSAS CITY, MO, 64108
PHONE: (816) 471-1239

SPINNAKER DESIGN COLLECTIVE
1215 DIELMAN INDUSTRIAL COURT
ST. LOUIS, 63132
PHONE: (314) 997-0202

SUFFICIENT GROUNDS
520 GILLIS
KANSAS CITY, 64106
PHONE: (816) 221-3685

STUDIOS/SOUND STAGES

HARDCASTLE FILM & VIDEO
7319 WISE AVE
ST. LOUIS, 63117
PHONE: (314) 647-4200

INNERVISION PRODUCTIONS, INC
11783 BORMAN DR
ST. LOUIS, 63117
PHONE: (314) 569-2500

KDNL-TV, CHANNEL 30
1215 COLE STREET
ST. LOUIS, 63106
PHONE: (314) 436-3030

KMOX-TV, CHANNEL 4
1 SOUTH MEMORIAL DR
ST. LOUIS, 36102
PHONE: (314) 621-2345

KPLR-TV, CHANNEL 11
4935 LINDELL BLVD
ST. LOUIS, 63108
PHONE: (314) 367-7211

KSDK-TV, CHANNEL 5
1000 MARKET ST
ST. LOUIS, 63101
PHONE: (314) 421-5055

LACLEDE COMMUNICATIONS
2675 SCOTT AVE., SUITE G
ST. LOUIS, 63102
PHONE: (314) 535-3999

MCC VIDEO PRODUCTIONS
14563 WEST 96TH TERRACE
LENEXA, KS., 66215
PHONE: (913) 888-7304

METRO VIDEO PRODUCTIONS INC
238 WEST 74TH ST
KANSAS CITY, MO.,64114
PHONE: (816) 444-7004

PREMIER FILM & RECORDING COMPANY
3033 LOCUST ST
ST. LOUIS, 63103
PHONE: (314) 531-3555

TECHNISONIC STUDIOS
1201 SOUTH BRENTWOOD BLVD
ST. LOUIS, 63117
PHONE: (314) 727-1055

VIDEO PUBLIC RELATIONS
1609 WEST 92ND ST
KANSAS CITY, MO, 64114
PHONE: (816) 444-8988

WAPT-TV
P.O. BOX 10297
JACKSON, MS ,39209
PHONE: (601) 922-1607

UNIONS AND ASSOCIATIONS

ACTORS EQUITY
1079 DARWICH
ST. LOUIS, 63132
PHONE: (314) 994-7738/ 367-4268

AFTRA, LOCAL 213
406 WEST 34TH ST, 206
KANSAS CITY, MO, 64111
PHONE: (816) 753-4557

AFTRA/SAG
906 OLIVE ST #1006
ST. LOUIS, 63101
PHONE: (314) 231-8410

CARPENTERS DISTRICT COUNCIL OF ST. LOUIS
1401 HAMPTON AVE
ST. LOUIS, 63139
PHONE: (314) 644-4800

IATSE, LOCAL 6
1611 SOUTH BROADWAY
ST. LOUIS, 63104
PHONE: (314) 621-5077
FAX: (314) 621-5709

IATSE, LOCAL 6
3460 HAMPTON AVE, ROOM 104
ST. LOUIS, 63139
PHONE: (314) 352-4432

IATSE, LOCAL 31
304 WEST 10TH ST, 102
KANSAS CITY, 64105
PHONE: (816) 842-5167

IATSE, LOCAL 421
P.O. BOX 1051
MARION, 62959
PHONE: (314) 334-6407

IBEW, LOCAL 4
4406 ST. VINCENT AVE
ST. LOUIS, 63119
PHONE: (314) 647-2288

INTERNATIONAL TELEVISION ASSOCIATION (ITVA)
C/O INNERVISION PRODUCTIONS
11783 BORMAN DR
ST. LOUIS, 63146
PHONE: (314) 569-2500/ 344-7275

KANSAS CITY FEDERATION OF MUSICIANS, LOCAL
2539 BROADWAY
KANSAS CITY, MO, 64108
PHONE: (816) 221-6934

PAINTERS UNION 1156
235 CLARK
ST. LOUIS, 64111
PHONE: (314) 946-4580

ST. LOUIS INDEPENDENT PRODUCTION PROFESSIONALS (SLIPP)
7536 FORSYTH BLVD #119
ST. LOUIS, 63105
PHONE: (314) 768-1197

TEAMSTERS JOINT COUNCIL 41
4501 VAN BRUNT BLVD
KANSAS CITY, 64130
PHONE: (816) 924-2000

TEAMSTERS JOINT COUNCIL 56
4501 VAN BRUNT BLVD
KANSAS CITY, 64130
PHONE: (816) 924-1650

TEAMSTERS, LOCAL 610
300 SOUTH GRAND
ST. LOUIS, 63103
PHONE: (314) 533-3517

NEBRASKA

MAKE UP SUPPLIERS

MANGELSENS
3457 SOUTH 84TH
OMAHA, 68124
PHONE: (402) 391-6225

PROPS

CHRISTENTEN LUMBER CO.
201 SOUTH MAIN
FREMONT, 68025
PHONE: (402) 721-3212/5239

DESIGNER ART COMPANY, THE
6306 CHARLES ST
OMAHA, 69132
PHONE: (402) 556-1625

FROM NEBRASKA
12TH & "O" ST - CENTRUM
LINCOLN, 68508
PHONE: (402) 476-2455/ 471-2122

METROPOLITAN STYLE
3421 "O" ST
LINCOLN, 68510
PHONE: (402) 377-4787

OMAHA SCENIC STUDIOS
4641 NORTH 82ND
OMAHA, 68134
PHONE: (402) 572-8014

PETERSON, TERRY
204 NORTH PLATTE
FREMONT, 68025
PHONE: (402) 721-6040/ 72709713

RAINBOW RECORDING STUDIOS, INC
2322 SOUTH 64TH AVE
RAINBOW PRODUCTIONS
OMAHA, 68106
PHONE: (402) 554-0123

STAUFFER, ED
705 SOUTH 40TH ST
LINCOLN, 68510
PHONE: (402) 472-2074/
488-4934

STUDIOS/SOUND STAGES

LODES/PETERSON PRODUCTIONS
4859 SOUTH 97TH ST
OMAHA, 68127
PHONE: (402) 592-7230/331-
5442

UNIONS AND ASSOCIATIONS

IATSE - LOCAL 151
6645 BALLARD
LINCOLN, 68507
PHONE: (402) 471-7166/466-2731
IATSE
2501 GINDY DR
OMAHA, 68147
PHONE: (402) 731-5907
**INTERNATIONAL BROTHER-
HOOD OF ELECTRICIANS #22**
8946 L STREET
OMAHA, 68127
PHONE: (402) 331-8147
**LINCOLN CENTRAL LABOR
UNION**
4625 "Y" ST
LINCOLN, 68503
PHONE: (402) 466-5080
NABET - LOCAL 45
1705 SOUTH 154TH CIRCLE
OMAHA, 68144
PHONE: (402) 333-2474
NEBRASKA STATE AFL - CIO
4660 SOUTH 60TH AVE
OMAHA, 68117
PHONE: (402) 734-1300
**OMAHA MUSICIANS
ASSOCIATION**
4535 LEAVENWORTH ST.,
SUITE #124
OMAHA, 68106
PHONE: (402) 553-5818
**TEAMSTERS LOCAL 70/LOCAL
554**
4359 SOUTH 90TH
OMAHA, 68127
PHONE: (402) 331-0550

NEVADA

SPECIAL EFFECTS

CAMEL PRODS
1105 TERMINAL WAY #126
RENO, 89502
PHONE: (702) 688-6250
**CINEMA SERVICES OF LAS
VEGAS**
4445 S VALLEY VIEW #708
LAS VEGAS, 89103
PHONE: (702) 876-4667
FOY ENTERPRISES
3275 E. PATRICK LANE
LAS VEGAS, 89120
PHONE: (702) 454-3300
LAS VEGAS VIDEO & SOUND
4221 THIRIOT ST
LAS VEGAS, 89103
PHONE: (702) 362-4660
MAGIC LATERN PRODS
P.O. BOX 93984
LAS VEGAS, 89193
PHONE: (702) 366-7934
THINGS COMPANY
3930 VANESSA DR
LAS VEGAS, 89103
PHONE: (702) 871-0035

MAKE-UP SUPPLIERS

DALE PRODUCTION SERVICES
P.O. BOX 4102
INCLINE VILLAGE, 89450
PHONE: (702) 831-2122
LENZ AGENCY
1591 E. DESERT INN RD #100
LAS VEGAS, 89109
PHONE: (702) 733-6888
**PHILIP E. SHELBURNE (SPECIAL
F/X MAKE-UP)**
904 N. 20TH ST
LAS VEGAS, 89127
PHONE: (702) 642-6461
THEATRE ARTS GROUP
1612 METROPOLITAN ST.
LAS VEGAS, 89102
PHONE: (702) 877-6463
VEGAS ENT. INTL
101 CONVENTION CTR. DR.
#1202
LAS VEGAS, 89109
PHONE: (702) 794-0052/732-
0995
WILD FX (SPECIAL F/X MAKE-UP)
534 BAYBERRY DR
LAS VEGAS, 89110
PHONE: (702) 438-6140/2800

PROPS

A & D SCENERY
3200 SIRIUS
LAS VEGAS, 89102
PHONE: (702) 362-9404
ANIMATED ELECTRONICS
1700 E DESERT INN DR #407
LAS VEGAS, 89109
PHONE: (702) 731-0148
ANTIQUES (OLD TOWN)
HWY. 95
GOLDFIELD, 89013
PHONE: (702) 485-3233
A. ANCIENT SLOTS & ANTIQUES
3127 INDUSTRIAL RD
LAS VEGAS, 89109
PHONE: (702) 796-7779/ (800)
322-SLOT
AVIATION CLASSICS
4825 TEXAS AVE
STEAD FIELD, 89506
PHONE: (702) 972-5540/5626
CAMRAC STUDIOS
1775 KUENZLI LN
RENO, 89502
PHONE: (702) 323-0965
DALE PRODUCTION SERVICES
P.O. BOX 4102
INCLINE VILLAGE, 89450
PHONE: (702) 831-2122
GAMBLERS GENERAL STORE
800 S. MAIN ST
LAS VEGAS, 89101
PHONE: (702) 382-9903/
(800) 322-CHIP
THE GUN STORE
2900 E. TROPICANA
LAS VEGAS, 89121
PHONE: (702) 454-1110
KST PROPS
4224 LOSSEE RD #B
N. LAS VEGAS, 89030
PHONE: (702) 642-8378
LANGWORTHY CASINO SUPPLY
611 N MAIN ST
LAS VEGAS, 89101
PHONE: (702) 382-9903
**LAS VEGAS THEATRICAL
DRAPERIES**
4350 ARVILLE #15
LAS VEGAS, 89103
PHONE: (702) 362-8408
LAS VEGAS VIDEO & SOUND
4221 THIRIOT ST
LAS VEGAS, 89103
PHONE: (702) 362-4660

OLD VEGAS
2440 BOULDER HIGHWAY
HENDERSON, 89015
PHONE: (702) 564-1311
THE PLASTIC MAN
3919 RENATE DR.
LAS VEGAS, 89103
PHONE: (702) 362-2133
STAGECOACH PRODS
2695, LINDELL
LAS VEGAS, 89102
PHONE: (702) 876-6697
STUDIO KAMINSKI, LTD
1040 MATLEY LANE
RENO, 89502
PHONE: (702) 786-2615
THE SURVIVAL STORE
3250 POLLUX AVE
LAS VEGAS, 89102
PHONE: (702) 871-7795
THINGS COMPANY
3930 VANESSA DR
LAS VEGAS, 89103
PHONE: (702) 871-0035
VEGAS ENT. INTL
101 CONVENTION CTR. DR.
#1202
LAS VEGAS, 89109
PHONE: (702) 794-0052/732-0995

PYROTECHNICS

DAVID J. BARKER
5855 NATURES DR
LAS VEGAS, 89122
PHONE: (702) 454-7007
KERBY BROS. PRODS
P.O. BOX 4575
LAS VEGAS, 89127
PHONE: (702) 648-7797
THE SURVIVAL STORE
3250 POLLUX AVE
LAS VEGAS, 89102
PHONE: (702) 871-7795

STUDIOS/SOUND STAGES

A-1/S.I.R. SOUNDSTAGE 1
3780 SCIPPS WAY
LAS VEGAS, 89103
PHONE: (702) 364-0203
ACT 1
1300 BRACKEN AVE
LAS VEGAS, 89104
PHONE: (702) 366-0836/367-
7371
AIRLIX
797 HARMON #44
LAS VEGAS, 89119
PHONE: (702) 796-9192
**ALTA HAM BLACK BOX
THEATER**
UNLV/DEPT. THEATRE ARTS
LAS VEGAS, 89154
PHONE: (702) 739-3535/3737
AXE-TRAX RECORDING STUDIO
1558 LINDA WAY
SPARKS, 89431
PHONE: (702) 358-9463
BILL DAGER
5867 NATURE DR
LAS VEGAS, 89122
PHONE: (702) 435-8733
CAMRAC STUDIOS
1775 KUENZLI LN
RENO, 89502
PHONE: (702) 323-0965
GBA MUSIC
1514 S. EASTERN AVE
LAS VEGAS, 89104
PHONE: (702) 334-9228
**GRANNYS HOUSE REC.
STUDIOS**
1515 PLUMAS ST
RENO, 89509
PHONE: (702) 736-2622

KINNEY/HANCOCK PRODS
560 E. OAKEY BLVD
LAS VEGAS, 89104
PHONE: (702) 731-1770
KAMINSKI STUDIOS
1040 MATLEY LN
RENO, 89502
PHONE: (702) 786-2615
KCRL (TV 4-NBC)
1790 VASSAR
RENO, 89502
PHONE: (702) 322-9145
KLAS (TV 8-CBS)
3228 CHANNEL 8 DR.
LAS VEGAS, 89114
PHONE: (702) 733-8850
KOLO (TV 8-ABC)
4850 AMPERE DR
RENO, 89502
PHONE: (702) 786-8880
KNPB (TV-5)
P.O. BOX 14730
RENO, 89503
PHONE: (702) 784-4555
KTNV (TV 2-CBS)
4925 ENERGY WAY
RENO, 89502
PHONE: (702) 786-2212
KTNV (TV 13-ABC)
3355 S. VALLEY VIEW
LAS VEGAS, 89102
PHONE: (702) 876-1313
KUPR FILM & TV PRODS
3111 S. VALLEY VIEW #A-112
LAS VEGAS, 89102
PHONE: (702) 364-1247
KVBC (TV 3-NBC)
1500 FOREMASTER LN
LAS VEGAS, 89101
PHONE: (702) 642-3333
LAS VEGAS VIDEO & SOUND
4221 THIRIOT ST
LAS VEGAS, 89103
PHONE: (702) 362-4660
MAGIC LANTERN PRODS
P.O. BOX 93984
LAS VEGAS, 89193
PHONE: (702) 366-7934/
368-5825
**NEVADA AUDIO VISUAL
SERVICES**
3062 SHERIDAN ST
LAS VEGAS, 89102
PHONE: (702) 876-6272
N.T.V. PRODS
5665 S. VALLEY VIEW #4
LAS VEGAS, 89118
PHONE: (702) 795-2688
OAKDALE STUDIOS
4549 OAKDALE
LAS VEGAS, 89121
PHONE: (702) 458-3133
THE PALLADIUM
3665 INDUSTRIAL RD
LAS VEGAS, 89109
PHONE: (702) 733-6366
SAMUEL J. SCOTT PRODS
2810 S. MARYLAND PKWY
LAS VEGAS, 89109
PHONE: (702) 792-9211
SCOT RAMMER MUSIC
5074 HIBBETTS CT
LAS VEGAS, 89103
PHONE: (702) 362-2462
**SOUNDWORKS RECORDING
STUDIOS**
2570 E TROPICANA #18
LAS VEGAS, 89121
PHONE: (702) 362-7943
STUDIO 96
3896 SWENSON AVE
LAS VEGAS, 89119
PHONE: (702) 735-8153

TAKE ONE
2708 HIGHLAND
LAS VEGAS, 89109
PHONE: (702) 871-5077

VEGAS VALLEY PRODS
25 TV-5 DR
LAS VEGAS, 89014
PHONE: (702) 435-5555

WAREHOUSE DIST. SERVICE
3920 S. EASTERN AVE
LAS VEGAS, 89109
PHONE: (702) 876-7723/5806

WILDERNESS LOCATIONS & PRODS
P.O. BOX 10260
S. LAKE TAHOE, 95731
PHONE: (916) 577-3008/
(800) 874-7488

UNIONS AND ASSOCIATIONS

I.A.T.S.E. LOCAL 363
P.O. BOX 5278
RENO, 89513
PHONE: (702) 786-2286

I.A.T.S.E. LOCAL 720
3000 SOUTH VALLEY VIEW BLVD
LAS VEGAS
PHONE: (702) 873-3451

NEVADA SCREENWRITERS GROUP (NSG)
101 CONESTOGA DR. #17
CARSON CITY, 89706
PHONE: (702) 885-8773/
826-0939

PAVCA
2375 E. TROPICANA #155
LAS VEGAS, 89119
PHONE: (702) 795-2424/
454-1067

SCREEN ACTORS GUILD OF NEVADA
3305 W. SPRING MTN RD. #60
LAS VEGAS, 89102
PHONE: (702) 367-8217

SEG OF NEVADA
2505 MASON AVE
LAS VEGAS, 89102
PHONE: (702) 878-5530

TEAMSTERS LOCAL 533
700 RYLAND, SUITE 11
RENO, 89502

TEAMSTERS LOCAL 995
300 SHADOW LANE
LAS VEGAS, 89106
PHONE: (702) 385-0995

NEW JERSEY

SPECIAL EFFECTS

AQUA-EFFECTS, INC
928 EAST HAZELWOOD AVE
RAHWAY, 07065
PHONE: (800) 451-2436
FAX: (201) 382-1651

3D/FX BY RALPH CIRELLA
75 PARKSIDE RD
BEDMINSTER, 07921
PHONE: (201) 781-6323

MATTHEW FEUER
12 VIBURNUM COURT
LAWRENCEVILLE, 08648
PHONE: (609) 396-4460

GEMINI LIGHTING TELEVISION PRODUCTIONS
68 MAPLE PLACE
KEYPORT, 07735
PHONE: (201) 264-5236

HILL THEATRE STUDIO SPECIAL SERVICES
35 WEST BROAD ST
PAULSBORO, 08066
PHONE: (609) 423-8910
FAX: (609) 224-0224

IMAGEFFECTS
21 RICHWOOD PLACE
DENVILLE, 07834
PHONE: (201) 586-2521

KIMBERLY THEATRICS
7C MARLEN DR
TRENTON, 08691
PHONE: (609) 587-7927/
(215) 389-7503
FAX: (609) 587-7867

PRO-SET
86 LACKAWANNA AVE
P.O. BOX 661
WEST PATERSON, 07424
PHONE: (201) 256-6626
FAX: (201) 256-4047

JACK N. RIDNER ASSOCIATES
112 WATER ST
TINTON FALLS, 07724
PHONE: (201) 542-3548

MAKE-UP SUPPLIERS

BAL-ZAC MISTY MORN STUDIO
263 MAIN ST
WEST ORANGE, 07052
PHONE: (201) 763-5072

EASTERN COSTUME
373 ROUTE 46 WEST
FAIRFIELD, 07006
PHONE: (201) 575-3503

HALLS COSTUMES & DANCE SUPPLY
680 HIGH ST
BURLINGTON, 08016
PHONE: (609) 386-5720

IMAGEFFECTS STUDIOS
21 RICHWOOD PLACE
DENVILLE, 07834
PHONE: (201) 586-2521

IRVINGS THEATRICAL MAKEUP
305 EAST RIDGEWOOD AVE
RIDGEWOOD, 07450

PROPS

AVIS RENT-A-CAR
500 HIGHWAY 34
MATAWAN, 07747

BEITEL DISPLAYS, INC
1880 PRINCETON AVE
LAWRENCEVILLE, 08648
PHONE: (609) 393-5512

CARS OF YESTERDAY RENTALS
US HIGHWAY 9 W
ALPINE, 07620
PHONE: (201) 784-0030

CHATTIN AWNING AND TENT COMPANY
601 NASSAU ST
NORTH BRUNSWICK, 08902
PHONE: (201) 214-9400

GARDEN STATE SIGN COMPANY
BOX 953
LAKEWOOD, 08701
PHONE: (201) 363-7645

HIBERNIA AUTO RESTORATIOS
MAPLE TERRACE
HIBERNIA, 97842
PHONE: (201) 627-1882

SUSAN HOGAN
331 MAITLAND AVE
TEANECK, 07666
PHONE: (201) 837-2144

ISLAND HEIGHTS STUDIO OF ART
CENTRAL & OCEAN AVENUES
P.O. BOX 552
ISLAND HEIGHTS, 08732
PHONE: (201) 929-0719

NAVY ARMS COMPANY INC
689 BERGEN BLVD
RIDGEGIELD, 07657
PHONE: (201) 945-2500

OF RARE VINTAGE
718 COOKMAN AVE
ASBURY PARK, 07712
PHONE: (201) 988-9459

PACKARD INDUSTRIES
76 MONROE ST
BOONTON, 07005
PHONE: (201) 334-2400

RELIABLE PLASTICS, INC
777 NORTH AVENUE EXT.
DUNELLEN, 08812
PHONE: (201) 752-1177

JACK RINDNER ASSOCIATES
112 WATER ST
TINTON FALLS, 07724
PHONE: (201) 542-3548

SUPERIOR SIGNAL COMPANY INC
WEST GREYSTONE RD
P.O. BOX 96
SPOTSWOOD, 08884
PHONE: (201) 251-0800

TRENCH IMPKO
1605 JOHN ST
FORT LEE, 07024
PHONE: (201) 461-6650

STUDIOS/SOUND STAGES

ALLSCOPE INC
P.O. BOX 4060
PRINCETON, 08540
PHONE: (212)925-4005/ (609)
799-4200

CNBC
2200 FLETCHER AVE
FORT LEE, 07024
PHONE: (201) 585-2622

HENRY CHARLES MOTION PICTURE STUDIOS
PLAINFIELD AVE
EDISON, 08818
PHONE: (201) 545-5104

COLOR LEASING, INC
330 ROUTE 46 EAST
FAIRFIELD, 07006
PHONE: (201) 575-1118

DAK PRODUCTIONS
41 JERSEY AVE
NEW BRUNSWICK, 08901
PHONE: (201) 247-4740

GORAJ LIGHTING AND STUDIO
2020 NEW JERSEY AVE
HADDON HEIGHTS, 08035
PHONE: (201) 546-6578

HILL THEATRE STUDIO
35 WEST BROAD ST
PAULSBORO, 08066
PHONE: (609) 423-891

INSTRUCTIVISION, INC
3 REGENT ST
LIVINGSTON, 07039
PHONE: (201) 992-9081

INTERNATIONAL MEDIA SERVICES
718 SHERMAN AVE
PLAINFIELD, 07060
PHONE: (201) 756-4060

LEWIS STUDIOS
344 KAPLAN DR
FAIRFIELD, 07006
PHONE: (201) 227-1234

R.J. MARTIN COMPANY
315 ROUTE 17
PARAMUS, 07652
PHONE: (201) 967-0005

LEO MEISTER PRODUCTIONS, INC
321 RIVER RD
NUTLEY, 07110
PHONE: (201) 667-2323

MIX PRODUCTIONS
81 MT. OLIVE ROAD
BUDD LAKE, 07828
PHONE: (201) 691-6919
FAX: (201) 366-9634

PHOTGRAPHIC HOUSE
P.O. BOX 825
158 WEST CLINTON ST
DOVER, 07801
PHONE: (201) 366-3000

PRESS BROADCASTING COMPANY
P.O. BOX 94, PRESS PLAZA
ASBURY PARK, 07712
PHONE: (201) 776-7744

THE PRUDENTIAL AUDIO AND VISUAL CENTER
213 WASHINGTON ST
NEWARK, 07101
PHONE: (201) 877-7831

RJO PRODUCTIONS, INC
102-108 MARYLAND AVE
PATERSON, 07503
PHONE: (201) 523-2500

THEATRE OF UNIVERSAL IMAGES
1020 BROAD ST
P.O. BOX 10036
NEWARK, 07101
PHONE: (201) 596-0407/
622-3458

ULTRA PHOTO WORKS, INC
468 COMMERCIAL AVE
PALISADES PARK, 07650
PHONE: (201) 592-7730

VIDEOCENTER OF NJ, INC
228 PARK AVE
EAST RUTHERFORD, 07073
PHONE: (201) 935-0900

VIDEO RESOURCE
140 CENTENNIAL AVE
PISCATAWAY, 08854
PHONE: (201) 457-8880

VIDEO STATION INC
374 SPRINGFIELD AVE
SUMMIT, 07901
PHONE: (201) 273-0024

UNIONS AND ASSOCIATIONS

IATSE LOCAL 21 - NEWARK
MILLBURN MALL
2933 VAUXHALL RD
VAUXHALL, 07088
PHONE: (201) 964-4033

IATSE LOCAL 49
1024 LIBERTY AVE
NORTH BERGEN, 07047
PHONE: (201) 864-5398

IATSE LOCAL 77, ATLANTIC CITY
190 C-1 OLD ZION RD
LINWOOD, 08221
PHONE: (609) 927-5959

IATSE LOCAL 408, CAMDEN
212 ARDMORE AVE
WESTMONT, 08108
PHONE: (201) 858-3854

IATSE LOCAL 895
(WARDROBE, ATLANTIC CITY)
21 SOUTH BENSON
MARGATE, 08402
PHONE: (609) 822-2854

NEW YORK

SPECIAL EFFECTS

A.P.A.
230 WEST 10TH ST
NYC, 10014
PHONE: (212)675-4894
ASSOCIATES & FERREN
PHONE: (516) 537-7800
BESTEK, INC
218 WEST HOFFMAN AVE
LINDENHURST, 11757
PHONE: (516) 225-0707
BROOKLYN MODEL WORKS
60 WASHINGTON AVE
BROOKLYN, 11205
PHONE: (718) 834-1944
DALE MALLIE & CO
34-31 35TH ST
ASTORIA, 11106
PHONE: (212)706-1233
DEVLIN VIDEOSERVICE
1501 BROADWAY, SUITE 408
NEW YORK, 10036
PHONE: (212) 391-1313
FAX: (212) 391-2744
DIRECT IMAGE
14 CAMERON ST
HUNTINGTON, 11743
PHONE: (516) 385-7056
EFEX SPECIALISTS, INC
35-39 37TH ST
LONG ISLAND CITY, 11101
PHONE: (718) 937-2417
FAX: 937-3920
FRENCH, ED
(SPECIAL MAKE-UP F/X)
70-78 WILLOUGHBY ST
BROOKLYN, 11201
PHONE: (718) 852-6149
THE MAKE-UP SHOPPE LTD
(SPECIAL MAKE-UP/CREATURE
F/X)
42 HURON RD
BELLEROSE VILLAGE, 11001
PHONE: (516) 437-4297/4929
PRODUCERS EAST MEDIA, INC
734 WALT WHITMAN DR
ROUTE 110, SUITE 204
MELVILLE, 11747
PHONE: (516) 421-4800
**THEATRICAL SERVICES AND
SUPPLIES, INC**
170 OVAL DR
CENTRAL ISLIP, 11722
PHONE: (516) 421-4800
**TIFFEN MANUFACTURING
COMPANY**
90 OSER AVE
HAUPPAUGE, 11788
PHONE: (516) 273-2500
VIDEO CENTER/CGL, INC
DESERVERSKY CONFERENCE
CENTER
NORTHERN BLVD
OLD WESTBURY, 11568
PHONE: (516) 626-3570
WEISS, PETER DESIGNS
32 UNION SQUARE EAST
NYC, 10003
PHONE: (212) 477-2659
WIZARDWORKS
39-40 21ST ST., L.I.C.
NYC, 11101
PHONE: (718) 786-8383
**ZELLER, GARY SPECIAL
EFFECTS**
MAIN STREET, P.O. BOX 2
DOWNSVILLE, 13755
PHONE: (212)627-7676/
(607) 363-7792

EQUIPMENT RENTALS

DEVLIN VIDEOSERVICE
1501 BROADWAY, SUITE 408
NEW YORK, 10036
PHONE: (212) 391-1313
FAX: (212) 391-2744
TEFEX SPECIALISTS, INC
35-39 37TH ST
LONG ISLAND CITY, 11101
PHONE: (718) 937-2417
FAX: 937-3920

MAKE-UP SUPPLIERS

**COLOR ME BEAUTIFUL,
CREATED IMAGES**
309 PHILADELPHIA AVE
MASSAPEQUA, 11762
PHONE: (516) 795-6274
**ENTERTAINMENT UNLIMITED
ARTISTS BUREAU CORP.**
64 DIVISION AVE
LEVITTOWN, 11756
PHONE: (516) 735-5550
FRENCH, ED
(SPECIAL MAKE-UP F/X)
70-78 WILLOUGHBY ST
BROOKLYN, 11201
PHONE: (718) 852-6149
HOUSE OF COSTUMES, LTD
166 JERICHO TURNPIKE
MINEOLA, 11501
PHONE: (516) 294-0170
KELLY, BOB COSMETICS INC
151 WEST 45TH ST
NYC, 10036
PHONE: (212)819-0030
**KRYOLAN PROFESSIONAL
MAKE-UP**
ALCONE COMPANY, INC.,
PARAMOUNT THEATRICAL
SUPPLY
575 8TH ST
NYC, 10018
**LONG BEACH THEATRE GUILD,
INC**
BOX 406
LONG BEACH, 11561
PHONE: (516) 889-1314
THE MAKE-UP SHOPPE LTD
(SPECIAL MAKE-UP/CREATURE
F/X)
42 HURON RD
BELLEROSE VILLAGE, 11001
PHONE: (516) 437-4297/4929
MAKING FACES LTD
46 SURREY LANE
HEMPSTEAD, 11550
PHONE: (516) 489-3291
**THEATRICAL SERVICES AND
SUPPLIES, INC**
170 OVAL DR
CENTRAL ISLIP, 11722
PHONE: (516)348-0262
FAX: (516) 348-4842
WILLIAM MATT SALON
83 WESTBURY AVE
CARLE PLACE, 11514
PHONE: (516) 7407-4433

MODELS/MINIATURES

LE DERNIER CRI, LTD
28 CHESTNUT ST
GREENVALE, 11548
PHONE: (718) 484-1737
MALLIE, DALE & COMPANY
35-30 38TH ST
ASTORIA, 11101
PHONE: (718) 706-1233
**MONTE SCOTT PEEPER
ARCHITECTS**
414 FOXHURST RD
OCEANSIDE, 11572
PHONE: (516) 678-1978

RMB DRAFTING SERVICES, INC
128 NORMAN DR
EAST MEADOW, 11554
PHONE: (516) 794-2198
SOUNDSMITH
755 NEW YORK AVE, SUITE 105
HUNTINGTON, 11743-4240
PHONE: (516) 673-5747

PROPS

ADVANCED RESTORATIONS
123 JERICHO TURNPIKE
FLORAL PARK, 11001
(516) 488-5222
ALTMANS LUGGAGE
135 ORCHARD ST
NYC, 10002
PHONE: (212) 254-7275
AMUSEMENT EMPORIUM, INC
2560 SUNRISE HIGHWAY
BELLMORE, 11710
PHONE: (516) 783-7840
FAX: (516) 826-7470
ANKRA
2200 JERICHO TURNPIKE
GARDEN CITY PARK, 10040
PHONE: (516) 747-5252
ANTIQUE CARRIAGES
942 CONNTEQUOT AVE
NORTH GREAT RIVER, 11722
PHONE: (516) 277-4967
ANTIQUES AT TRADERS COVE
230 TRADERS COVE
PORT JEFFERSON, 11777
PHONE: (516) 331-2261
BALDWIN CHIROPRACTIC
2108 GRAND AVE
BALDWIN, 11510
PHONE: (516) 378-6800
BARGAIN SPOT
64 3RD AVE
NYC, 10003
PHONE: (212)674-1188
**BAYVIEW FLORIST AND
MONTAGE**
4644 MERRICK RD
MASSAPEQUA, 11758
PHONE: (516) 799-7222
CAMEO GALLERY, INC
5-09 BURNS AVE
HICKSVILLE, 11801
PHONE: (516) 9699
CAPTAIN HOWARD UNGER
80 BEACH RD
GREAT NECK, 11020
PHONE: (516) 829-9800/
(212) 639-3578
CHATEAU STABLES INC
608 W 48TH ST
NYC, 10036
PHONE: (212)246-0520
**CHESTER A. FULTON & SON
FUNERAL HOME, INC**
49 W. MERRICK RD
FREEPORT, 11520
PHONE: (516) 378-3401
CLASSIC CONVERTIBLES, LTD
74 FIREPLACE RD
EAST HAMPTON, 11520
PHONE: (516) 324-1018
COLLECTORS COMICS
3247 SUNRISE HIGHWAY
WANTAGH, 11793
PHONE: (516) 783-8700
FAX: (516) 783-6372
**CRADLE OF AVIATION
MUSEUM**
MUSEUM LANE, MITCHEL FIELD
GARDEN CITY, 11530
PHONE: (516) 222-1191
DALE MALLIE & CO
34-31 35TH ST
ASTORIA, 11106
PHONE: (212)706-1233

DESIGNED LABORATORIES
ONE GARDEN ST
GARDEN CITY, 11530
PHONE: (516) 746-2586
DIVE SHOP
1152 BROADWAY
HEWLETT, 11557
PHONE: (516) 374-7756
**EAST MEADOW BUSINESS
MACHINES INC**
2410 HEMPSTEAD TURNPIKE
EAST MEADOW, 11554
PHONE: (516) 795-1332/
(800) 637-7721
FAX: (516) 796-1375
ECLECTIC PROPERTIES INC
620 WEST 26TH ST
NYC, 10001
PHONE: (212)645-8380
F.E.S. AUCTIONS INC
9 PARK LANE
ROCKVILLE CENTRE, 11570
PHONE: (516) 764-7459
FOLLOW YOUR ART INC
28 EAST PARK AVE
LONG BEACH, 11561
PHONE: (516) 431-6262
**FRANKLIN FUNERAL HOME,
INC**
42 NEW HYDE PARK RD
FRANKLIN SQUARE, 11010
PHONE: (516) 775-9491
FRITZ, HENRY J
2480 PARK AVE
BALDWIN, NY, 11510
PHONE: (516) 223-4799
GEMCO WARE, INC
1 GEMCO PLACE
FREEPORT, 11520
PHONE: (516) 432-4196
G. GLESSMAN INC
58 NEWMANS COURT
HEMPSTEAD, 11550
PHONE: (516) 481-9561
FAX: (516) 481-9562
GILBERT, ROBERT P.
2169 DECKER AVE
MERRICK, 11566
PHONE: (516) 623-5553
GRAFFITI UNLIMITED
739 FRANLK N AVE
GARDEN CITY, 11530
PHONE: (516) 741-5539
**GUY & COMPANY HAIR
DSIGNERS**
629 EAST PARK AVE
LONG BEACH, 11561
PHONE: (516) 889-9333/
432-9222
ISLAND HELICOPTER
NORTH AVE
GARDEN CITY, 11530
PHONE: (516) 228-9355
**J. MORRISON BETTER CARS,
INC**
117 3RD ST
GARDEN CITY, 11530
PHONE: (516) 741-1322
KENMORE FURNITURE CO. INC
352 PARK AVENUE SOUTH
NYC, 10010
PHONE: (212) 683-1888
KIDZART, INC
347 GLEN COVE AVE
SEA CLIFF, 11579
PHONE: (516) 671-6613
**LEWIS & VALENTINE LAND-
SCAPING, INC**
627 CEDAR SWAMP RD
GLEN HEAD, 11545
PHONE: (516) 671-1200
LIGHTER THAN AIR ENT. INC
235 JERICHO TURNPIKE
FLORAL PARK, 11001
PHONE: (516) 488-3121

LIN-DEL ASSOCIATES
964 EAST BROADWAY
WOODMERE, 11598
PHONE: (516) 374-0173

LIST, DANIEL ANTIQUE CARS
45 CHRISTOPHER ST
NYC, 10014
PHONE: (212) 255-6068

LONG BEACH TEATRE GUILD, INC
BOX 406
LONG BEACH, 11561
PHONE: (516) 889-1314

LUCCI AUTO PROPS
749-757 HICKS ST
BROOKLYN, 11231
PHONE: (212) 624-6050

MEAN STREET CUSTOM AUTOMOTIVE
4188 AUSTIN BLVD
ISLAND PARK, 11558
PHONE: (516) 889-1375

MEMORY LANE RECORDS
1321 GRAND AVE
BALDWIN, 11510
PHONE: (516) 623-2247

MERCY AMBULANCE-AMBULETTE
P.O. BOX 194
ISLAND PARK, 11558
PHONE: (516) 431-0444

MITCHEL ALAN FURS
1798 MERRICK RD
MERRICK, 11566
PHONE: (516) 546-4419

OCALLAGHAN, KEVIN
PHONE: (516) 751-1673

OBSOLETE FLEET
45 CHRISTOPHER ST
NYC, 10014
PHONE: (212)255-6068

OLDE KRAFT CO, LTD
60 PEBBLE LANE
ROSLYN HEIGHTS, 11577

ORIGINAL ST. JAMES GENERAL STORE
516 MORICHES RD
ST. JAMES, 11780
PHONE: (516) 862-8333

OYSTER BAY FRAME SHOP
95 AUDREY AVE
OYSTER BAY, 11771
PHONE: (516) 922-5332

PROP HOUSE INC, THE
653 11TH AVE
NYC, 10036

PROPS AND ANTIQUES
3123 LEE PL
BELLMORE, 11710
PHONE: (516) 785-4764

RENT A WRECK
85 GLEN HEAD RD
GLEN HEAD, 11545
PHONE: (516) 676-6520

SAFES INCORPORATED
3426 MERRICK RD
SEAFORD, 11783
PHONE: (516) 781-6982

SAY IT IN NEON
434 HUDSON ST
NYC, 10014
PHONE: (212)691-7977

SHOWROOM OUTLET
625 W 55TH ST
NYC, 10019
PHONE: (212)581-0470

SKYSHOTS
95 BRADLEY PL, SUITE 2 S
MINEOLA, 11501
PHONE: (516) 248-3617

SOUTH STREET PARTY, INC
78 SOUTH STREET
OYSTER BAY, 11771
PHONE: (516) 922-4066

STATE SUPPLY EQUIPMENT CO.
210 11TH AVE
NYC, 10001
PHONE: (212)675-1430

SUFFOLK MARINE MUSEUM
P.O. BOX 184, 86 WEST AVE
WEST SAYVILLE, 11796
PHONE: (516) 567-1733

THINGS ANTIQUE INC
483 AMSTERDAM AVE
NYC
PHONE: (212)873-4655

THOMPSON, ROBERT
176 GERHARD RD
PLAINVIEW, 11803
PHONE: (516) WE8-4540

TRI-COUNTY FIRE EQUIPMENT
110 MAPLE AVE
CEDARHURST, 11516
PHONE: (516) 569-0030

ULTIMATE CLASS LIMOUSINE SERVICE
3698 COLLECTOR LANE
BETHPAGE, 11714
PHONE: (516) 735-5801

WALSH, RICHARD W.
118 LINDEN ST
BELLMORE, 11710
PHONE: (718) 217-2988

WANTAGH 5 & 10 EMPORIUM
1901 WANTAGH AVE
WANTAGH, 11793
PHONE: (516) 785-1259

WIZARDWORKS
39-40 21ST ST., L.I.C.
NYC, 11101
PHONE: (718) 786-8383

PYROTECHNICS

DEVLIN VIDEOSERVICE
1501 BROADWAY, SUITE 408
NEW YORK, 10036
PHONE: (212) 391-1313
FAX: (212) 391-2744

TEFEX SPECIALISTS, INC
35-39 37TH ST
LONG ISLAND CITY, 11101
PHONE: (718) 937-2417
FAX: 937-3920

FIREWORKS BY GRUCCI
ONE GRUCCI LANE
BROOKHAVEN, 11719
PHONE: (516) 286-0088
FAX: (516) 286-9036

STUDIOS/SOUND STAGES

ANS INTERNATIONAL VIDEO LTD
386 FIFTH AVE, 3RD FLOOR
NYC, 10018
PHONE: (212)736-1007

A.P.A.
230 WEST 10TH ST
NYC, 10014
PHONE: (212)675-4894

ARIES PRODUCTIONS
415 OSER AVE
HAUPPAUGE, 11788
PHONE: (516) 435-8282
FAX: (516) 435-8290

BESTEK THEATRICAL PRODUCTIONS
218 WEST HOFFMAN AVE
LINDENHURST, 11757
PHONE: (516) 225-0707

BLOOMCREST FABRICS, INC
1874 GRAND AVE
BALDWIN, 11510
PHONE: (516) 867-9880

BOKEN STUDIO
513 WEST 54TH ST
NYC, 10019
PHONE: (212)581-5507

BRUCE MORGAN PHOTOGRAPHY, INC
15 WEST SUNRISE HIGHWAY
FREEPORT, 11520

CABLEVISION
ONE MEDIA CROSSWAYS DR
WOODBURY, 11797
PHONE: (516) 364-8450

CABLEVISION
1600 MOTOR PARKWAY
HAUPPAUGE, 11788
PHONE: (516) 348-6800

CAMERA MART INC., THE
456 WEST 55TH ST
NYC, 10019
PHONE: (212)757-6977
TELEX: 275619
FAX: (212)582-2498

CECO INTERNATIONAL CORP.
440 W 15TH ST
NYC, 10011
PHONE: (212)206-8280

CESTARE STUDIOS
188 HERRICKS RD
MINEOLA, 11501
PHONE: (516) 742-5550

CINE STUDIO
241 WEST 54TH ST
NYC, 10019
PHONE: (212)581-1916

DIMENSION CABLE
201 OLD COUNTRY ROAD
RIVERHEAD, 11901
PHONE: (516) 727-7170

DSI/VERITAS STUDIOS
527 W 45TH ST
NY, 10036
PHONE: (212)581-2050

DEVLIN VIDEOSERVICE
1501 BROADWAY, SUITE 408
NEW YORK, 10036
PHONE: (212) 391-1313
FAX: (212) 391-2744

TEMPIRE STAGES
50-20 25TH ST
LONG ISLAND CITY, 11101
PHONE: (718) 392-4747

FARKAS
385 3RD AVE
NY, 10016
PHONE: (212)679-8212

FIVE TOWNS COLLEGE
2165 SEAFORD AVE
SEAFORD, 11783
PHONE: (516) 783-8800

GROUP W CABLE
565 EAST MAIN ST
BAY SHORE, 11706
PHONE: (516) 666-2263

HBO STUDIO PRODUCTION
120A EAST 23RD ST
NY, 10010
PHONE: (212)512-7800
FAX: (212)512-7980

HOFSTRA UNIVERSITY
HEMPSTEAD, 11550
PHONE: (516) 560-6817

INMAC (INTER-MEDIA ART CENTER, INC)
370 NEW YORK AVE
HUNTINGTON, 11743
PHONE: (516) 549-9666

INTERNATIONAL PRODUCTION CENTRE
514 W 57TH ST
NYC, 10019
PHONE: (212) 582-6530

KAUFMAN ASTORIA STUDIOS
34-12 36TH ST
ASTORIA, 11106
PHONE: (718) 392-5600
TELEX: 6971038

LIBERTY STUDIOS
238 EAST 26TH ST
NY, 10010
PHONE: (212)532-1865

LIFETIME TELEVISION STUDIOS
34-12 36TH ST
ASTORIA, 11106
PHONE: (718) 706-3513
FAX: (718) 706-3589

LONG BEACH PHOTO CENTER, LTD
255 WEST PARK AVE
LONG BEACH, 11561
PHONE: (516) 431-6900

LTV CHANNEL 27
211 SPRINGS FIREPLACE RD
EAST HAMPTON, 11937
PHONE: (516) 324-3315

MALIBU RESORTS INTERNATIONAL, LTD
LIDO BLVD
LIDO BEACH, 11561
PHONE: (516) 432-1600

MOLLOY COLLEGE
1000 HEMPSTEAD AVE
ROCKVILLE CENTRE, 11570
PHONE: (516) 678-5000
EXT. 200

MTI
885 2ND AVE
NYC, 10017
PHONE: (212)355-0510

MINISTAGE
304 EAST 45TH ST
NY, 10017
PHONE: (212)697-8388

MOTHERS SOUNDSTAGE
210 EAST 5TH ST
NY, 10003
PHONE: (212)529-5097

MUSICAL ADVENTURES
7 THE PLAZA
LOCUST VALLEY, 11560
PHONE: (516) 759-2872

N.E.P. PRODUCTIONS INC
56 W 45TH ST
NY, 10036
PHONE: (212) 382-1100

NEW COMMUNITY CINEMA
423 PARK AVE, P.O. BOX 498
HUNTINGTON, 11743
PHONE: (516) 423-7619/0

PATHE STUDIOS-M.T.I. TV CITY
105 EAST 106TH ST
1443 PARK AVE
NY, 10029
PHONE: (212) 722-1818

PORT WASHINGTON UNION FREE SCHOOL DISTRICT
100 CAMPUS DR
PORT WASHINGTON, 11050
PHONE: (516) 883-4000

PRIMA LUX VIDEO
30 WEST 26TH ST
NY, 10010
PHONE: (212) 206-1402
FAX: (212) 206-1826

PRODUCERS EAST MEDIA, INC
734 WALT WHITMAN RD
ROUTE 110, SUITE 204
MELVILLE, 11747
PHONE: (516) 421-4800

REEL TYME PRODUCTIONS/ REEL TYME RECORDING INC.
282 JERICHO TURNPIKE
FLORAL PARK, 11001
PHONE: (516) 354-0246

SILVERCUP STUDIOS INC
42-25 21ST ST
LONG ISLAND CITY, 11101
PHONE: (718) 784-3390

SUNY FARMINGDALE
COLLEGE OF TCHNOLOGY
FARMINGDALE, 11735
PHONE: (516) 420-2400
FAX: (516) 420-2693

TELETIME VIDEO PRODUCTIONS
37-39 WATERMILL LANE
GREAT NECK, 11021
PHONE: (516) 466-3882

TIME SQUARE STUDIOS
1481 BROADWAY
NY, 10036
PHONE: (212) 704-9700

UNITEL - NEW YORK
515 W 57TH ST
NYC, 10019
PHONE: (212) 265-3600

VIDEO CENTER/CGL, INC
DESEVERSKY CONFERENCE CENTER
NORTHERN BLVD
OLD WESTBURY, 11568
PHONE: (516) 626-3570

VERITAS PRODUCTIONS INC
527 W. 45TH ST
NY, 10036
PHONE: (212) 581-2050

WINDSOR TOTAL VIDEO
8 WEST 38TH ST
NYC, 10018
PHONE: (212) 944-9090

WLIW CHANNEL 21
1425 OLD COUNTRY RD
PLAINVIEW, 11803
PHONE: (516) 454-8866

WOODBURY COUNTRY CLUB
884 JERICHO TURNPIKE
WOODBURY, 11797
PHONE: (516) 692-6200

UNIONS AND ASSOCIATIONS

AICP
100 EAST 42ND ST
NYC, 10017
PHONE: (212) 867-5720

AIVF
625 BROADWAY
NEW YORK, 10012
PHONE: (212) 473-3400

ACTORS EQUITY ASSOCIATION
165 WEST 46TH ST
NYC, 10036
PHONE: (212) 869-8530

AMERICAN FEDERATION OF TV & RADIO ARTISTS
260 MADISON AVE, 7 FLOOR
NY, 10016
PHONE: (212) 532-0800

CINEMATOGRAPHERS (IATSE LOCAL 644)
505 8TH AVE, 16 FLOOR
NY, 10018
PHONE: (212) 244-2121

DIRECTORS GUILD OF AMERICA
110 WEST 57TH ST
NYC, 10019
PHONE: (212) 581-0370

IATSE
1515 BROADWAY
NYC, 10036
PHONE: (212) 730-1770

IATSE - LOCAL 764
THEATRICAL WARDROBE UNION
1501 BROADWAY, ROOM 1313
NEW YORK, 10036
PHONE: (212) 221-1717
FAX: (212) 302-2324

IBTC TEAMSTERS, LOCAL 817
ONE HOLLOW LANE
LAKE SUCCESS, NY, 11042
PHONE: (516) 365-3470

LAB TECHNICIANS, MOTION PICTURE (IATSE LOCAL 702)
165 WEST 46TH ST
NYC, 10036
PHONE: (212) 869-5540

MAKE UP ARTISTS & HAIR STYLISTS (IATSE LOCAL 798)
31 WEST 21ST ST
NYC, 10016
PHONE: 627-0660

NABET 16
322 8TH AVE, 5 FLOOR
NY, 10001
PHONE: (212) 633-9292

PHOTOGRAPHERS INTERNATIONAL
505 8TH AVE
NEW YORK, 10018
PHONE: (212) 244-2121

RADIO AND T.V. BROADCAST ENGINEERS
230 WEST 41ST ST
NEW YORK, 10036
PHONE: (212) 354-6770

S.M.P.T.E.
595 W HARTSDALE AVE
WHITE PLAINS, 10607
PHONE: (914) 761-1100

SCENIC ARTISTS IBPAT
575 8TH AVE
NY, 10018
PHONE: (212) 736-4498

SCREEN ACTORS GUILD
1515 BROADWAY, 44 FLOOR
NY, 10036
PHONE: (212) 944-1030

STUDIO MECHANICS, MOTION PICTURE (IATSE LOCAL 52)
326 WEST 48TH ST
NY, 10036
PHONE: (212)399-0980

STAGE DIRECTORS & CHOREOGRAPHERS
1501 BROADWAY
NY, 10036
PHONE: (212) 391-1070

STUDIO MECHANICS LOCAL 52
326 WEST 48TH ST
NY, 10036
PHONE: (212) 399-0980

UNITED SCENIC ARTIST (IATSE LOCAL 829)
575 89TH AVE
NY, 10018
PHONE: (212) 736-4498

VIDEO AND STAGE MECHANICS
LOCAL 340 IATSE
P.O. BOX 434
MEDFORD, NY, 11763
PHONE: (516) 289-9499

WRITERS GUILD OF AMERICA EAST INC
555 WEST 57TH ST
NYC, 10019
PHONE: (212) 245-6180

NORTH CAROLINA

SPECIAL EFFECTS

MANWARING STUDIOS
HORSE SHOE, 28742
PHONE: (704) 891-9442

MAKE-UP SUPPLIERS

ASHEVILLE COMMUNITY THEATRE
35 WALNUT STREET EAST
ASHEVILLE, 28801
PHONE: (704)253-4931

CAROLINA SENIC STUDIO
200 WAUGHTON ST
WINSTON-SALEM, 27106
PHONE: (919) 784-7170

DANCERS SHOP
2211 NEW HOPE CHURCH ROAD
RALEIGH, 27604
PHONE: (919) 876-9800

MOBLEYS RALEIGH ART CENTER
113 SOUTH SALISBURY ST
RALEIGH, 27601
PHONE: (919) 832-4775

MONROE COSTUMES & THEATRICAL SUPPLIES
3108 MONROE RD
CHARLOTTE, 28205
PHONE: (704)333-4653

MONTFORD PARK PLAYERS
BOX 2663
ASHEVILLE, 28802
PHONE: (704)254-4540

WESTERN CAROLINA UNIVERSITY
CENTER FOR IMPROVING MOUNTAIN LIVING
CULLOWHEE, 28723
PHONE: (704) 227-7492

OHIO

SPECIAL EFFECTS

CLASSIC VIDEO
2690 STATE ROAD/SUITE 100
CUYAHOGA FALLS, 44223
PHONE: (216) 928-7773
FAX: (216)923-6803

STUDIOS/SOUND STAGES

BRIGHT LIGHT PRODUCTION COMPANY
420 PLUM ST
CINCINNATI, 45202
PHONE: (513) 721-2574

CREATIVE TECHNOLOGY INC
853 COPLEY RD
AKRON, 44320
PHONE: (216) 535-5778

FACET COMMUNICATIONS
1223 CENTRAL PARKWAY
CINCINNATI, 45214
PHONE: (513) 381-4033

MEDIA GROUP, THE
1480 DUBLIN RD
COLUMBUS, 4325
PHONE: (614) 488-0621

S O S PRODUCTIONS
753 HARMON AVE
COLUMBUS, 43223
PHONE: (614) 221-0966

SPECTRUM VIDEO
6888 ALPHA DR
CLEVELAND, 44143
PHONE: (216) 449-0522

TELECATION INC
278 NORTH FIFTH ST
COLUMBUS, 43215
PHONE: (614) 224-4400

VALDHERE INC
360 VALLEYWOOD DR
DAYTON, 45429
PHONE: (513) 293-2191

VANGUARD PRODUCTIONS
7084 HUNTLY RD
COLUMBUS, 43229
PHONE: (614) 436-4610

VIDEO GENESIS INC
24000 MERCANTILE RD
BEACHWOOD, 44122
PHONE: (216) 464-3635

WVIZ-TV CHANNEL 25
4300 BROOKPARK RD
CLEVELAND, 44134
PHONE: (216) 398-2800

OKLAHOMA

SPECIAL EFFECTS

CAIN, JAMES F.
729 TURTLE CREEK RD
MOORE, 73160
PHONE: (405) 691-5712

CARVER, RICHARD
3500 WEST APACHE
TULSA, 74127
PHONE: (918) 583-7455/583-4814

CLARK, JIM F.
5737 E. 26 PLACE
TULSA, 74129
PHONE: (918) 836-4145

EATON, ROD K.
1535 E. 62ND ST.
TULSA, 74136
PHONE: (918) 747-1629

HENLEY, MIKE
P.O.BOX 54848
TULSA, 74155
PHONE: (918) 627-4345/599-5733

MAKE-UP SUPPLIERS

HAZELS COSTUMES
1515 NORTH PORTLAND
OKLAHOMA CITY, 73107
PHONE: (405) 942-9960

MODELS/MINIATURES

HENLEY, MIKE
P.O. BOX 54848
TULSA, 74155
PHONE: (918) 627-4345/599-5733

PROPS

CROOK, ROBERT M. (MILITARY)
RT. 2, BOX 98
POTEAU, 74953
PHONE: (918) 658-3963

ECLIPSE CREATIVE SERVICES (HOT AIR/HELIUM BALLOONS)
2727 E. 21ST ST #600
TULSA, 74114
PHONE: (918) 742-9526/742-9525

MCCARTNEY, STEPHEN B. (MILITARY, WEAPONS, WESTERN)
5518 EAST 36TH ST
TULSA, 74135
PHONE: (918) 665-3146/663-4751

SHELL, HARVEY
4320 W. 43 ST
TULSA, 74107
PHONE: (918) 446-9273/592-1857

PYROTECHNICS

CLARK, JIM F.
5737 E. 26 PLACE
TULSA, 74129
PHONE: (918) 836-4145

STUDIOS/SOUND STAGES

COLE INC
P.O. BOX 2803, 615 E 4TH ST
TULSA, 74101
PHONE: (918) 585-9119/582-9601

CSU PUBLIC SERVICE TV
100 NORTH UNIVERSITY
EDMOND, 73034
PHONE: (405) 341-2980
EXT. 589

KETA
7403 NORTH KELLEY ST
OKLAHOMA CITY, 73111
PHONE: (405) 848-8501
KOCO-TV
1300 EAST BRITTON RD
OKLAHOMA CITY, 73113
PHONE: (405) 478-3000
KTVY
500 EAST BRITTON RD
OKLAHOMA CITY, 73114
PHONE: (405) 478-1212
KWTV
7401 NORTH KELLEY ST
OKLAHOMA CITY, 73111
PHONE: (405) 843-6641
LONG BRANCH STUDIOS
6314 E 13TH ST
TULSA, 74112
PHONE: (918) 832-7640/7654
OMNI LIGHTING
212 NORTH MAIN
TULSA, 74103
PHONE: (918) 583-6464/
584-6593
STUDIO 212
212 NORTH MAIN
TULSA, 74103
PHONE: (918) 583-4814/7455
STUDIO VII
417 NORTH VIRGINIA
OKLAHOMA CITY, 73106
PHONE: (405) 236-0643

UNIONS AND ASSOCIATIONS

OKLAHOMA STUNTMAN ASSOCIATION
3444 SO. 132ND E. AVE
TULSA, 74134
PHONE: (918) 622-1780
IATSE
6823 E. 65TH PL
TULSA, 74133
PHONE: (918) 496-7722
IATSE LOCAL 112
P.O. BOX 112
OKLAHOMA CITY, 73107
PHONE: (405) 232-4793/
278-0239
TEAMSTERS (INTERNATIONAL BROTHERHOOD OF)
P.O. BOX 1836
TULSA, 74101
PHONE: (918) 587-3358/
496-8600
3528 W. RENO
OKLAHOMA CITY, 73125
PHONE: (405) 947-2333

OREGON

SPECIAL EFFECTS

FEKE, JETHRO
652 B AVE
LAKE OSWEGO, 97034
PHONE: (503) 635-4217
HOLLYWOOD LIGHTS, INC
0625 S.W. FLORIDA
PORTLAND, 97219
PHONE: (503) 244-5808/(800)
826-9881
FAX: (503) 244-6045
HYBRID IMAGES
1505 S.E. OXFORD LANE
MILWAUKIE, 97222
PHONE: (503) 653-8293
THE ILLUSION WORKS-MOTION PICTURE SPECIAL EFFECTS
7969 S.E. 8TH
PORTLAND, 97202
PHONE: (503) 235-6046

LUNDELL, DON
SUNRISE AND SONS SETS
3221 N.E. HOYT
PORTLAND, 97232
PHONE: (503) 238-1432/
(206) 745-9557
JOE K. MARKS FILM & VIDEO PRODUCTION
3728 S.W. 55TH DR
PORTLAND, 97221
PHONE: (503) 292-8626/
250-2994
FAX: (503) 292-0656
WILLIAM A. MCINTIRE ENTERPRISES
P.O. BOX 4244
PORTLAND, 97203
PHONE: (5030 286-4193
NORTHWEST THEATRICAL
P.O. BOX 3687
SUNRIVER, 97707
PHONE: (503) 593-7240
OREGON SCENIC & LIGHTING
11853 S.E. HWY. 212
CLACKAMAS, 97015
PHONE: (503) 656-2234
FAX: (503) 656-7223
PACIFIC GRIP & LIGHTING
1239 S.E. 12TH
PORTLAND, 97214
PHONE: (503) 233-4747
FAX: (503) 233-5830
TROMANELLI, WILL
4318 S.W. CONDOR AVE
PORTLAND, 97201
PHONE: (503) 241-1082

EQUIPMENT RENTALS

MCINTIRE, WILLIAM A. ENTERPRISES
BOX 4244
PORTLAND, 97208
PHONE: (503) 286-4193

EXPENDABLES

HOLLYWOOD LIGHTS, INC
0625 S.W. FLORIDA
PORTLAND, 97219
PHONE: (503) 244-5808/(800)
826-9881
FAX: (503) 244-6045
NORTHWEST THEATRICAL
P.O. BOX 3687
SUNRIVER, 97707
PHONE: (503) 593-7240
PACIFIC GRIP & LIGHTING
1239 S.E. 12TH
PORTLAND, 97214
PHONE: (503) 233-4747
FAX: (503) 233-5830
PORTLAND CINE CO.
519 N.E. HANCOCK
PORTLAND, 97212
PHONE: (503) 284-2717
FAX: (503) 282-5622

MINIATURES/MODELS

ACTION SCENES, INC
P.O. BOX 10733
PORTLAND, 97210
PHONE: (503) 771-5283
KING, LARRY
P.O. BOX 14882
PORTLAND, 97214
PHONE: (503) 771-5283
JOSE SOLIS CREATIVE SERVICES
625 S.E. 76TH AVE
PORTLAND, 97215
PHONE: (503) 255-9279/
780-0497
FAX: (503) 232-5615

PROPS

CLASSIC CARRIAGE SERVICE, INC
38940 CAMP CREEK RD
SPRINGFIELD, 97478
PHONE: (503) 726-9461
PETER CORVALLIS PRODUC-TIONS, INC
237 S.W. FRONT
PORTLAND, 97204
PHONE: (503) 222-1664
END OF THE TRAIL COLLECTIQUES
5937 N. GREELY
PORTLAND, 97217
PHONE: (503) 283-0419
HABROMANIA, INC
316 S.W. 9TH
PORTLAND, 97205
PHONE: (503) 223-0767
HARTS REPTILE WORLD
11264 S. MACKSBURG RD
CANBY, 97013
PHONE: (503) 266-7236
HIPPO HARDWARE & TRADING CO.
1040 E. BURNSIDE
PORTLAND, 97214
PHONE: (503) 231-1444
PORTLAND REPERTORY THEATER
TWO WORLD TRADE CENTER
25 S.W. SALMON
PORTLAND, 97204
PHONE: (503) 224-4491
THE PROP SHOP
1019 N.W. EVERETT
PORTLAND, 98209
PHONE: (503) 227-1487
REJUVENATION HOUSE PARTS CO
901 N. SKIDMORE
PORTLAND, 97217
PHONE: (503) 249-2038
STAR RENTAL AND SALES
1735 S.E. UNION
PORTLAND, 97214
PHONE: (503) 231-7300

PYROTECHNICS

WILLIAM A. MCINTIRE ENTERPRISES
P.O. BOX 4244
PORTLAND, 97208
PHONE: (5030 286-4193
PATRICK CORPORATION
1600 N. 28TH
SPRINGFIELD, 97478
PHONE: (503) 746-7528
STUDIOS/SOUND STAGES
CORPORATE MEDIA SERVICES, LTD
417 N. WATER
SILVERTON, 97381
PHONE: (503) 873-2899/
(800) 452-7007
IMAGE SPACE STUDIO
333 N.W. PARK AVE
PORTLAND, 97209
PHONE: (503) 223-6794
KPDX-TV PRODUCTION CENTER
910 N.E. MARTIN LUTHER KING BLVD
PORTLAND, 97232
PHONE: (503) 239-4949
FAX: (503) 239-6184
MIRA FILM & VIDEO
116 N. PAGE ST
PORTLAND, 97227
PHONE: (503) 464-0630
FAX: (503) 464-9782

NORTHWEST STAGE
GRAY ADAMS PRODUCTIONS, INC
2580 N.W. UPSHUR
PORTLAND, 97210
PHONE: (503) 224-3800
PARAGON CABLE
LOCAL ORIGINATION PRODUCTION
3075 N.E. SANDY BLVD
PORTLAND, 97232
PHONE: (503) 230-2099
PORTLAND CINE CO.
519 N.E. HANCOCK
PORTLAND, 97212
PHONE: (503) 230-2099
THE SOUND STAGE
5253 N.E. SANDY
PORTLAND, 97213
PHONE: (503) 287-5387

UNIONS AND ASSOCIATIONS

AFL - CIO OF OREGON
1900 HINES ST. SE
SALEM, 97032
PHONE: (503) 224-3169
AFTRA
516 S.E. MORRISON M-3
PORTLAND, 97214
PHONE: (503) 238-6914
AMERICAN FEDERATION OF MUSICIANS - LOCAL 99
325 N.E. 20TH
PORTLAND, 97232
PHONE: (503) 235-8791
IATSE LOCAL 28
P.O. BOX 1728
PORTLAND, 97207
PHONE: (503) 295-2828
IATSE LOCAL 675
100 HOWARD
EUGENE, 97404
PHONE: (503) 344-6306/
688-6645
INTERNATIONAL TELEVISION ASSOCIATION (ITVA)
P.O. BOX 12379
PORTLAND, 97212
PHONE: (503) 294-4200
MID OREGON PRODUCTION ARTS NETWORK (MOPAN)
P.O. BOX 11008
EUGENE, 97440
PHONE: (503) 341-9118
NORTHWEST FILM & VIDEO CENTER/OREGON ART INSTITUTE
1219 S.W. PARK
PORTLAND, 97205
PHONE: (503) 221-1156
FAX: (503) 226-4842
OREGON MEDIA PRODUCTION ASSOC. (OMPA)
P.O. BOX 2784,1612 S.W. JEFFERSON
PORTLAND, 97208
PHONE: (503) 228-8822
PORTLAND ADVERTISING FEDERATION (PAF)
P.O. BOX 14067
PORTLAND, 97214
PHONE: (503) 771-4033
SOUTHERN OREGON FILM & VIDEO ASSOC. (SOFVA)
1102 W. 10TH
MEDFORD, 97501
PHONE: (503) 779-4033

PENNSYLVANIA

SPECIAL EFFECTS

AQUA-EFFECTS, INC
928 EAST HAZELWOOD AVE
RAHWAY, NJ , 07065
PHONE: (800) 451-2436
FAX: (201) 382-1651

CINEKYD ENTERPRISES INC
129 TERWOOD ROAD
WILLOW GROVE, 19090
PHONE: (215) 659-4696

COLOR FILM SERVICE INC
46 GARRETT ROAD
UPPER DARBY, 19082
PHONE: (215)528-6747

HILL THEATRE STUDIO SPECIAL SERVICES
35 WEST BROAD ST
PAULSBORO, NJ , 08066
PHONE: (609) 423-8910
FAX: (609) 224-0224

ILLUSION HOUSE
2617 HERR ST
HARRISBURG, 17103
PHONE: (717) 233-8848

IMPOSSIBLE PRODUCTIONS
711 VIRGINIA AVE
PITTSBURGH, 15215
PHONE: (412) 781-5166/
782-2381

KIMBERLY THEATRICS
7C MARLEN DR
TRENTON, NJ , 08691
PHONE: (609) 587-7927/
(215) 389-7503
FAX: (609) 587-7867

LOPATIN PRODUCTION INC
26 ROCK HILL ROAD
BALA CYNWYD, 19004
PHONE: (215)667-4144

MCMANUS ENTERPRISES/THE COMPLETE PRODUCTION SERVICE INC
P.O. BOX 780, 111 UNION AVE
BALA CYNWYD, 19004
PHONE: (215)664-8605

METROPOLIS STUDIOS
1410-1418 SOUTH DARIEN ST
PHILADELPHIA, 19147
PHONE: (215) 453-3000
FAX: (215) 463-1322

MODERN VIDEO PRODUCTIONS
1600 MARKET ST, 33RD FLOOR
PHILADELPHIA, 19103
PHONE: (215)569-4100

PRICE, JOHN M. FILMS
BOX 81
RADNOR, 19087
PHONE: (215) 687-6699

RHODES, LEON S.
2960 KING RD
BRYN ATHYN, 19009
PHONE: (215)947-4044

ROSE, PETER PRODUCTIONS
4372 FLEMING ST
PHILADELPHIA, 19128
PHONE: (215)483-7660/
893-3140

SAPSIS RIGGING INC
233 NORTH LANDSOWNE AVE
LANDSDOWNE, 19050
PHONE: (215) 849-6660
FAX: (215) 849-8010

THE SENTRY POST
1434 BYWOOD AVE
UPPER DARBY, 19082
PHONE: (215) 352-3788/1649
FAX: (215) 352-1649

SETWORKS, INC
28 SUNSET DR
PAOLI, 19301
PHONE: (215) 296-0882/0668

SKELETONS IN MY CLOSET PRODUCTIONS
112 EAST WASHINGTON LANE
PHILADELPHIA, 19144
PHONE: (215) 848-2231/
299-1154

STEWART, E.J. INC
525 MILDRED AVE
PRIMOS, 19018
PHONE: (215)626-6500

TRACES
10 PRESIDENTIAL BLVD.,
SUITE 115
BALA CYNWYD, 19004
PHONE: (215) 660-9699
FAX: (215) 660-0915

WPHL PRODUCTIONS
5001 WYNNEFIELD AVE
PHILADELPHIA, 19131
PHONE: (215)878-1700

EQUIPMENT RENTALS

PERFORMANCE LIGHTING & PRODUCTION SERVICES, INC.
4853 CAMPBELLS RUN RD
PITTSBURGH, 15205
PHONE: (800) 722-1134/
(412) 747-0501

SAFEGUARD LIGHTING SYSTEMS, INC
112 WEST EAGEE RD
HAVERTOWN, 19083
PHONE: (215) 853-3030/
876-2800

SKYWORKS, INC
1600 MARKET ST. SUITE 2601
PHILADELPHIA, 19103
PHONE: (215) 564-3962

MAKE-UP SUPPLIERS

DEFEO DESIGN STUDIOS
SPRING RD, R.D. #4
MALVERN, 19355
PHONE: (215) 644-2814

DONNA MOYER - MAKE-UP & HAIR SERVICES
258 LYCEUM AVE
PHILADELPHIA, 19128
PHONE: (215) 487-7616/
(212) 753-2310

IMAGEFFECTS STUDIOS
21 RICHWOOD PL
DENVILLE, NJ , 07834
PHONE: (201) 586-2521

WANIELISTA, ANNETTE
1430 ALABAMA AVE
PITTSBURGH, 15216
PHONE: (412) 343-7906

MODELS/MINIATURES

ANIVISION
981 WALNUT ST
PITTSBURGH, 15234
PHONE: (412) 563-2221

METROPOLIS STUDIOS
1410-1418 SOUTH DARIEN ST
PHILADELPHIA, 19147
PHONE: (215) 453-3000
FAX: (215) 463-1322

THE OBJECT WORKS
12 EIGHTH ST
PITTSBURGH, 15222
PHONE: (412) 261-3513/3798

RILEIGH'S, INC
P.O. BOX 504, 1701 UNION BLVD
ALLENTOWN, 18105
PHONE: (215) 432-0242/
437-5493

PROPS

ACADEMY OF NATURAL SCIENCES MUSEUM
19TH AND BEN FRANKLIN PARKWAY
PHILADELPHIA, 19103
PHONE: (215) 269-1012

AIR ATLANTIC AIRLINES, INC
R.D. #2 - BOX 165
CENTRE HALL, 16828
PHONE: (814) 364-1477/1664

ARCHITECTURAL ANTIQUES EXCHANGE
715 NORTH 2ND ST
PHILADELPHIA, 19123
PHONE: (215) 922-3669

BRENNER CAR AND TRUCK RENTAL
P.O. BOX 1955
HARRISBURG, 17105
PHONE: (717) 232-4271

BUCKSHIRE CORPORATION - ANIMALS
P.O. BOX 155, 2025 RIDGE ROAD
PERKASIE, 18944
PHONE: (215) 257-0116

BUFFALO ENTERPRISES
308 WEST KING ST
EAST BERLIN, 17316
PHONE: (717) 259-9081

CINEMA SUPPLY COMPANY, INC
502 S. MARKET ST. (L)
MILLERSBURG, 17061
PHONE: (717) 692-4744/3708

CLASSIC LIMOUSINE SERVICE
403 WOODBINE AVE
FEASTERVILLE, 19047
PHONE: (215)355-4576

COAST TO COAST CATERING
722 FULTON ST
PHILADELPHIA, 17147
PHONE: (215) 922-7864

DESIGN DISCOVERY
782 HILLVIEW RD
MALVERN, 19355
PHONE: (215) 353-2683/0474

FIVE STATE DEALERS ADVERTISING ASSOCIATION
INDEPENDENCE MALL WEST
PHILADELPHIA, 19106
PHONE: (215)922-3945

GARDEN OF EDEN
100 BROWNSVILLE RD
PITTSBURGH, 15120
PHONE: (412) 431-9337

GREAT ADVENTURE BALLOON CLUB
BOX 1172
LANCASTER, 17603
PHONE: (717) 397-3623

IKOS DESIGN
59 S. CHARLOTTE #14
MANHEIM, 17545
PHONE: (717) 665-5579

IMPERIAL CARRIAGES LTD
P.O. BOX 522
CHESTER SPRINGS, 19425
PHONE: (215) 827-9343/7783

JAMES & SON COSTUMIERS AND MILITARY CLOTHIERS
1230 ARCH ST, 8TH FLOOR
PHILADELPHIA, 19107
PHONE: (215)922-7409

KELSO RESTORATIONS
536 E. POPLAR ST
MCCONNELLSBURG,
PHONE: (717) 485-4344/4693

MAIN LINE PARTY RENTALS
400 FEHELEY DR
KING OF PRUSSIA, 19406
PHONE: (215) 227-5757

METROPOLIS STUDIOS
1410-1418 SOUTH DARIEN ST
PHILADELPHIA, 19147
PHONE: (215) THE-SHOW
FAX: (215) 463-1322

M D ATLANTIC AIR MUSEUM
R.D. #9 - BOX 9381
READING, 19605
PHONE: (215) 372-7333

MILEY TRUCK RENTAL
23 CHESTNUT ST
CARNEGIE, 15106
PHONE: (800) 822-8118/
(412) 279-6200

NEW HOPE-IVYLAND RAILROAD
P.O. BOX 196
PENNDEL, 19047
PHONE: (215)757-3790

THE OBJECT WORKS
12 EIGHTH ST
PITTSBURGH 15222
PHONE: (412) 261-3513/3798

PALOMBO SERVICES INC.
123 EDGWOOD AVENUE
PITTSBURGH 15218
PHONE: (412) 241-2431

PHOTRONIX/ED WOLFKILL ASSOCIATES
BOX 78
SHADY GROVE, 17255
PHONE: (717) 762-5825

RAIL TOURS, INC
P.O. BOX 285
JIM THORPE, 18229
PHONE: (717) 325-4606

RENN-ART ASSOCIATES
120 COLLEGE AVE
LANCASTER, 17603
PHONE: (717) 393-4755

RILEIGH'S, INC
P.O. BOX 504, 1701 UNION BLVD
ALLENTOWN, 18105
PHONE: (215) 432-0242/
437-5493

ROYAL LIMOUSINE OF PITTSBURGH, INC
2215 GROVELAND ST
PITTSBURGH, 15234
PHONE: (412) 884-3306/
(800) 369-7995

THE SENTRY POST
1434 BYWOOD AVE
UPPER DARBY, 19082
PHONE: (215) 352-3738/1649

SETWORKS, INC
28 SUNSET DR
PAOLI, 19301
PHONE: (215) 296-0882/430-
1545

TAYLOR RENTAL
2407 OLD GETTYSBURG RD
CAMP HILL, 17011
PHONE: (717) 763-1233

TYBURN RAILROAD COMPANY
152 MONROE AVE
P.O. BOX 196
PENNDEL, 19047
PHONE: (215) 757-3793/
(717) 392-1129

VALLEY MEDICAL SERVICES, INC
35 EAST ELIZABETH AVE
P.O. BOX 469
BETHLEHEM, 19018
PHONE: (215) 861-8188/
252-0286

VIDEOTONE PRODUCTIONS
208 SYCAMORE DR
CORAOPOLIS, 15108
PHONE: (412) 269-0080/0306

WINGSN WHEELS OF PITTS-BURGH, INC
BOX 56 GLADE MILL ROAD,
ROUTE 228 EAST
VALENCIA, 16059
PHONE: (215)898-3513

WILSON STUDIO
818 LIBERTY AVE, SUITE 500
PITTSBURGH, 15222
PHONE: (412) 391-7499/78

ZAP PRODUCTION SERVICES
P.O. BOX 539
LANSDOWNE, 19050
PHONE: (215) 380-8995/622-2215

STUDIOS/SOUND STAGES

ATLANTIC TELEPRODUCTION
201 FOURTEENTH ST
PITTSBURGH, 15222
PHONE: (412) 456-1300

CENTER CITY FILM & VIDEO
1503-05 WALNUT ST
PHILADELPHIA, 19102
PHONE: (215) 568-4134

CINEKYD ENTERPRISES INC
129 TERWOOD ROAD
WILLOW GROVE, 19090
PHONE: (215) 659-4696

CONTINENTAL FILM GROUP
PARK ST AND WENGLER AVE
SHARON, 16146-3090
PHONE: (412) 981-3456

DIGITAL VIDEO PRODUCTIONS
62 SOUTH FRANKLIN ST
WILKES-BARRE, 18773
PHONE: (717) 823-2828

KENNEDY/LEE INC
RD 12
YORK, 17406
PHONE: (717) 757-4666

LAIRD PRODUCTIONS
2153 MARKET ST
CAMP HILL, 17011
PHONE: (717) 737-1556

LOPATIN PRODUCTION INC
26 ROCK HILL ROAD
BALA CYNWYD, 19004
PHONE: (215)667-4144

MPTV VIDEO PRODUCTION UNIT
2116 NOBLE ST
PITTSBURGH, 15218-2514
PHONE: (412) 271-6788/
(800) 648-6788

METROPOLIS STUDIOS
1410-1418 SOUTH DARIEN ST
PHILADELPHIA, 19147
PHONE: (215) 463-3000
FAX: (215) 463-1322

**OCONNOR, WALTER G.
ASSOCIATES**
658 OLD YORK RD
ETTERS, 17319
PHONE: (717) 938-6466

**PENN MUTUAL VIDEO
COMMUNICATIONS**
INDEPENDENCE SQUARE
PHILADELPHIA, 19172
PHONE: (215)625-6101

PHICO VIDEO CENTER
ONE PHICO DR - P.O. BOX 85
MECHANICSBURG, 17055
PHONE: (717) 776-1122/5120

THE PRODUCTION BLOCK
2833 N. FRONT ST
HARRISBURG, 17110
PHONE: (717) 233-4155/6

STEWART, E.J. INC
525 MILDRED AVE
PRIMOS, 19018
PHONE: (215)626-650

TELE-IMAGE INC
225 CITY LINE AVE
BALA CYNWYD, 19004
PHONE: (215) 667-1004

**TPC COMMUNICATIONS, INC./
CHANNEL ONE LTD**
PRODUCTION PLAZA, 79 N IND.
PARK
SEWICKLEY, 15143
PHONE: (412) 741-4000/
(800) 331-3735

VITECH PRODUCTIONS
610 MELWOOD AVE
PITTSBURGH, 15213
PHONE: (412) 481-1110

WHYY, INC
INDEPENDENCE MALL EAST
PHILADELPHIA, 19106
PHONE: (215)351-1200

WJAC-TV CREATIVE SERVICES
HICKORY LANE
JOHNSTOWN, 15905
PHONE: (814) 255-7600

WPHL PRODUCTIONS
5001 WYNNEFIELD AVE
PHILADELPHIA, 19131
PHONE: (215)878-1700

WQED
4802 5TH AVE
PITTSBURGH, 15213
PHONE: (412) 622-1300

WILSON STUDIO
818 LIBERTY AVE, SUITE 500
PITTSBURGH, 15222
PHONE: (412) 391-7499/78

UNIONS AND ASSOCIATIONS

AFTRA
THE PENHOUSE SUITE 2007, 625
STANWIX ST
PITTSBURGH, 15222
PHONE: (412) 281-6767
230 SOUTH BROAD ST,
10TH FLOOR
PHILADELPHIA19102
PHONE: (215)545-3150

**GENERAL TEAMSTERS LOCAL
UNION #249**
4701 BUTLER ST
PITTSBURGH, 15201
PHONE: (412) 682-3700/33

IATSE LOCAL 3
101 6TH ST, 633 FULTON
BUILDING
PITTSBURGH, 15222
PHONE: (412) 281-4568

IATSE LOCAL 8
1720 DELANCY ST
PHILADELPHIA, 19103
PHONE: (215)732-3316

**INTERNATIONAL TV & VIDEO
ASSOCIATION (ITVA) -
PHILADELPHIA**
2032 MT. VERNON ST
PHILADELPHIA, 19130
PHONE: (215) 235-2646/
628-9818
FAX: (215) 523-3710

**PHILADELPHIA INDEPENDENT
FILM/VIDEO ASSOCIATION
(PIFVA)**
3701 CHESTNUT ST
PHILADELPHIA, 19104
PHONE: (215) 895-6594
FAX: (215) 895-6562

**SCREEN ACTORS GUILD OF
PHILADELPHIA**
203 SOUTH BROAD ST,
10TH FLOOR
PHILADELPHIA, 19102
PHONE: (215)545-3150

**STEADICOM OPERATORS
ASSOCATION**
100 SPRING GARDEN ST
PHILADELPHIA, 19123
PHONE: (215) 225-5226/
592-9747

TEAMSTERS LOCAL 107
107 SPRING GARDEN ST
PHILADELPHIA, 19123
PHONE: (215)923-1480

TEAMSTERS LOCAL 249
PHONE: 4701 BUTLER ST
PITTSBURGH, 15201
PHONE: (412) 682-3700

SOUTH CAROLINA

SPECIAL EFFECTS

JONES, RANDY
115 FOREST CIRCLE
SUMMERVILLE, 29483
PHONE: (803) 871-3653

PHILLIPS, DAVID
3012 HILLDALE RD
WEST COLUMBIA , 29169
PHONE: (803) 796-3352

SCHEETZ, CHRIS
1215-A SESSIONS ST
MYRTLE BEACH, 29577
PHONE: (803) 626-7259

MAKE-UP SUPPLIERS

COSTUME ASSOCIATES, INC
701 EAST MCBEE AVE
GREENVILLE, 29601
PHONE: (803) 271-4260

**HOUSE OF FABRICS-CHEZ
FABRIC**
1312 MAIN ST
COLUMBIA, 29201
PHONE: (803) 765-2485

THEATRICS UNLIMITED, INC
19 MAGNOLIA RD
CHARLESTON, 29407
PHONE: (803) 763-2326

PROPS

LEWIS AND CLARK
1231 LINCOLN ST
COLUMBIA, 29201
PHONE: (803) 765-2405

**SPOLETO CONSTRUCTION
SHOP**
62 BRIGADE ST
CHARLESTON, 29403
PHONE: (803) 723-6524/
883-3920

STUDIOS/SOUND STAGES

**EARL OWENSBY STUDIOS -
GAFFNEY, S.C.**
1 MOTION PICTURE BLVD
P.O. BOX 184
SHELBY,28150
PHONE: (704) 482-0611
FAX: (704) 487-4763

**COASTAL STUDIOS -
WALTERBORO, S.C.**
500 MEMORIAL AVE
WALTERBORO, 29488
PHONE: (803) 538-5000

**CAROLINA SOUTH STUDIOS -
CHARLESTON, S.C.**
192 E. BAY ST, SUITE 303
CHARLESTON, 29401
PHONE: (803) 723-1075/65
FAX: (803) 723-9454

UNIONS AND ASSOCIATIONS

IATSE REPRESENTATIVE
LOCAL 333
1713 BRANTLEY DR
JAMES ISLAND, 29412
PHONE: (803) 795-7228

IATSE REPRESENTATIVE
LOCAL 347
1825 HATFIELD ST
COLUMBIA,29204
PHONE: (803) 256-1553

IATSE REPRESENTATIVE
LOCAL 697
11 HUNT ST
TRAVELERS REST, 29690
PHONE: (803) 834-4411

TEAMSTERS REPRESENTATIVE
LOCAL 509
1213 STATE ST
CAYCE, 29033
PHONE: (803) 796-6172

TENNESSEE

PROPS

BUFFALOE, BOB
3965 PIKES PEAK
MEMPHIS, 38108
PHONE: (901) 386-4681

COOK, ANNA LUE
7892 CROSS PIKE DR
MEMPHIS
PHONE: (901) 755-4338

FLASHBACK
2304 CENTRAL AVE
MEMPHIS, 38104
PHONE: (901) 272-2304

GILBERT, THOMAS
5338 SHADY GROVE TERRACE
MEMPHIS
PHONE: (901) 683-2325

GYPSYS VINTAGE FASHIONS
2613 BROAD AVE
MEMPHIS, 38112
PHONE: (901) 454-0386

LEBOVITZ, STEPHEN I.
P.O. BOX 17084
MEMPHIS
PHONE: (901) 683-8967/761-0626

MEMPHIS SCENIC, INC
504 CUMBERLAND AVE
MEMPHIS, 38112
PHONE: (901) 458-8171
FAX: (901) 458-4527

STUDIOS/SOUND STAGES

API PHOTOGRAPHERS, INC
3111 STONEBROOK CIRCLE
MEMPHIS, 38116
PHONE: (901) 396-8650

BULLET VIDEO
49 MUSIC SQUARE WEST
NASHVILLE, 37203
PHONE: (615) 327-4621

IMAGEMAKER INC
220 GREAT CIRCLE RD, SUITE 118
NASHVILLE, 37228
PHONE: (615)244-1700

KAREM MITCH STUDIO, INC
2803 C FOSTER AVE
NASHVILLE, 37210
PHONE: (615)331-2228

KINGSWOOD PRODUCTIONS
810 12TH AVE SOUTH
NASHVILLE, 37203
PHONE: (615)256-0530

**MOTION PICTURE PRODUC-
TIONS, INC (FILM &TAPE)**
220B TERRACE VIEW RD
LOUISVILLE, 37777
PHONE: (615)970-2192

OPRYLAND PRODUCTIONS
2806 OPRYLAND DR
NASHVILLE, 37214
PHONE: (615)889-6840

**OWENS, JIM ENTERTAINMENT,
INC (TAPE)**
1525 MCGAVOCK
NASHVILLE, 37203
PHONE: (615)256-7700

PRISM STUDIOS
1027 ELM HILL PIKE
NASHVILLE, 37210
PHONE: (615)225-1919

ROXY PRODUCTION CENTER
827 MERIDIAN ST
NASHVILLE, 37207
PHONE: (615)226-1122

SAMS & MAY/ THE STUDIO, INC
145 12TH AVE NORTH
NASHVILLE, 37203
PHONE: (615)256-4960
SHOE PRODUCTIONS INC
P.O. BOX 12025, 485 NORTH
HOLLYWOOD
MEMPHIS, 38112
PHONE: (901) 458-4496
STUDIO SIX
1711 POPLAR AVE
MEMPHIS
PHONE: (901) 278-8970
TENNESSE PERFORMING ARTS CENTRE
505 DEADRICK ST
NASHVILLE, 37219

UNIONS AND ASSOCIATIONS

IATSE LOCAL 69
6818 STORNAWAY DR
MEMPHIS, 38119
PHONE: (901) 754-7054
IATSE LOCAL 825 WARDROBE
6818 STORNAWAY DR
MEMPHIS, 38119
PHONE: (901) 754-7054
INTERNATIONAL TELEVISION ASSOCIATION (ITVA)
MEMPHIS CHAPTER
781 SOUTH MAIN ST
MEMPHIS, 38106
PHONE: (901) 774-4944/
(800) 444-4MPL
MEMPHIS & SHELBY COUNTY COUNTY FILM, TAPE AND MUSIC COMMISSION
245 WAGNER PL, SUITE 4
MEMPHIS, 38103-3815
PHONE: (901) 527-8300
NABET
1776 BROADWAY, SUITE 1900
NEW YORK, NY, 10019
PHONE: (212) 633-9292
SAG/AFTRA
1108 17TH AVE SOUTH, P.O.
BOX 121087
NASHVILLE, 37212
PHONE: (615) 327-2958
TEAMSTERS (LOCAL 984)
3020 SANDBROOK ST
MEMPHIS, 38116
PHONE: (901) 398-2329
TENNESSEE FILM, ENTERTAIN-MENT & MUSIC COMMISSION
320 6TH AVE. NORTH, 7TH
FLOOR
NASHVILLE, 37219-5308
PHONE: (615)741-FILM
(800) 342-8470 - IN TN
(800) 251-8594 - OUTSIDE TN
TENNESSEE ORGANIZATION OF PRODUCER SERVICES (T.O.P.S.)
P.O. BOX 11717
MEMPHIS, 38111
PHONE: (901) 725-1928

TEXAS

SPECIAL EFFECTS

ART F/X
HOUSTON
PHONE: (713) 861-5660
THIGH TICH STUNTS & SPECIAL FX
4615 GREEN TRAIL DR
HOUSTON, 77084
PHONE: (713) 855-3820
FAX: (713) 579-6569

STAGE PRODUCTION SERVICES
10915 SILKWOOD
HOUSTON, 77031-1706
PHONE: (713) 778-1040

EXPENDABLES

DUNCAN, VICTOR INC
6305 N. O'CONNOR, BLDG 4
IRVING, 75039
PHONE: (214) 869-0200
FAX: (214) 869-98'0
TEXCAM INC
3226 IRVING BLVD
DALLAS, 75247-6007
PHONE: (214) 630-1487
FAX: (214) 630-1761
TEXCAM INC
3263 BRANARD
HOUSTON, 77098
PHONE: (713) 524-2774
FAX: (713) 524-2779

MAKE-UP SUPPLIERS

GIBSON COSTUME SHOP
SAN ANTONIO
PHONE: (512) 826-7811
INCOGNITO COSTUMES INC
7001 FAIR OAKS, SUITE 510
DALLAS, 75231
PHONE: (214) 696-8901
NOROSTCO/TEXAS COSTUME
2607 ROSS
DALLAS, 75201
PHONE: (214) 953-1255/880-8650
PERFORMING ARTS SUPPLY CO
HOUSTON
PHONE: (713) 666-2787
POSNICK, SUSAN INC., DBA SURFACE
COMPLES 105
IRVING, 75039
PHONE: (214) 869-1144
RANDY MAZE MAKE-UP CENTER
12906 VILLAWOOD LANE
HOUSTON, 77072
PHONE: (713) 933-3139
STARLINE COSTUME PRODUCTS
1286 BANDERA RD
SAN ANTONIO, 78228
PHONE: (512) 435-3535
TEXAS SCENIC CO. INC.
5423 JACKWOOD
SAN ANTONIO, 78238
PHONE: (512) 684-0091/(800)

PROPS

4TH U.S. MEMORIAL CAVALRY REGIMENT
23490 I-10 WEST
SAN ANTONIO, 78257
PHONE: (512) 698-2842
AMELIAS RETRO-VOGUE & RELICS
2024 S. LAMAR
AUSTIN, 78704
PHONE: (512) 442-4446/
444-1289
COOL CAR CLUB - AUSTIN
225 CONGRESS #294
AUSTIN, 78701
PHONE: (512) 440-8191
CROSSWALK DELI
121-C ALAMO PLAZA
SAN ANTONIO, 78205
PHONE: (512) 228-0880/
222-8583
DESIGN DIMENSIONS
7208 MCNEIL RD ,
SUITES 103/104
AUSTIN , 78729
PHONE: (512) 258-8596/331-5193

DIXIE DUDE RANCH
P.O. BOX 548
BANDERA, 78003
PHONE: (512) 796-4481/3217
THE EXECUTICE BRANCH
6460 HART LANE
AUSTIN, 78731
PHONE: (512) 343-1575/837-6154
GREAT FX
2611 N. BELTLINE #126
SUNNYVALE, 75182
PHONE: (214) 226-6226
CLIFF HANGER
18002 EASY ST.
JONESTOWN, 78645
PHONE: (512) 267-9740/9150
HOLT HELICOPTERS, INC
P.O. BOX 1669
UVALDE, 78802-1669
PHONE: (512) 278-9463/1866
JIFASCO INTERNATIONAL INC.
P.O. BOX 180415
AUSTIN, 78718-0415
PHONE: (512) 335-7563/258-8826
JUBILEE TENTS & AWNINGS
P.O. BOX 26831
AUSTIN, 78755-0831
PHONE: (512) 990-1209
LONE STAR LIVING HISTORY ASSOCIATION, INC
130 PASEO ENCINAL
SAN ANTONIO, 78212
PHONE: (512) 735-6051/
658-1267
JOE MECHANIC
5015 DUVAL
AUSTIN, 78751
PHONE: (512) 450-1075/
389-0472
MUZZLEFLASH
P.O. BOX 200842
AUSTIN, 78720
PHONE: (512) 835-6102/
478-6754
PEGASUS MOTORCARS
1719 ALLEN GENOA
HOUSTON, 77502
PHONE: (713) 473-3933
FAX: (713) 473-2001
PRODUCTION SERVICES
6016 G-4 DONIPHAN
EL PASO, 79932
PHONE: (915) 584-6903
PROP CITY
1136 EUCLID
HOUSTON, 77009
PHONE: (713) 880-1954/353-5667
PROPS & DROPS
1735 OJEMAN
HOUSTON, 77055
PHONE: (713) 984-9514/
528-1860
PROPS OF TEXAS
2930 CANTON ST
DALLAS, 75226
PHONE: (214) 748-7767
RAGTOPS OF TEXAS, INC
HOUSTON,
PHONE: (713) 358-9126
ROCKIN J RANCH
3600 KIPHEN RD
ROUND ROCK, 78664
PHONE: (512) 244-1252
SHOWGUNS
5540 MERCEDES
DALLAS, 75206
PHONE: (214) 821-2101
SIGN BUILDERS OF TEXAS, INC
4125 TODD LANE
AUSTIN, 78744
PHONE: (512) 447-3147/
445-5757
SIGNATURE WORKS
WESTERN
SAN ANTONIO
PHONE: (512) 646-0227

STUNTS GALORE
13430 LOUISVILLE
HOUSTON, 77015
PHONE: (800) 621-4793/
(713) 451-4826
TERLINGUA GHOST TOWN
P.O. BOX 362
TERLINGUA, 79852
PHONE: (915) 371-2234/424-3234
TEXAS BULLET PRODUCTIONS
P.O. BOX 1026
FLORESVILLE, 78114-1026
PHONE: (512) 393-3250/8166
THE TEXAS STATE RAILROAD
P.O. BOX 39
RUSK, 75785
PHONE: (214) 683-2561
THE TEXAS TRADER
1208 SLAUGHTER LANE
AUSTIN, 78748
PHONE: (512) 282-1215
WEAPONS MASTERS
P.O. BOX 830648
RICHARDSON, 75033
PHONE: (214) 681-3600
WEST TEXAS FILM SERVICES
P.O. BOX 1180
MARFA, 79843
PHONE: (915) 729-4472/2386

PYROTECHNICS

CRECE, RICHARD
SAN ANTONIO
PHONE: (512) 656-8131
HIGH TICH STUNTS & SPECIAL FX
4615 GREEN TRAIL DR
HOUSTON, 77084
PHONE: (713) 855-3820
FAX: (713) 579-6569
STAGE PRODUCTION SERVICES
10915 SILKWOOD
HOUSTON, 77031-1706
PHONE: (713) 778-1040

STUDIOS/SOUND STAGES

ASHE-BOWIE PRODUCTIONS
SAN ANTONIO
PHONE: (512) 690-8283
CONTINENTAL PRODUCTIONS
3900 HARRY HINES BLVD
DALLAS, 75219
PHONE: (214) 521-3900
DALLAS COMMUNICATIONS COMPLEX
6301 NORTH O'CONNOR RD
IRVING, 75039
PHONE: (214) 869-0700/
869-7739
F.P.S. INC.
11250 PAGEMILL RD
DALLAS, 75243
PHONE: (214) 340-8585
FIRST VIDEO PRODUCTIONS, INC
4235 CENTERGATE
SAN ANTONIO, 78217
PHONE: (512) 655-1111
FAX: (512) 655-9249
HOUSTON STUDIOS
P.O. BOX 1461
HOUSTON, 77251
PHONE: (713) 528-1318
JOHNSON PRODUCTIONS
SAN ANTONIO
PHONE: (512) 220-1790
KENS-TV PRODUCTION
SAN ANTONIO
PHONE: (512) 366-5000
KIDD, RICHARD PRODUCTIONS INC
5610 MAPLE
DALLAS, 75235
PHONE: (214) 638-5433

NEW AGE RECORDING
SAN ANTONIO
PHONE: (512) 641-9988
PANDA ENTERTAINMENT GROUP, INC
P.O. BOX 57457
WEBSTER, 77598
PHONE: (713) 333-2236
FAX: (713) 332-7701
PEARLMAN PRODUCTIONS
2401 W. BELLFORT
HOUSTON, 77054
PHONE: (713) 668-3601
PINNACLE STUDIOS
2410 FARRINGTON ST
DALLAS, 75207
PHONE: (214) 637-2748/
630-4500
SALSA PRODUCTIONS
SAN ANTONIO
PHONE: (512) 822-3963
SOUTH COAST VIDEO INC
5234 ELM ST
HOUSTON, 77081
PHONE: (713) 661-3550
SOUTHWEST TELEPRODUCTIONS INC
2649 TARNA DR
DALLAS, 75229-2222
PHONE: (214) 243-5719
FAX: (214) 243-5800
SPINDLETOP PRODUCTIONS
6311 NORTH O'CONNOR RD,
SUITE 125
IRVING, 75039
PHONE: (214)869-0222
STARS OVER TEXAS INC
7021 JOHN CARPENTER FWY
DALLAS, 75247
PHONE: (214) 638-6200/
(800) 322-7009
BILL STOKES ASSOCIATES
5642 DYER
DALLAS, 75206
PHONE: (214) 363-0161
STUDIOS AT LAS COLINAS/ DALLAS
6301 N. O'CONNOR RD., LB 19
IRVING, 75039
PHONE: (214) 869-0700/7661
TELE-IMAGE
6305 NORTH O'CONNOR LB6
IRVING, 75039-3510
PHONE: (214) 869-0060/
(800) 882-0060
THIRD COAST STUDIOS
501 NORTH I-H 35
AUSTIN, 78702
PHONE: (512) 476-8905
W.F.P. & ASSOCIATES
717 E. IRVING BLVD
IRVING, 75060
PHONE: (214) 579-1630/0649
WHITE LION PICTOGRAPH
SAN ANTONIO
PHONE: (512) 826-3615
WILMING REAMS ANIMATION
SAN ANTONIO
PHONE: (512) 342-214

UNIONS AND ASSOCIATIONS

IATSE LOCAL #51
P.O. BOX 403
HOUSTON, 77001
PHONE: (713) 861-5453
IATSE LOCAL #76
201 E. PARK, SUITE 107
SAN ANTONIO, 78212
PHONE: (512) 223-3911
IATSE LOCAL #126
P.O. BOX 1175
FORT WORTH, 76101
PHONE: (817) 685-6363/
834-3838

IATSE LOCAL #127
5409 N. JIM MILLER #213
DALLAS, 75227
PHONE: (214) 388-4741
IATSE LOCAL #480
615 ROBINSON AVE
EL PASO, 79902
PHONE: (915) 544-6818/
533-4131
SCREEN ACTORS GUILD/AFTRA - DALLAS
TWO DALLAS
COMMUNICATIONS COMPLEX
6309 N. O'CONNOR RD.,
SUITE 111, LB25
IRVING, 75039-3510
PHONE: (214) 363-8300
SCREEN ACTORS GUILD/AFTRA - HOUSTON
2650 FOUNTANVIEW, SUITE 326
HOUSTON, 77057
PHONE: (713) 972-1806
TEAMSTERS LOCAL #657
8214 ROUGH RIDER
SAN ANTONIO, 78239
PHONE: (512) 590-2013
TEAMSTERS LOCAL #745
P.O. BOX 17270
DALLAS, 75217
PHONE: (214) 398-0661
TEAMSTERS LOCAL #1111
6001 GULF FREEWAY,
SUITE A117
HOUSTON, 77023
PHONE: (713) 926-8311

UTAH

SPECIAL EFFECTS

STS
5181 AMELIA EARHART DR
SALT LAKE CITY, 84116
PHONE: (801) 537-1427
FAX: (801)-596-1194

EQUIPMENT RENTALS

CTM STUDIOS
210 WEST 1500 SOUTH
PROVO, 84601
PHONE: (801) 375-3066
LARSEN, ROLF
1099 WINDSOR ST
SALT LAKE CITY, 84105
PHONE: (801) 486-1755
MENZEL, MIKE
1472 EAST LONGDALE DR
SANDY, 84092
PHONE: (801) 571-5555
ROSE & COMPANY
1375 WEST 500 NORTH #13
PROVO, 84601
PHONE: (801) 374-6995

PROPS

CRAFTSMAN UPHOLSTERY & INTERIORS
354 NORTH STATE
OREM, 84057
PHONE: (801) 226-6606/ (800) 367-6221
DENNYS WIGWAM
48 EAST CENTER
KANAB, 84741
PHONE: (801) 644-2452/5036
ELKHORN MOVIE RANCH
776 WEST VINE ST
TOOELE, 84074
PHONE: (801) 882-5666/3863

GENERAL MERCANTILE
P.O. BOX 70
PARK CITY, 84060
PHONE: (801) 654-3859
HIGHLAND FILM PRODUCTION SERVICES, INC
11401 WILLOW HILL DRIVE
SANDY, 84092
PHONE: (801) 571-1790/1962
ILLUSION TECH
P.O. BOX 161
BOUNTIFUL, 84010
PHONE: (801) 295-6510
L.S. PRODUCTIONS, INC
2669 SOUTH HIGHLAND DR.,
APT. F
SALT LAKE CITY
PHONE: (801) 466-1664
PERFORMANCE FABRICATION
5325 FERNCREST CIRCLE
SALT LAKE CITY, 84118
PHONE: (801) 966-5417
RIGHT SIDE PRODUCTION SERVICES
5835 SOUTH WATERBURY DR.,
SUITE H
SALT LAKE CITY, 84121
PHONE: (801) 272-5060
SCENIC SERVICE SPECIALISTS
1240 EAST 800 NORTH
OREM, 84057
PHONE: (801) 224-4293

UNIONS AND ASSOCIATIONS

DIRECTORS GUILD OF AMERICA
7950 SUNSET BLVD
LOS ANGELES, 90046
PHONE: (213) 656-1220
IATSE LOCAL 648
3189 NAVAJO LANE
PROVO, 84604
PHONE: (801) 465-3389
IATSE LOCAL 99
100 SOUTH WEST TEMPLE
SALT LAKE CITY, 84101
PHONE: (801) 359-0513/
582-8711
SCREEN ACTORS GUILD - REGIONAL OFFICE
950 SOUTH CHERRY, SUITE 502
DENVER,80222
PHONE: (800) 527-7517
SCREEN ACTORS GUILD - UTAH CHAPTER
13734 SOUTH 2550 WEST
RIVERTON
PHONE: (801) 254-1446/
(800) 527-7517
TEAMSTERS LOCAL 222
2614 SOUTH 3270 WEST
SALT LAKE CITY, 84119
PHONE: (801) 972-1898
UNITED STUNTMENS ASSOCIATION
P.O. BOX 744
PAROWAN, 84761
PHONE: (801) 477-8212/
(206) 542-1649
UTAH STUNTMEN ASSOCIATION
10010 NORTH 6000 WEST
AMERICAN FORK, 84003
PHONE: (801) 756-6254
WRITERS GUILD OF AMERICA
1800 SOUTH WEST TEMPLE,
SUITE 103
SALT LAKE CITY, 84115
PHONE: (801) 485-9253

VERMONT

STUDIOS/SOUND STAGES

BURLINGTON STUDIOS INC
444 SOUTH UNION STREET
BURLINGTON, 05401
PHONE: (802) 863-1115

VIRGINIA

SPECIAL EFFECTS

CHURCH HILL PRODUCTION SERVICES
2605 E. FRANKLIN ST
RICHMOND, 23223
PHONE: (804) 649-8421/
648-7909
FAX: (804) 358-9263
COMMON MAN MOTION PICTURE CORP
TAYLOR MADE IMAGES, INC
5042 SOUTH 22ND ST
ARLINGTON, 22206
PHONE: (703) 931-3431
FAX: (703) 931-3474
D & E MINIATURE
835 HOLLY HEDGE LANE
VIRGINIA BEACH, 23452
PHONE: (804) 468-4687
FAX: (804) 468-3724
DEN PRODUCTIONS
5308 PINELAND COURT
RICHMOND, 23234
PHONE: (804) 275-1355
DR. BOB'S THEATRICITY
424 INVESTORS PLACE,
SUITE 101
VIRGINIA BEACH, 23452
PHONE: (804) 499-0720/2723
PARK AVENUE ARMORERS
16 PARK AVE
NEWPORT NEWS,23607
PHONE: (804) 380-0844

PROPS

ARCHITECTURAL RENOVATIONS, INC
2877 GUARDIAN LANE,
SUITE 111
VIRGINIA BEACH, 23452
PHONE: (804) 498-9364/
463-4662
AIR-RAGEOUS
P.O. BOX 1127, 12104
HARROWGATE RD
CHESTER, 23831-1127
PHONE: (804) 748-2830
FAX: (804) 796-9457
AUTOMOTIVE CUSTOM SERVICES
1104 ELLIOT DR
MECHANICSVILLE, 23111
PHONE: (804) 746-5771
BACKSTAGE, INC
310 WEST BROAD ST
RICHMOND, 23220-4258
PHONE: (804) 644-1433
FAX: (804) 644-4913
THE BLANKENSHIP COMPANY
4221 BROOK RD
RICHMOND, 23227
PHONE: (804) 262-6367
BREAK OF DAWN ANIMAL TALENTS
4077 DARBYTOWN RD
RICHMOND, 23231
PHONE: (804) 795-2998/5933

CHURCH HILL PRODUCTION SERVICES
2605 E. FRANKLIN ST
RICHMOND, 23223
PHONE: (804) 649-8421/
648-7909
FAX: (804) 358-9263

COMMON MAN MOTION PICTURE CORP
TAYLOR MADE IMAGES, INC
5042 SOUTH 22ND ST
ARLINGTON, 22206
PHONE: (703) 931-3431
FAX: (703) 931-3474

THE COSTUME WORKS
5021 BROOK RD, SUITE 117
RICHMOND, 23227
PHONE: (804) 266-9287

CUSTOM GLASS ETCHERS
9206 VENTURE CT., C-6
MANASSAS PARK, 22111
PHONE: (703) 335-2569

D & E MINIATURE
835 HOLLY HEDGE LANE
VIRGINIA BEACH, 23452
PHONE: (804) 468-4687
FAX: (804) 468-3724

THE DAMN HORSETRADERS
6490 OSBORNE TP.
RICHMOND, 23231
PHONE: (804) 222-6805

DEN PRODUCTIONS
5308 PINELAND COURT
RICHMOND, 23234
PHONE: (804) 275-1355

DU JOUR
5806 GROVE AVE
RICHMOND, 23226
PHONE: (804) 285-1301

EAST COAST GUN & SUPPLY, INC
T/A DISCOUNT GUN EXCHANGE
3924 OLD FARM LANE
VIRGINIA BEACH, 23452
PHONE: (804) 486-7936

ENGRAVING SERVICE & DESIGNM, INC
11010 TRADE RD
RICHMOND, 23236
PHONE: (804) 379-2812

FIXTURES AND FORMS
317 N. 17TH ST, P.O. BOX 5606
RICHMOND, 23220
PHONE: (804) 648-8300
FAX: (804) 648-8306

FOUR SEASONS SPECIAL EVENTS
1200 W. MAIN ST
RICHMOND, 23220
PHONE: (804) 355-4584

HALCYON: VINGTAGE CLOTHING
117 NORTH ROBINSON ST
RICHMOND, 23220
PHONE: (804) 358-1311

JACK DE TREVILLE & CO.
1613 W. MAIN ST
RICHMOND, 23220
PHONE: (804) 358-6618/
323-1018

KEYSTONE PROPS
11070 BOTTOM CREEK RD
BENT MOUNTAIN, 24059
PHONE: (703) 929-4906
FAX: (703) 929-5005

LIVING HISTORY ASSOCIATES, LTD.
P.O. BOX 4914, 1100 WEST
FRANKLIN ST
RICHMOND, 23220
PHONE: (804) 353-8166/
264-9451

LIVING HISTORIANS UNLIMITED
RT. 2, BOX 1300
WOODFORD, 22580
PHONE: (703) 899-6464

LOVEABLE LLAMAS, HISTORICAL HORSES & ETC.
5042 CHARLES CITY RD
RICHMOND, 23231-6542
PHONE: (804) 795-5641

MCPROPS
400 BALDWIN RD
RICHMOND, 23229
PHONE: (804) 288-4793

P. GRAYSON ENTERPRISES
RTE. 1, BOX 188-9
LINVILLE, 22834
PHONE: (703) 833-5593/
433-7137

PLEASANTS HARDWARE
2024 W. BROAD ST
RICHMOND, 23220
PHONE: (804) 359-9381
FAX :(804) 355-5928

PREMIERE, INC
3339 WEST CARY ST
RICHMOND, 23221
PHONE: (804) 355-3887

SCECON SCENIC SERVICES
P.O. BOX 6689
CHESAPEAKE, 23323
PHONE: (804) 487-8418

SIGNS OF IMAGINATION
3921 DEEP ROCK RD
RICHMOND, 23233
PHONE: (804) 346-3032
FAX: (804) 346-5062

STEIN'S THEATRICAL & DANCE SUPPLY CENTER
3100 CLARENDON BLVD
ARLINGTON, 22201
PHONE: (703) 522-2660

SYSTEMS WIRELESS, LTD
465 HERNDON PARKWAY
HERNDON, 22070
PHONE: (703) 471-7887

THEATRE IV
114 WEST BROAD ST
RICHMOND, 23220
PHONE: (804) 783-1688

WALKER DESIGN
2341 PURPLE MARTIN LANE
VIRGINIA BEACH, 23455
PHONE: (804) 460-4616

MAKE-UP SUPPLIERS

STEIN'S THEATRICAL & DANCE SUPPLY CENTER
3100 CLARENDON BLVD
ARLINGTON, 22201
PHONE: (703) 522-2660

STUDIOS/SOUND STAGES

ATLANTIC FILM STUDIOS
1000 FILMWAY
SUFFOLK, 23434
PHONE: (804) 934-7000
FAX: (804) 7010

ATLANTIC VIDEO, INC
150 SOUTH GORDON ST
ALEXANDRIA, 22304
PHONE: (703) 823-2800

CENTRAL VIRGINIA TELECOMMUNICATIONS CORP
23 SESAME ST
RICHMOND, 23235
PHONE: (804) 320-1301

CLASSIC ENTERTAINMENT PRODUCTIONS
5241 CLEVELAND ST, SUITE 113
VIRGINIA BEACH, 23462
PHONE: (804) 499-9243
FAX: (804) 499-6178

CREATIVE VIDEO & FILM PRODUCTION
808 LIVE OAK DR., SUITE 101
CHESAPEAKE, 23320
PHONE: (804) 420-3605
FAX: (804) 366-0919

ERNST & YOUNG VISUAL SERVICES GROUP
1950 ROLAND CLARKE PL
RESTON, 22091
PHONE: (703) 648-2301
FAXL (703) 620-3335

LONGWORTH COMMUNICATIONS, INC
230 SOUTH CRATER RD
PETERSBURG, 23803
PHONE: (804) 748-2512
FAX: (804) 733-0305

MEDIA GENERAL CABLE OF FAIRFAX
2917 ESKRIDGE RD
FAIRFAX, 22031
PHONE: (703) 849-8430
FAX: (703) 207-9458

METRO COMMUNICATIONS, INC
424 DUKE OF GLOUCESTER ST
WILLIAMSBURG, 23185
PHONE: (804) 253-0050
FAX: (804) 253-8558

METRO VIDEO PRODUCTIONS, INC
8-A SOUTH PLUM ST
RICHMOND, 23220
PHONE: (804) 358-2500
FAX: (804) 354-1552

NORTHSTAR ENTERTAINMENT GROUP, INC
1000 CENTERVILLE TURNPIKE
VIRGINIA BEACH, 23463
PHONE: (804) 523-7293
FAX: (804) 523-7180

PARK AVENUE TELEPRODUCTIONS
3500 MAYLAND COURT
RICHMOND, 23233
PHONE: (804) 346-3232

R.L. STUDIO
6812 EVERGLADES DR
RICHMOND, 23225
PHONE: (804) 674-4103

SIGNET BANK VIDEO FACILITY
11013 WEST BROAD ST. RD.
GLEN ALLEN, 23060
PHONE: (804) 747-2064/2278
FAX: (804) 747-2985

STUDIO THEATRE OF RICHMOND, INC
P.O. BOX 149
RICHMOND, 23201
PHONE: (804) 780-1356

V.M. PRODUCTION
RICHMOND SOUND STAGE
2100 BREMO RD
RICHMOND, 23230
PHONE: (804) 282-8273

WARNER CABLE OF RESTON, INC
397 HERNDON PARKWAY,
SUITE 25
HERNDON, 22070
PHONE: (703) 471-1749/4205

WOLF TRAP FOUNDATION
1624 TRAP RD
VIENNA, 22180
PHONE: (703) 255-1902

WRLH-TV
1925 WESTMORELAND ST
RICHMOND, 23230
PHONE: (804) 358-3535/1495

WVPT TV
298 PORT REPUBLIC RD
HARRISONBURG, 22801
PHONE: (703) 434-5391
FAX: (703) 434-7481

ZUZU FILMS, INC
13 WEST MAIN ST
RICHMOND, 23220
PHONE: (804) 780-1990
FAX: (804) 780-2331

UNIONS AND ASSOCIATIONS

AMERICAN FEDERATION OF TV AND RADIO ARTISTS
AND SCREEN ACTORS GUILD
5480 WISCONSIN AVE,
SUITE 201
CHEVY CHASE, 20815
PHONE: (301) 657-2560

C 644 EAST COAST-SOUTHERN REP
1406 NEWTON ST
WASHINGTON, D.C., 20017
PHONE: (202) 269-4607

I.A. LOCAL 531
2602 LYNCHBURG ST
HOPEWELL, 23860
PHONE: (804) 458-0226

IATSE LOCAL 22
1247-B LOCKWOOD DR
SILVER SPRING, MD 20901
PHONE: (301) 593-4650

IATSE LOCAL 72/NORFOLK, VIRGINIA BEACH, PORTSMOUTH,
CHESAPEAKE
P.O. BOX 3431
NORFOLK, 23514
PHONE: (804) 485-0876

IATSE LOCAL 264/NEWPORT NEWS, HAMPTON, WILLIAMSBURG
P.O. BOX 9124
HAMPTON, 23670
PHONE: (804) 826-9191/
838-9045

IATSE/TBSE L819
12302 PERSIMMON PL
WOODBRIDGE, 22192
PHONE: (703) 491-4725/
(202) 244-5151

IASTE LOCAL 370/RICHMOND
P.O. BOX 1681
RICHMOND, 23213
PHONE: (804) 288-9068

IATSE M LOCAL 572 STAUNTON
ROUTE 1
LINVILLE, 22834
PHONE: (703) 833-5593/
433-7137

INTERNATIONAL TELEVISION ASSOCIATION
P.O. BOX 11375
RICHMOND, 23230
PHONE: (804) 747-3031

M 711 CHARLOTTESVILLE (IATSE)
P O. BOX 611
CHARLOTTESVILLE, 22902
PHONE: (804) 293-7495

M 563 DANVILLE (IATSE)
2134 BAXTER ST
DANVILLE, 24540
PHONE: (804) 793-8208

M 55 ROANOKE (IATSE)
520 WALNUT AVE, S.W.
ROANOKE, 24016
PHONE: (703) 981-1133

NATIONAL ASSOC. OF BROADCASTING EMPLOYEES & TECHNICIANS
NABET LOCAL 31
5034 WISCONSIN AVE, N.W.,
SUITE 202
WASHINGTON, D.C. 20016
PHONE: (202) 966-4073

TEAMSTERS
3705 CAROLINA AVE
RICHMOND, 23222
TT 868
5308 WESTPORT RD
CHEVY CHASE , MD ,20815
PHONE: (301) 657-9194
**VIRGINIA PRODUCTION
SERVICES ASSOCIATION**
P.O. BOX 7419
RICHMOND, 23221-0419
PHONE: (804) 287-5070
**VIRGINIA SCREENWRITER'S
FORUM**
105 BIRCH CIRCLE
MANAKIN-SABOT, 2313
PHONE: (804) 784-5835

WASHINGTON

SPECIAL EFFECTS

RIGGS, BOBBY
5717 S. MUTINY BAY ROAD
FREELAND, 98249
PHONE: (206) 321-4558

PYROTECHNICS

DON DUMAS
318 1ST AVE S.
SEATTLE, 98104
PHONE: (206) 343-7045
**KNOTT LIMITED SPECIAL
EFFECTS**
1192 EUCLID AVE
BELLINGHAM 98226
PHONE: (206) 738-9598

EXPENDABLES

OPPENHEIMER CAMERA, INC
666 S. PLUMMER ST
SEATTLE, 98134
PHONE: (503) 294-0762/
(206) 467-8666
FAX: (206) 467-9165
**RANIER PHOTOGRAPHIC
SUPPLY**
8730 RAINIER AVE. S.
SEATTLE, 98118
PHONE: (206) 255-3456
SEATTLE CINE RENTALS
433 8TH AVE. N
SEATTLE, 98109
PHONE: (206) 622-8540/
(800) 367-0085
FAX: (206) 292-2919
SPOKANE CINE CO.
EAST 9514 MONTGOMERY,
BAY 32
SPOKANE, 99212
PHONE: (509) 926-5001
FAX: (509) 926-4999

UNIONS AND ASSOCIATIONS

**WASHINGTON FILM & VIDEO
ASSOC. (WFVA)**
P.O. BOX 84588
SEATTLE, 98124
PHONE: (206) 441-4569

WEST VIRGINIA

SPECIAL EFFECTS

JUNE BINDER
HEARTWOOD
BIG BEND, 26136
PHONE: (304) 354-7874
GLEN DALE MORTON, JR
198 FRAME RD
ELKVIEW, 25071
PHONE: (304) 965-0080

PYROTECHNICS

BRADFORD D. BOLL
800 CEDAR RD., APT. C
CHARLESTON, 25314
PHONE: (304) 342-0108/
344-4572

STUDIOS/SOUND STAGES

WBOY-TV
904 W. PIKE ST
P.O. BOX 1590
CLARKSBURG, 26302
PHONE: (304) 623-3311
WCHS-TV
1301 PIEDMONT RD
CHARLESTON, 25301
PHONE: (304) 346-5358
**WEST VIRGINIA EDUCATIONAL
NETWORK**
WEST VIRGINIA STATE COLLEGE
CAMPUS BOX 195
INSTITUTE, 25112
PHONE: (304) 766-2068
WOWK-TV
555 5TH AVE
HUNTINGTON, 25701
PHONE: (304) 525-7661
WPBY-TV
THIRD AVE
HUNTINGTON, 25701
PHONE: (304) 348-2160/696-
6630
**WSAZ TELEVISION 3
PRODUCTIONS**
645 FIFTH AVE
HUNTINGTON, 25701
PHONE: (304) 697-4780
WSWP-TV
P.O. BOX AH
BECKLEY, 25802
PHONE: (304) 255-1501/2
WVVA TELEVISION 6, INC
P.O. BOX 1930
BLUEFIELD, 24701-6930
PHONE: (304) 325-5487

WISCONSIN

SPECIAL EFFECTS

ALLEZ, GEORGE
309 N. HILLSIDE TERR.
MADISON, 53705
PHONE: (608) 238-192

PYROTECHNICS

**STURMS SPECIAL EFFECTS
INTL**
P.O. BOX 691
LAKE GENEVA, 53147
PHONE: (414) 245-6594

PROPS

ACME PRODUCTION SERVICES
529 WEST NATIONAL AVE
MILWAUKEE, 53204
PHONE: (414) 645-7030
FAX: (414) 645-7118
ANOTHER DISPLAY COMPANY
3828 NORTH HUBBARD
MILWAUKEE, 53212
PHONE: (414) 332-6093
BCA INC.
6192 HOLY HILL RD
HARTFORD, 53027
PHONE: (414) 964-1111/628-
3724
BADGER EXPOSITION SERVICE
440 WEST VLIET ST
MILWAUKEE, 53212
PHONE: (414) 276-8757
FAX: (414) 276-8912
**CREATIVE SERVICES
INTERNATIONAL**
1310 W. BURNHAM ST
MILWAUKEE, 53204
PHONE: (414) 645-0700
GREAT BIG PICTURES
1444 EAST WASHINGTON AVE
MADISON, 53703
PHONE: (608) 257-7071
GREAT BIG PICTURES, INC
240 NORTH MILWAUKEE ST
MILWAUKEE, 53202
PHONE: (414) 224-9070
**MAINSTAGE THEATRICAL
SUPPLY, INC**
129 WEST PITTSBURGH AVE
MILWAUKEE, 53204
PHONE: (414) 278-0878
FAX: (414) 278-0986
MARK GUBIN, INC
2893 SOUTH DELAWARE
MILWAUKEE, 53207
PHONE: (414) 482-0640
FAX: (414) 481-9320
PM ASSOCIATES
13435 WEST COURTLAND AVE
BROOKFIELD, 53005
PHONE: (414) 781-4051

STUDIOS/SOUND STAGES

BETA 1 VIDEO PRODUCTIONS
W140 N5910 LILLY RD
MENOMONEE FALLS, 53051
PHONE: (414) 252-3833
LOGAN PRODS INC
8035 N PT. WASHINGTON RD.
MILWAUKEE, 53217
PHONE: (414) 352-9691
FAX: (414) 352-4993
MARX PRODUCTION CENTER
3100 WEST VERA AVE
MILWAUKEE, 53209
PHONE: (414) 351-5060
FAX: (414) 351-4652
MPI FILM & VIDEO
MADISON
PHONE: (608) 233-8588
NELSON PRODUCTIONS INC
1533 NORTH JACKSON ST
MILWAUKEE, 53202
PHONE: (414)271-5211
PROVIDEO
2302 WEST BADGER RD
MADISON, 53713
PHONE: (608) 273-2507
PYRAMID PICTURES, INC
6812 WEST CALUMET RD
MILWAUKEE, 53223
PHONE: (414) 355-5114
FAX: (414) 355-3373

RW VIDEO INC
4902 HAMMERSLEY RD
MADISON, 53711
PHONE: (608)274-4000
FAX: (608)274-9357
**TAYLOR VIDEO
COMMUNICATIONS INC**
10838 WEST WISCONSIN AVE
WAUWATOSA, 53226
PHONE: (414) 778-0362
FAX: (414) 778-2067
VIDEO WISCONSIN
18110 W. BLUEMOUND RD
BROOKFIELD, 53005
PHONE: (414) 785-1110
WINDSOR LAKE STUDIOS LTD
STUDIO DR
EAGLE RIVER, 54521
PHONE: (715) 479-6666
FAX: (715) 479-6241
WISCONSIN CABLE ADS, INC
623 CLERMONT ST,
P.O. BOX 502
ANTIGO, 54409
PHONE: (715) 627-2001
FAX: (715) 623-6445

INTERNATIONAL

ARGENTINA

FILM COMMISSIONS

INSTITUTO NACIONAL DE CINEMATOGRAFIA
LIMA 319,(1073)
BUENOS AIRES
ARGENTINA
PHONE: (37)8429 EXT.
PHONE: TX 21104 EXT.

SECRETARIA DE CULTURA DE LA NACION
AV. ALVEAR 1690,(1041)
BUENOS AIRES
ARGENTINA
PHONE: (42)8551 EXT.

SUBSECRETARIA DE TURISMO
SUIPACHA 1111
(1368)BUENOS AIRES
ARGENTINA
PHONE: (312)8856 EXT.

UNIONS AND ASSOCIATIONS

ASOCIACION ARGENTINA DE ACTORES (ACTORS GUILD)
ALSINA 1762,(1088) BUENOS AIRES
ARGENTINA
PHONE: (49)1781 EXT.

ASOCIACION ARGENTINA DE PRODUCTORES DE CINE INDEPENDIENTES
GUATEMALA 5188
14250BUENOS AIRES
ARGENTINA
PHONE: (774)9480 EXT.

ASOCIACION DE CRONISTAS CINEMATOGRAFICOS DE LA ARGENTINA
MAIPU 621
(1006) BUENOS AIRES
ARGENTINA
PHONE: (392)6625 EXT.

ASOCIACION DE PRODUCTORES CINEMATOGRAFICOS DE LA ARGENTINA
LAVALLE 1860
(1051) BUENOS AIRES
ARGENTINA
PHONE: (40) 3438/3439 EXT.

ASOCIACION DE VIDEOEDITORES IDEPENDIENTES DE LA ARGENTINA
BOLIVAR 391,PISO 6
(1066) BUENOS AIRES
ARGENTINA
PHONE: (34)4876 EXT.
PHONE: TX 18650 BONES EXT.

DIRECTORES ARGENTIONOS CINEMATOGRAFICOS
JUNCAL 2029
(1116) BUENOS AIRES
ARGENTINA
PHONE: (84)8774 EXT.
PHONE: (84)7544 EXT.
FAX: (84)0208

SINDICATO DE INDUSTRIA CINEMATOGRAFICA ARGENTINA
JUNCAL 2029
(1116) BUENOS AIRES
ARGENTINA
PHONE: (84)8774 EXT.
PHONE: (84)7544EXT.
FAX: (84)0208

SOCIEDAD DE AUTORES DE LA ARGENTINA
PACHECO DE MELO 1820
(1126) BUENOS AIRES
ARGENTINA
PHONE: (42)2838 EXT.

AUSTRALIA

SPECIAL EFFECTS

4FX/PETER MCKIE
7 QUIRK STREET,ROZELLE
NSW 2043
AUSTRALIA-NEW SOUTH WALES
PHONE: (02)818 5466 EXT.
FAX: (02)810 2219

AUSTRALIAN EFEX COMPANY PTY LTD
178 GEORGE STREET,ERSKINVILLE
NSW 2043
AUSTRALIA-NEW SOUTH WALES
PHONE: (02)516 3033 EXT.
FAX: (02) 512 486

COURTLEY,STEVE
C/O TOP TECHNICIANS
AUSTRALIA-NEW SOUTH WALES
PHONE; (02)981 1622 EXT.

CREATEK ENTERPRISES PTY LTD
17 LACKEY STREET,ST. PETERS
NSW 2044
AUSTRALIA-NEW SOUTH WALES
PHONE: (02)550 1477 EXT.

DESIGN FIELD PTY LTD.
1 CALEDONIA STREET,PADDINGTON
NSW 2021
AUSTRALIA-NEW SOUTH WALES
PHONE: (2)328 7366 EXT.

EFFECTS ENGINEERING
LOT 46,LAITOI ROAD,TERRY HILLS
NSW 2084
AUSTRALIA-NEW SOUTH WALES
PHONE: (02)450 2956 EXT.

MAXWELL, ALAN SPECIAL EFFECTS
LOT 46,LAITOKI RD ,TERREY HILLS
NW 2084
AUSTRALIA-NEW SOUTH WALES
PHONE: (02)450 1648 EXT.

POLLARD, ANDY
233 NELSON LANE,ANNANDALE
NSW 2038
AUSTRALIA-NEW SOUTH WALES
PHONE: (02)660 7172 EXT.

PRIDE EFFECTS STUDIOS
19 HIGGINBOTHAM RD.,GLADESVILLE
NSW 2111
AUSTRALIA-NEW SOUTH WALES
PHONE: (02)807 3826 EXT.

CRANES/CHERRY PICKERS

MOBILE LIGHTS
57 SEVEN HILLS ROAD NORTH
SEVEN HILLS
NSW 2147
AUSTRALIA-NEW SOUTH WALES
PHONE: (02)624 3699 EXT.
PHONE: TX AA25106 EXT.
FAX: (02)674 6351

SAMUELSON FILM SERVICE PTY LTD
1 MCLACHLAN AVENUE
ARTARMON
NSW 2064
AUSTRALIA-NEW SOUTH WALES
PHONE: (02)436 1844 EXT.
PHONE: TX 25189 EXT.
FAX: (02)438 2585

EXPENDABLES & MISC. EQUIPMENT

AVTRONICS PTY LTD
33 HIGGINBOTHAM ROAD
GLADESVILLE
NSW 2111
AUSTRALIA-NEW SOUTH WALES
PHONE: (02) 807 1444 EXT.
PHONE: TX AA 25629 EXT.
FAX: (02) 809 7136

FILMTRONICS (AUSTRALIA) PTY LTD
33 HIGINBOTHAM ROAD
GLADESVILLE
NSW 2111
AUSTRALIA-NEW SOUTH WALES
PHONE: (02)807 1444 EXT.
PHONE: TX AA 25629 EXT.
FAX: (02)809 7136

JOHN BARRY GROUP PTY LTD
1 MCLACHLAN AVENUE
ARTARMON
NSW 2064
AUSTRALIA-NEW SOUTH WALES
PHONE: (02)439 6955 EXT.
PHONE: TX AA 25188 EXT.
FAX: (02)439 2375

LEMAC FILM & VIDEO EQUIPMENT
1/33 COLLEGE STREET
GLADESVILE
SYDNEY
NSW 2111
AUSTRALIA-NEW SOUTH WALES
PHONE: (02)816 4266 EXT.
FAX: (02)816 2584
277 HIGHETT STREET, RICH-MOND
VIC 3121
AUSTRALIA-VICTORIA
PHONE: (03) 429-3588 EXT.
FAX: (03) 428-3336

LEROY JONES MOTIONS PICTURES SERVICES
133 MULLENS STREET
ROZELLE
NSW 2039
AUSTRALIA-NEW SOUTH WALES
PHONE: (02)818 1844 EXT.

STANMART FILM SERVICES
39 JONES STREET
ULTIMO
NSW
AUSTRALIA-NEW SOUTH WALES
PHONE: (02)660 2270 EXT

FILM COMMISSIONS

AUSTRALIAN FILM COMMISSION
8 WEST STREET
NORTH SYDNEY
NSW 2060
AUSTRALIA-NEW SOUTH WALES
PHONE: (02)925 7333 EXT.
PHONE: TX 25157 EXT.
FAX: (02)959 5403

AUSTRALIAN FILM INSTITUTE
213 PALMER STREET
DARLINGHURST
NSW 2010
AUSTRALIA-NEW SOUTH WALES
PHONE: (02)332 2111 EXT.
TX 27286

FILM AUSTRALIA
EATON ROAD, LINDFIELD.
N.S.W 2070
PHONE 02 467 9777
TX AA 22734
FAX 02 465672

AUSTRALIAN FILM & TELEVISION AND RADIO SCHOOL
P O. BOX 126 NORTH RYDE ,
N.S.W. 2113
PHONE 02 805 6611
FAX 02 8871030
TX. AA27575

FILM AUSTRALIA
ETON ROAD
LINDFIELD
NSW 2070
AUSTRALIA-NEW SOUTH WALES
PHONE: (02)467 9777 EX
FAX 465 672

FILM CENSORSHIP BOARD
7TH FLOOR, PICCADILLY COURT
222 PITT STREET
SYDNEY
NSW 2000
AUSTRALIA-NEW SOUTH WALES
PHONE: (02)267 2711 EXT

NATIONAL FILM ARCHIVE
G.P.O. BOX 2002
CANBERRA
ACT 2601
AUSTRALIA-NEW SOUTH WALES
PHONE: (062)671 711 EXT.
FAX: (02)251 3740

NEW SOUTH WALES FILM/TV OFFICE
45 MACQUARIE STREET
SYDNEY
NSW 2000
AUSTRALIA-NEW SOUTH WALES
PHONE: (02)251 7233 EXT.
FAX: (02)251 3740

QUEEN ISLAND FILM INDUSTRY DEVELOPMENT OFFICE
BRISBANE,QLD 4000
AUSTRALIA-QUEENSLAND
PHONE: (07)229 1233 EXT.
PHONE: TX 43048 EXT.
FAX: (07)229 1535

SOUTH AUSTRALIAN FILM CORP
113 TAPLEY'S HILL ROAD,
HENDON
S.A. 5014
AUSTRALIA-SOUTH AUSTRALIA
PHONE: (08)452277 EXT.
PHONE: TX AA 88206 EXT.
FAX: (08)347 0385

PROPS

CRAWFORD PRODUCTIONS PTY LTD
259 MIDDLEBOROUGH
ROAD,BOX HILL
VIC 3128
AUSTRALIA-VICTORIA
PHONE: (03)895 2211 EXT.
PHONE: TX AA 31604 EXT.
FAX: (03)890 5732

STUDIOS & SOUND STAGES

AUSTRALIAN BROADCASTING CORP.
BOX 4444,CROWS NEST
NSW 2065
AUSTRALIA-NEW SOUTH WALES
PHONE: (02)430 3999 EXT.
PHONE: TX AA 177143 EXT.
FAX: (02)430 3888

AUSTRALIAN EFEX COMPANY PTY LTD
178 GEORGE
STREET,ERSKINVILLE
NSW 2043
AUSTRALIA-NEW SOUTH WALES
PHONE: (02)516 3033 EXT.
FAX: (02) 512 486

AUSTRALIAN FILM STUDIOS LTD
1-15 BARR STREET,BALMAIN
NSW 2041
AUSTRALIA-NEW SOUTH WALES
PHONE: (02)555 1555 EXT.
FAX: (02)810 6173

MAX STUDIOS
19-25 BIRMINGHAM STREET
ALEXANDRIA
NSW 2018
AUSTRALIA-NEW SOUTH WALES
PHONE: (02)317 5799 EXT.
FAX: (02)669 5975

AUSTRIA

SPECIAL EFFECTS

NEUNER SFX
THEODOR SICKELGASSE 16/8/3
A-1100 WIEN
AUSTRIA
PHONE: 0222-8892745 EXT.
FAX: 0222-8892744

FILM COMMISSIONS

ORF
WURZBURGGASSE 30
A-1136 WIEN
PHONE: (222) 87878/2228
TELEX: 133601
FAX: (222) 87878/238

ORF
AUSTRIAN BROADCASTING
CORPORATION RADIO &
TELEVISION
WURZBURGGASSE 30
1136 VIENNA
PHONE: 87878/2228
TELEX: 133601
FAX: 222/87878/2387

UNIONS AND ASSOCIATIONS

AKM (PERFORMING RIGHTS SOCIETY)
BAUMANNGASSE 10
1030 WIEN
AUSTRIA
PHONE: 757679 EXT.
FAX: 757679/27

AUSTRIAN FILM COMMISION
NEUBAUGASSE 36/2/17A
1070 WIEN
AUSTRIA
PHONE: 963323 EXT.
FAX: 933436

DETLEF KREJCI
WEBGASSE 43/12
A-1060 WIEN
AUSTRIA
PHONE: 78 35 76 EXT.
FAX: 0222-792171

FACHVERBAND DER AUDIOVISIONS
UND FILMINDUSTRIE
OSTERREICHES
BUNDESWIRTSCHAFTSKAMMER
WIEDNER HAUPTSTRASSE 63
1045 WIEN
AUSTRIA
PHONE: 6505/3010 EXT.
FAX: 0222/5057007

WIENER FILMBURO
MA 53
RATHAUSSTRABE 9
STIEGE 2, HALBSTOCK
A-1082 VIENNA, AUSTRIA

BELGIUM

SPECIAL EFFECTS

APIGRAPHIC ET VISUALS
ROUTE DES RARNIERES 1
1328 OHAIN
PHONE: 02-633 19 26

BERTIAUX & MARCFILM SPRL
SQUARE PLASKY 86
1040 BRUXELLES
PHONE: 02-736 30

CAMPENS, JACQUES
RUE FONTAINE D'AMOUR 24
1030 BRUXELLES
PHONE: 02-241 68 82

DE BROUCKERE, THIERRY
RUE FRANZ MERJAY 28
1060 BRUXELLES
PHONE: 02-345 48 05

GERARD CHAPELLE PRODUCTIONS
WOLSENSTRAAT 30
1710 DILBEEK
PHONE: 02-4666 76 06

GUERIG, MARC-PHILIPPE
RUE OTLET 28
1070 BRUXELLES
PHONE: 02-520 99 76

IMAGE +
RUE DE L'ABRE BENIT 93
1050 BRUXELLES
PHONE: 02-649 06 13

KEY GRIP SYSTEMS SPRL
RUE DE L'ETE 72
1050 BRUXELLES
PHONE: 32-2-646 25 26
FAX: 32-2-646 25 24

MARC FILMS STUDIO
SQUARE EUGENE PLASKY 86
1050 BRUXELLES
PHONE: 02-736 01 3

MICHEL, PHILIPPE
RUE FRANS HALS 43
1070 BRUXELLES
PHONE: 02-520 90 28
PHOTOGRAFILM
RUE DES ACCACIAS 21
1950 KRAAINER
PHONE: 02-731 70

S.E.T.E.C.T.S.A.
SQUARE PLASKY 86
1040 BRUXELLES
PHONE: 02-736 01 30

EQUIPMENT RENTALS

AMPEX S.A.
RUE DE L'INDUSTRIE 37
1400 NIVELLES
PHONE: 067-22 49 21
TELEX: 57432 CABLE BELGIUM

MAGNA-TECH ELECTRONIC CO. INC - AGENTS: A.R.C.
RUE DU BOIS DE LINHOUT 45
1200 BRUXELLES
PHONE: 02-735 75 99
TELEX: 846 63765

FILM COMMISSIONS

COMMISSION CONSULTATIVE DE L'AUDIO-VISUEL
RUE JEAN STEVENS 7
1000 BRUXELLE
PHONE: (02) 518123

COMMISSION DE CONTROLE DU FILM
PLACE POELART 3
1000 BRUXELLES
PHONE: 02-511 42 00

COMMISSION DE SELECTION DES FILMS
GALERIE RAVENSTEIN 3-4
1000 BRUXELLES
PHONE: 02-513 94 40

CONSEIL SUPERIEUR DE L'AUDIO-VISUEL
GALERIE RAVENSTEIN 4
1000 BRUXELLES
PHONE: 02-513 94 40

DIRECTION D'ADMINISTRATION DE L'AUDIO-VISUEL
GALERIE RAVENSTEIN 28
1000 BRUXELLES
PHONE: (02) 5139440

FILMDIENST MINISTERIE VOOR NEDERLANDSE CULTUUR
RUE DE COLONIES 29/31
1000 BRUXELLES
PHONE: (02) 5737465

MINISTERE DE LA JUSTICE
COMMISSION DE CONTROLE
DE FILMS
PLACE PLELAERT 4
1000 BRUXELLES
PHONE: (02) 5114200

MINISTERE DES AFFAIRES ECONOMIQUES
COMMISSION DU FILM,
AVENUE J.A. DE MOT, 26
1040 BRUXELLES
PHONE: (02) 233611

MINISTERES DES RELATIONS EXTERIEURES
SERVICE CINEMA
RUE DES 4 BRAS 2
1000 BRUXELLES
PHONE: (02) 5168111

MAKE-UP

KRYOLAN PROFESSIONAL MAKE UP
JEAN AVONDSTONDT ET FRES,
S.P.R.L., RUE MELSENS 13
1000 BRUXCXELLES
PHONE: 02-511 54 44

STUDIOS/SOUND STAGES

BEL EQUIPMENT BVBA
AVENUE LEON MAHILLON
LAAN, 102
1040 BRUXELLES
PHONE: 02-736 06 36

CITY FILMS SC
RUE DE VERREWINKEL 95
1180 BRUXELLES
PHONE: 02-375 54 50
02-375 44 89
TELEX: 23332
FAX: 02-3753234

GLOBE SHOW CENTER
FABRIEDSTRAAT 6
9470 DENDERLEEUW
PHONE: 32 (0) 53/663466
TELEX: 12015
FAX: 053/663945

NEW WAVE TV
111 RUE DU COLLEGE
1050 BRUXELLES
PHONE: 02 648 77 97

SODEP
RUE RASSON 21
1040 BRUXELLES
PHONE: 02-736 54 37

SOFEDI-FILMS
AVENUE LOUISE 485
1050 BRUXELLES
PHONE: 02-720 71 80

VANDAM-KH S.A.
CHAUSSEE DE BLEURGAT 195
1050 BRUXELLES
PHONE: 02-6492075

ZI & A FILM & TV PRODUCTIONS
50 AVENUE VANDERAEY
1180 BRUXELLES
PHONE: 02-374 90 63
FAX: 02-375 41 30

UNIONS AND ASSOCIATIONS

ACTION CINE-JEUNESSE
CAMPUS IHECS, CHAUSSEE DE
BINCHE
7000 MONS
PHONE: 065-317351 (EXT. 208)

ASSOC. BELGE DES AUTEURS DE FILMS ET DE TELEVISION
AVENUE EVERARD, 55
1190 BRUXELLES
PHONE: 02-3457478

ASSOC. EUROPENNE DU FILM D'ANIMATION
RUE DE FACQZ 50
1050 BRUXELLES
PHONE: 02-537 62 17

CENTRE FRANCOPHONE DU FILM PUR L'ENFANCE ET LA JEUNESSE
CHAUSSEE D'AUDERGHEM 189
1040 BRUXELLE

CENTRE UNIVERSITAIRE DU FILM SCIENTIFIQUE
AVENUE FRANKLIN ROOSEVELT
50
1050 BRUXELLES
PHONE: 02-649 00 30

CHAMBRE PROFESSIONNELLE DE LA CINEMATOGRAPHIE
RUE ROYALE 288
1210 BRUXELLES
PHONE: 02-2181267

ECOLE NATIONALE SUPERIEURE DES ARTS VISUALS (ENSAV)
INSTITUT DE CINEMATOGRAPHIE,
EXPERIMENTALE D'ANIMATION
21 ABBAYE DE LA CAMBRE
1050 BRUXELLES
PHONE: 02-6489619

ENTREPRENEURS DE L'AUDIOVISUEL EUROPEEN
RUE THERESIENNE 8
1000 BRUXELLES
PHONE: 02-511 98 39

MEDIA (EUROPEAN PROGRAMS)
RUE DE LA LOI 200 (2 ETAGE, LOCAL 64)
1049 BRUXELLES
PHONE: 02-236 07 18

PARALLAX (STUDIO D'ACTEURS)
30 BOULEVARD DE WATERLOO
1000 BRUXELLES
PHONE: 02-5127385

SERVICE NATIONAL DES CINE-CLUBS (FILM SOCIETIES)
GALERIE RAVENSTEIN 72
1000 BRUXELLES
PHONE: 02-513 41 55

UNIBELFILM
RUE ROYALE 266 - BOITE 6
1030 BRUXELLES
PHONE: 02-218 43 01

UNION DES PRODUCTEURS DE FILMS FRANCOPHONES
C/O M. GILLON, RUE G. EVERESTI, 20
1350 LIMAL
PHONE: 010 41 30 35

UNION PROFESSIONNELLE DES PRODUCTEURS DE TELEVISION
RUE RASSON, 21
1040 BRUXELLES
PHONE: 02-736 54 37

BRAZIL

FILM COMMISSIONS

CONSELO NACIONAL DO CINEMA
RUE MAYRINK VEIGA 28,
RIO DE JANEIRO
BRAZIL

TEMPRESA BRASILIERA DE FILMS SA
RUA ALVARO ALVIM 37
SALA 518,RIO DE JANEIRO
CEP 20031
BRAZIL
PHONE: 240-7135 EXT.

UNIONS AND ASSOCIATIONS

ASSOCIACAO BRASILIEIRA
CINEMATOGRAFICA
RUN MEXICO 31/603
BRAZIL
PHONE: 240-2276 EXT.
FAX: BRAZIL

ASSOCIACO BRASILIERA DE PRODUCTORES INDEPENDENTES
RUA CONDE DE IRAJA
53,BOTOFOGO
RIO DE JANEIRO
BRAZIL

SINDICATO NACIONAL DA INDUSTRIA CINEMATOGRAFICA
RIA JOAQUIM SILVA 11,
SALA 20241,RIO DE JANEIRO
BRAZIL
PHONE: 23204102 EXT.

CANADA

FILM COMMISSIONS

ALBERTA

ALBERTA FILM INDUSTRY DEVELOPMENT OFFICE
9940 106TH ST., STERLING PL, 10TH FLOOR
EDMONTON, ALBERTA T5K 2P6
PHONE: (403) 427-2005

ALBERTA MOTION PICTURE DEVELOPMENT CORPORATION
SUITE 690, THE PHIPPS-MCKINNON BLDC, 10020-101A AVE
EDMONTON T5J3G2
PHONE;(403)424-3855
FAX; (403) 424-7669

CALGARY ECONOMIC DEVEL-OPMENT AUTHORITY
FILM DEVELOPMENT OFFICE,
P.O. BOX 2100, POSTAL STATION M
CALGARY, ALBERTA T2P 2M5
PHONE: (403) 268-2771
FAX: (403) 268-1946

EDMONTON MOTION PICTURE & TV BUREAU
9797 JASPER AVE , SUITE 108
EDMONTON, ALBERTA T5J 1N9
PHONE: (403) 424-7870/ (800) 661-6965
FAX: (403) 426-0535
NATIONAL FILM BOARD
120-9700 JASPER AVE
EDMONTON, AB T5J4C3
PHONE; 9403) 495-3013

TELEFILM CANADA.
5525 ARTILLARY PL, 2ND FLOOR
HALIFAX, NOVA SCOTIA
PHONE: (902) 426-8425
FAX: (902) 426-4445

BRITISH COLUMBIA

B.C. FILM
555 BROOKSBANK AVE
N. VANCOUVER, V7J 3S5
PHONE: (604) 983-5400
FAX: (604) 983-5401

BRITISH COLUMBIA FILM COMMISSION
750 PACIFIC BLVD., S., 3RD FLOOR
VANCOUVER, B.C.V6B 5E7
PHONE: (604) 660-2732

BRITISH COLUMBIA FILM COMMISSION
3RD FL 601 WEST CORDOVA ST
VANCOUVER, V6B 1G1
PHONE: (604) 660-2732
FAX: (604) 660-4790

BURNABY FILM OFFICE
4949 CANADA WAY
BURNABY, B.C. V5G 1M2
PHONE: (604) 294-7231
FAX: (604) 294-7710

KNOWLEDGE NETWORK/OPEN LEARNING AGENCY
300-475 W. GEORGIA
VANCOUVER, V6B 4M9
PHONE: (604) 660-2000/ (800) 663-1678
FAX: (604) 660-2048

NATIONAL FILM BOARD ENGLISH PROGRAM - PACIFIC CENTRE
300 - 1045 HOWE ST
VANCOUVER, V6Z 2B1
PHONE: (604) 666-3838

TELEFILM CANADA
350-375 WATER ST
VANCOUVER, B.C. V6B 5C6
PHONE: (604) 666-1566
FAX: (604) 666-7754

THOMPSON-NICOLA REGIONAL DISTRICT
2079 FALCON RD
KAMLOOPS, B.C. V2C 4J2
PHONE: (604) 372-9336/ 573-4671
FAX: (604) 372-5048

VANCOUVER CITY HALL
453 WEST 12TH AVE
VANCOUVER, V5Y 1VA
PHONE: (604) 873-7339

VICTORIA/VANCOUVER ISLAND FILM & VIDEO COMMISSION
525 FORT ST
VICTORIA, B.C.V8W 1E8 CN
PHONE: (604) 386-3976

MANITOBA

FILM MANITOBA
177 LOMBARD AVE, 4TH FLOOR
WINNIPEG R38 2W5
PHONE: (204) 945-8827

LOCATION MANITOBA
100-93 LOMBARD AVE
WINNIPEG, MANITOBA
MB R3B 3B1
PHONE: (204) 947-2040
FAX: (204) 956-5261

NATIONAL FILM BOARD ENGLISH PROGRAM-PRAIRIE CENTRE
245 MAIN ST'W NNIPEG, MB R3C 1A7
PHONE: (204) 983-3160

NEWFOUNDLAND & LABRADOR

DEPARTMENT OF DEVELOPMENT AND TOURISM
P.O. BOX 8700
ST. JOHN'S, NEWFOUNDLAND A1B 4J6
PHONE: (709) 576-4079
TELEX: 016-4949
FAX: (709) 576-5936

NEW BRUNSWICK

GOVERNMENT FILM SERVICES
NEW BRUNSWICK TOURISM, RECREATION & HERITAGE, FILM &VIDEO COMMISSION
P.O.BOX 12345
FEDERICTON E3B 5C3
PHONE: (506)453-2553

NEW BRUNSWICK FILM AND VIDEO COMMISSION
P.O. BOX 12345, SUITE 208
FEDERICTON, NB E3B 5C3
PHONE: (506) 453-2553
FAX: (506) 453-2416

NOVA SCOTIA

FILM NOVA SCOTIA
P.O. BOX 2287, STATION M
HALIFAX, N.S. B3J 3C8
PHONE: (902) 422-3402

NATIONAL FILM BOARD-ATLANTIC CENTRE
1571 ARGYLE STREET
HALIFAX, N.S. E3L1K1
PHONE: (902) 426-6000/8901

TELEFILM CANADA
TOUR DE LA BANQUE NATIONALE, 500 DE LA GAUCHETIERE STREET WEST, 14TH FLOOR
MONTREAL, QUEBEC H3B 4L2
PHONE: (514) 283-6363
TELEX: 055-60998
FAX: (514) 283-8212

NATIONAL FILM BOARD-ATLANTIC CENTRE
1571 ARGYLE STREET
HALIFAX, N.S. B3L1K1
PHONE: (902) 426-5000/8901

ONTARIO

CANADIAN BROADCASTING CORPORATION
1500 BORNSON AVE
OTTAWA, K1G 3J5
PHONE: (613) 724-1200
TELEX: 05-34260
FAX: (613) 738-6887

CANADIAN GOVERNMENT EXPOSITIONS & AUDIO VISUAL CENTRE
(FILM & VIDEO GROUP)
450 COVENTRY RD.
OTTAWA, ON K1A0T1
PHONE: (613) 993-4014

CITY OF OTTAWA DEPT. OF ECONOMIC DEVELOPMENT
111 SUSSEX DR
OTTAWA, ON K1N 5A1 CN
PHONE: (613) 564-4147
FAX: (613) 564-8070

LONDON FILM COMMISSION
300 DUFFERIN AVE, EOX 5035
LONDON, ON N6A 4L9
PHONE: (519) 661-4595
FAX: (519) 661-5331

NATIONAL FILM, TV & SOUND ARCHIVES
344 WELLINGTON ST
OTTAWA, K1A 0N3
PHONE: (613) 995-1311
TELEX: 053-4311

NIAGARA REGION DEVELOP-MENT CORPORATION
P.O. BOX 1042, 2201 ST. DAVID'S RD
THOROLD, ON L2V 4T7
PHONE: (416) 685-1308
FAX: (416) 637-497

ONTARIO FILM DEVELOPMENT CORPORATION
81 WELLESLEY ST EAST
TORONTO, M4Y 1H6
PHONE: (416) 965-8393
FAX: (416) 965-0329

OTTAWA-HULL FILM & TELEVISION ASSOCATION
P.O. BOX 142, STN B
OTTAWA, ONT K1P 6C3
PHONE: (613) 233-3836
FAX :(613) 233-0698

TELEFILM CANADA
2 BLOOR ST WEST, 22ND FLOOR
TORONTO, ONT M7W 3E2
PHONE: (416) 973-6436
TELEX: 06-218344
FAX: (416) 973-8606

TORONTO FILM LIAISON
18TH FLOOR, EAST TOWER, CITY HALL
TORONTO, ONT M5H 2N2
PHONE: (416) 392-7570
FAX: (416) 392-067

PRINCE EDWARD ISLAND

DEPARTMENT OF TOURISM & PARKS
P.O. BOX 2000, 3RD FLOOR
SHAW BLDG
CHARLOTTETOWN, PRINCE
EDWARD ISLAND, C1A 7N8
PHONE: (902) 368-550

QUEBEC

INSTITUT QUEBECOIS DU CINEMA
80 RUE DE BRESOLES
MONTREAL, H2Y 1V5
PHONE: (514) 288-7655
TELEX: 055-62171 MTL

MONTREAL FILM COMMISSION
425 PLACE JACQUES CARTIER,
SUITE 300
MONTREAL, QUEBEC, H2Y 3B1
PHONE: (514) 872-2883
FAX: (514) 872-1153

NATIONAL FILM BOARD
FRENCH PROGRAM BRANCH:
PHONE: (514) 283-9285
SERVICES:
PHONE: (514) 283-9149
ENGLISH PROGRAM BRANCH:
PHONE: (514) 283-9501

NATIONAL FILM BOARD OF CANADA
3155 COTE DE LIESSE RD, P.O.
BOX 6100,
MONTREAL, QUEBEC, H3C 3H5
PHONE: (514) 283-9000/9437
FAX: (514) 496-1646

QUEBEC FILM OFFICE
1755 EAST RENE-LEVESQUE
BLD., SUITE 200
MONTREAL, QUEBEC, H2K 4P6
PHONE: (514) 873-7768
FAX: (514) 873-4388

TELEFILM CANADA
TOUR DE LA BANQUE
NATIONALE, 600 DE LA
GAUCHETIERE
STREET WEST, 14TH FLOOR
MONTREAL, H3B 4L2
PHONE: (514) 283-6363
TELEX: 055-60998
FAX: (514) 283-8212

SASKATCHEWAN

SASKATCHEWAN FILM DEVELOPMENT OFFICE
1942 HAMILTON ST
REGINA, SASKATCHEWAN, S4P 2C5
PHONE: (306) 787-8148
SASKFILM
1840 MCINTYRE ST
REGINA, SK S4P 2P9
PHONE: (306) 347-3456
FAX: (306) 359-7768

YUKON

YUKON FILM PROMOTION OFFICE
P.O. BOX 2703
WHITEHORSE, YK, Y1A 2C6
PHONE: (403) 667-5400
FAX: (403) 667-2634

CANADA - ALBERTA

SPECIAL EFFECTS

MTM EQUIPMENT RENTALS
525 MCKNIGHT BLVD. N.E.
CLAGARY T2E 5S9
PHONE: (403) 276-1505
FAX: (403) 277-0554

STARDOM STUDIOS INC
47 SUN CASTLE BAY S.E.
CALGARY T2X 2E9
PHONE: (403) 256-7396

STUDIOS/SOUND STAGES

ITV PRODUCTIONS
5325-104TH ST
EDMONTON T6H 5B8
PHONE: (403) 436-1250
FAX: (403) 438-8448

STUDIO III
P.O. BOX 7060, STATION E
CALGARY T3C 3L9
PHONE: (403) 246-7140

UNIONS AND ASSOCIATIONS

ALBERTA MOTION PICTURE INDUSTRIES ASSOCIATION (AMPIA)
209 MCLEOD BLDG,
10136-100TH ST
EDMONTON T5J 0P1
PHONE: (403) 423-0709
FAX: (403) 426-3057

ALLIANCE OF CANADIAN CINEMA, TELEVISION & RADIO ARTISTS
1414 - 8TH ST. S.W.,
MOUNT CALGARY PLACE, SUITE 260
CALGARY T2R 1J6
PHONE: (403) 228-3123

ALLIANCE OF CANADIAN CINEMA, TELEVISION & RADIO ARTISTS
302 LIBERTY BUILDING,
10506 JASPER AVE.
EDMONTON T5J 2W9
PHONE: (403) 423-0669

CALGARY SOCIETY OF INDEPENDENT FILMMAKERS
P.O. BOX 30089,
STATION B
CALGARY T2M 4N7
PHONE: (403) 277-1741

DIRECTORS GUILD OF CANADA
ALBERTA DISTRICT
COUNCIL, 524 - 11TH AVE
CALGARY T2R 0C8
PHONE: (403) 237-0689
FAX: (403) 266-1598

MOTION PICTURE THEATRE ASSOCIATION OF CANADA
C/O THEATRE
AGENCIES, 522 - 11TH AVE. S.W.
CALGARY T2R 0C8
PHONE: (403) 264-4660
FAX: (403) 264-6571

CANADA-BRITISH COLUMBIA

SPECIAL EFFECTS

ACTION STUNT SYSTEMS, INC
3835 FIR ST
BURNABY, V5G 2A6
PHONE: (604) 437-8137
FAX: (604) 645-7905

ROB BOULES/ SPFX ARTIST
21861 DEWDENY TRUNK RD
MAPLE RIDGE, V2X 3G8
PHONE: (604) 463-8996

CINETRONICS TECHNOLOGIES, INC
1144 W. 22ND AVE
N VANCOUVER, V7P 2E8
PHONE: (604) 986-8677

H. DASKEN MECHANICAL & TECHNICAL INNOVATIONS
65 - 3031 WILLIAMS RD
RICHMOND, V7E 1H9
PHONE: (604) 271-3121

DIGITAL POST & GRAPHICS
1921 MINOR AVE
SEATTLE, WA, 98101
PHONE: (206) 623-3444
FAX: (206) 340-1548

FINALE POST PRODUCTION INC
225 W. 2ND AVE
VANCOUVER, V5Y 1C7
PHONE: (604) 876-7678
FAX: (604) 876-3299

G & W RAIN MAKERS
237 WOOD ST
NEW WESTMINSTER, V3M 5D=K5
PHONE: (604) 526-3480

GALAIRE PRODUCTIONS INC
P.O. BOX 2445 , 349 W.
GEORGIA ST
VANCOUVER, V6B 3W7
PHONE: (604) 435-6134

GASTOWN POST & TRANSFER
50 W. 2ND AVE
VANCOUVER, V5Y 1B3
PHONE: (604) 872-7000
FAX: (604) 872-2106

DAVID HINKS
312 - 163 W. 5TH ST
N. VANCOUVER, V7M 1J6
PHONE: (604) 986-8850

KITCHEN SYNC TECHNOLOGIES INC
1144 W. 22ND AVE
N. VANCOUVER, V7P 2E8
PHONE: (604) 986-8677

KNOTT LIMITED SPECIAL EFFECTS
1192 EUCLID AVE
BELLINGHAM, WA ,98226
PHONE: (206) 738-9598
FAX: (206) 671-2851
6919 TREASURE TRAIL
LOS ANGELES, CA, 90068
PHONE: (213) 876-9724
FAX: (213) 876-2356

LASERHOUSE PRODUCTIONS INC
3-1438 ARBUTUS ST
VANCOUVER, V6J 3W8
PHONE: (604) 738-4918

LEES SPECIAL EFFECTS LTD
SITE 13, BOX 13, S.S. 1
CALGARY, AB, T4B 2B4
PHONE: (403) 286-6949
MURRAYS WATER HAULING
10866 272ND ST
WHONNOCK, V2X 8X8
PHONE: (604) 462-8430

PALLER SPECIAL EFFECTS
919 ROBINSON ST
COQUITLAM, V3J 4G8
PHONE: (604) 931-3993
FAX: (604) 931-3633

POST HASTE VIDEO
A DIVISION OF TEGRA
INDUSTRIES INC
1177 2 8TH AVE
VANCOUVER, V6H 1C6
PHONE: (604) 734-7727

STARFIRE SPECIAL EFFECTS
1554 E 13TH AVE
VANCOUVER, V5N 2B8
PHONE: (604) 879-0725
FAX: (604)650-3984

STUDIO POST & TRANSFER
5305 104 ST
EDMONTON, AB, T6H 5B8
PHONE: (403) 436-4444
FAX: (403) 438-8495

TASMAN INDUSTRIES LTD
7469 HUME AVE. TILBURY
INDUSTRIAL PARK
DELTA, V4G 1C3
PHONE: (604) 946-0455
FAX: (604) 946-7839

THOMAS SPECIAL EFFECTS LTD
140 RIVERSIDE DR
N. VANCOUVER, V7H 1T9
PHONE: (604) 929-5455
FAX: (604) 929-6653

TRINITY ELECTRONICS SYSTEMES LTD
10708 181 ST
EDMONTON, AB , T5S 1K9
PHONE: (403) 489-3199

UNREEL EFFECTS
BOX 178
ROCKYFORD, AB, T0J 2R0
PHONE: (403) 934-5048

MICHAEL WALLS
109-3755 W 6TH
VANCOUVER, V6R 1T9
PHONE: (604) 224-6442

WESTPOST
25 E. 2ND AVE
VANCOUVER, V5T 1B3
PHONE: (604) 875-0444
FAX: (604) 875-1715

EQUIPMENT SALES/ RENTALS/SERVICES

DUBBERLEYS ON DAVIE LTD.
920 DAVIE ST
VANCOUVER, V6Z 1B8
PHONE: (604) 684-5981

INTELLIPROMPT
3705 W. 18TH AVE
VANCOUVER, V6S 1B3
PHONE: (604) 224-4471
FAX: (604) 224-7010

KERRISDALE CAMERA (INDUSTRIAL DIVISION)
2170 W. 41ST AVE
VANCOUVER, V6M 1Z5
PHONE: (604) 263-0715

TELAV AUDIO VISUAL SERVICES INC
125 W. 3RD AVE
VANCOUVER, V5Y 1E6
PHONE: (604) 879-1999

PROPS

ARTABRAC
303 1226 HOMER ST
VANCOUVER, V6B 2Y5
PHONE: (604) 687-7098
FAX: (604) 736-3798

JIMMY CHOW
4652 VICTORY ST
BURNABY, V5J 1R9
PHONE: (604) 438-2245

CREATIVE PROPS & PARAPHERNALIA
5262 GRANVILLE ST
VANCOUVER, V6M 3B8
PHONE: (604) 261-4338

CUSTOMCOLOR PHOTO & A/V RENTALS
1114 ROBSON ST
VANCOUVER, V6E 1B2
PHONE: (604) 669-1574

DEBBIE ERHARDT
314 W. PENDER ST
VANCOUVER, V6B 1T3
PHONE: (604) 734-7111

HOLLYWOOD NORTH PROPS & LOCATIONS
3RD FLOOR - 322 WATER ST
VANCOUVER, V6B 1B6
PHONE: (604) 0150

BRUCE A. HOSICK
3605 14TH ST. NW
CALGARY, AB , T2K 1J3
INTERNATIONAL MOVIE SERVICES
148 E. 15TH ST
N. VANCOUVER, V7L 2R1
PHONE: (604) 985-4020
JOHNS JUKES LTD
2343 MAIN ST
VANCOUVER, V5T 3C9
PHONE: (604) 872-5757
FAX: (604) 872-8073
KELLY DEYOUNG SOUND SERVICES
271 E. 2ND AVE
VANCOUVER, V5T 1B8
PHONE: (604) 873-3841
KENS ILLUSIONARIUM
3288 MAIN ST
VANCOUVER, V3V 3M5
PHONE: (604) 875-9712
LANCASTER MEDICAL SUPPLIES & PRESCRIPTIONS
20-601 WEST BROADWAY
VANCOUVER, V5Z 4C2
PHONE: (604) 873-8585
FAX: (604) 873-2381
D. BRENT LANE
4112 PUGET DR
VANCOUVER, V5L 2V6
PHONE: (604) 731-6830
LINDSAY MODELS & DESIGNS CONSULTANTS LTD
831 W. 1ST
N. VANCOUVER, V7P 1A4
PHONE: (604) 988-5751
FAX: (604) 980-2361
MMI PRODUCT PLACEMENT
14 - 130 INDUSTRY ST
TORONTO, ON, M6M 4M8
PHONE: (416) 769-5000
FAX: (416) 762-8180
MOUNTAIN CARRIAGE TOUR CO
BOX 12 SITE 17, RR #2
CALGARY, AB, T2P 2G5
PHONE: (403) 242-9796
WAYNE MCLAUGHLIN
BOX 2626
VANCOUVER, V7L 1G8
PHONE: (604) 736-1035
FRANK PARKER
762 E. 3RD ST
N. VANCOUVER, V7L 1G9
PHONE: (604) 988-0412
PPC THE PRODUCT PLACEMENT COMPANY
1733 NAPIER ST
VANCOUVER, V5L 2N1
PHONE: (604) 251-2225
FAX: (604) 251-2225
PETLEY JONES GALLERY
1838 W. 57TH AVE
VANCOUVER, V6P 1T7
ROBERT J. SCOTT PRODUCTIONS/RENTALS
316 E. 1ST AVE
VANCOUVER, V5T 1A9
PHONE: (604) 872-2766
DAN SISSONS
6059 WALKER AVE
BURNABY, V5E 3B5
PHONE: (604) 524-2584/290-5859
STAMPEDE TACK
17982 56TH AVE
CLOVERDALE, V3S 1C7
PHONE: (604) 574-7427
BILL THUMM
3998 MARINE DR
W. VANCOUVER, V7V 1N3
PHONE: (604) 926-0484
VISIONS SET DECORATORS LTD
2734 AVE. NW
CALGARY, AB, T2N 0P8
PHONE: (403) 283-7958

PROPS - DESIGN & BUILD

BLUE CELLO SCENICS LTD
5374 KEITH RD
W. VANCOUVER, V7W 2Y7
PHONE: (604) 921-7800
FAX: (604) 921-7377
ROB BOULE/SPFX ARTIST
21861 DEWDNEY TRUNK RD
MAPLE RIDGE, V2X 3G8
PHONE: (604) 463-8996
COLORIFIC
195 W. 7TH AVE
VANCOUVER, V5Y 1L8
PHONE: (604) 879-1511
FAX: (604) 879-6643
F & D SCENE CHANGES LTD.
2B 803 24TH AVE
CALGARY, AB, T2G 1P5
PHONE: (403) 233-7633
FAX: (403) 266-7597
LOUS RENT-ALL SERVICES LTD
12160 88 AVE
SURREY, V3W 3J2
PHONE: (604) 596-1588
FAX: (604) 596-7914
NORQUIP SERVICES LTD
2110 FRONT ST
N. VANCOUVER, V7H 1A3
PHONE: (604) 929-6219
FAX: (604) 251-3232
PACIFIC URETHANE SYSTEMS
4134 VIRGINIA CR
N. VANCOUVER, V7R 3Z6
PHONE: (604) 985-3565
PLASTI-FAB LTD
679 ALDFORD AVE
ANNACIS ISLAND, V3M 5P5
PHONE: (604) 526-2771/
(800) 663-1388
FAX: (604) 526-8519
R N B SCENIC PRODUCTIONS LTD
8286 ONTARIO ST
VANCOUVER, V5X 3E3
PHONE: (604) 327-9481
RONCO POLE LTD
10354 120 ST
SURREY, V3V 4G2
PHONE: (604) 585-6088
SET DESIGN & DECOR
407-3717 HAMBER PL
N. VANCOUVER, V7G 2J7
PHONE: (604) 929-1943
FAX: (604) 929-8068
SETSPLUS DESIGN LTD
29 BRAID ST
NEW WESTMINSTER, V3L 3P2
PHONE: (604) 521-2206
FAX: (604) 521-3252
STELLAR PRODUCTION SERVICES
L330-560 BEATTY ST
VANCOUVER, V6B 2L3
PHONE: (604) 684-1886
FAX: (604) 688-2404
TASMAN INDUSTRIES LTD
7469 HUME AVE. TILBURY
INDUSTRIAL PARK
DELTA, V4G 1C3
PHONE: (604) 946-0455
FAX: (604) 946-7839
TRANS PEACE CONSTRUCTION (1987) LTD
9637 116TH ST
GRAND PRAIRIE, AB, T8V 5W3
PHONE: (403) 539-6855
FAX: (403) 539-3158
WESTSUN
108 GARDEN AVE
N. VANCOUVER, V7P 3H2
PHONE: (604) 984-3251
FAX: (604) 984-8236

STUDIOS/SOUND STAGES

ARTRAY PRODUCTIONS
BOX 4700
VANCOUVER, V6B 4A3
PHONE: (604) 420-2288
ALLARCOM STUDIOS
5305 104 ST
EDMONTON, AB, T6H 5B8
PHONE: (403) 436-1250
FAX: (403) 438-8495
ATN PRODUCTIONS
BOX 4700
VANCOUVER, V6J 1R2
PHONE: (604) 736-9990
FAX: (604) 736-9982
BRIDGE STUDIOS
2400 BOUNDARY RD
BURNABY, V5M 3Z3
PHONE: (604) 291-0650
FAX: (604) 291-0493
THE BURNABY STUDIO
3737 NAPIER ST
BURNABY, V5C 3E4
PHONE: (604) 299-1050
CABLE REGINA
2250 PARK ST
REGINA, SK, S4N 7K7
PHONE: (306) 569-3510
FAX: (306) 757-3262
NORTH SHORE STUDIOS
555 BROOKSBANE AVE
N. VANCOUVER, V7J 3S5
PHONE: (604) 983-5555
FAX: (604) 983-5554
NORTHSTAR INTERNATIONAL STUDIOS IND
C/O 400-375 WATER ST
VANCOUVER, V6B 5C6
PHONE: (604) 687-1144
FAX: (604) 687-6020
PANAVISION CANADA LTD, VANCOUVER
3999 EAST 2ND AVE
BURNABY, V5C 3W9
PHONE: (604) 291-7262
FAX: (604) 291-0422

UNIONS AND ASSOCIATIONS

A.C.T.R.A.
1622 W 7 AVE
VANCOUVER, V6J 1S5
PHONE: (604) 734-1414
FAX: (604) 734-1417
A.M.P.I.A.
ALBERTA MOTION PICTURE
INDUSTRIES ASSOC.
209-10136 100 ST.
EDMONTON AB, T5J 0P1
PHONE: (403) 423-0709
FAX: (403) 426-3057
ACADEMY OF CANADIAN CINEMA, TELEVISION & RADIO ARTISTS
911-525 SEYMOUR ST
VANCOUVER, V6B 3H7
PHONE: (604) 681-1101
FAX: (604) 681-0466
ACADEMY OF CANADIAN CINEMA & TELEVISION
1102-207 W. HASTINGS
VANCOUVER, V6B 3L6
PHONE: (604) 684-4528
FAX: (604) 684-4979
P.O. BOX 4250
SUNLAND, CA.,91040
PHONE: (818) 951-2842
ALLIANCE OF CANADIAN REGIONAL MOTION PICTURE INDUSTRY
ASSOCS. (ACRMPIA)
C/O B.C. MOTION PICTURES
ASSOC., 204-111 WATER ST
VANCOUVER, V6B 1A7
PHONE: (604) 688-1420
FAX: (606) 688-1425

THE ASSOC. OF CANADIAN FILM CRAFTSPEOPLE (ACFC)
1395 N. GRANDVIEW HWY
VANCOUVER, V5N 1N2
PHONE: (604) 254-2232
FAX: (604) 254-7790
B.C. COUNCIL OF FILM UNIONS
#303-1209 JERVIS ST.
VANCOUVER, V6E 1H7
PHONE: (604) 682-1930
FAX: (604) 682-1959
BRITISH COLUMBIA FILM INDUSTRY ASSN.
770 PACIFIC BLVD., 3RD FLOOR
VANCOUVER,
PHONE: (604) 660-2732
BRITISH COLUMBIA MOTION PICTURE ASSOCIATION
204-111 WATER ST
VANCOUVER, V6B 1A7
PHONE: (604) 688-1420
FAX: (604) 688-1425
COMPOSERS, AUTHORS & PUBLISHERS ASSOCIATION OF CANADA LTD
(CAPAC)
1155 ROBSON ST, SUITE 703
VANCOUVER, V6E 1B9
PHONE: (604) 689-8871
FAX: (604) 688-1142
CALGARY SOC. OF INDEPENDENT FILM MAKERS
BOX 30089, STN. B
CALGARY, AB, T2N 4N7
PHONE: (403) 277-1741
CANADIAN FILM& TV PRODUCTION ASSOC. (CFTPA)
663 YONGE ST., SUITE 401
TORONTO, ON, M4Y 2A4
PHONE: (416) 927-8942
FAX: (416) 922-4038
CANADIAN SOCIETY OF CINEMATOGRAPHERS
89 PINEWOOD TRAIL
MISSISSAUGA, ON, L5G 2L2
PHONE: (416) 271-4634
FAX: (416) 271-7360
CANADIAN INDEPENDENT FILM CAUCUS
ENT. BUS. CENTER 387 BLOOR
5TH SL
TORONTO, ON, M4W 1H7
PHONE: (416) 962-0624
FAX: (416) 362-3608
4371 ESPLANADE
MONTREAL, PQ H2W 1T2
CINEWORKS INDEPENDENT FILMMAKERS SOCIETY
300-1131 HOWE ST
VANCOUVER, V6Z 2L7
PHONE: (604) 685-3841
DIRECTORS GUILD OF CANADA B.C. DISTRICT COUNCIL
130-1152 MAINLAND ST.
VANCOUVER, V6B 4X2
PHONE: (604) 688-2976
EDUCATIONAL VIDEO PROJECT SOCIETY
328 W. 4TH ST
N. VANCOUVER, V7M 1.1
PHONE: (604) 986-3270
FAVA - FILM AND VIDEO ARTS SOCIETY OF ALBERTA
9722 102 ST.
EDMONTON, AB, T5K 0X4
PHONE: (403) 429-1671
GREATER VANCOUVER MEDIA ASSOCIATION
695 COMBIE ST
VANCOUVER, V6B 2P1
PHONE: (604) 681-1420
IATSE (ATLANTIC REGION)
PHONE: 9020 455-5016

IATSE (LOCAL 210)
EDMONTON, AB
PHONE: (403) 423-1863
FAX: (403) 426-0307

IATSE (LOCAL 212)
2A-3704 6TH ST. NE
CALGARY, AB, T2E 6K5
PHONE: (403) 230-8304/
295-3206
FAX: (403) 230-5922

IATSE (LOCAL 295)
REGINA, SK
PHONE: (306) 545-0454
FAX: (306) 569-8649

IATSE (LOCAL 300)
SASKATOON, SK
PHONE: (306) 343-7933
FAX: (306) 664-6423

IATSE (LOCAL 873)
TORONTO, ON
PHONE: (416) 252-5025
FAX: (416) 252-3923

IATSE (LOCAL 63)
BOX 394
WINNIPEG, MB, R3C 2H6
PHONE: (204) 944-0511
FAX: (204) 488-4306

IATSE (LOCAL 928-CANADIAN ART DIRECTORS)
TORONTO, ON
PHONE: (416) 759-9489
FAX: (416) 759-7111

IATSE (LOCAL 891)
1640 BOUNDARY RD.
BURNABY, V5K 4V4
PHONE: (604) 294-8422
FAX: (604) 298-3456

INTERIOR FILM PROMOTION ASSN.
612 BAY AVE
KELOWNA, V1Y 7J9
PHONE: (604) 762-2391

MANITOBA SOC. OF INDEPENDENT ANIMATORS
245 MAIN ST.
WINNIPEG, MB, R3C 1A7
PHONE: (204) 983-1276
FAX: (204) 774-3246

NABET 800
923 WEST 8TH AVE
VANCOUVER, V5Z 1E4
PHONE: (604) 736-0300
FAX: (604) 736-9023

QUICKDRAW ANIMATION SOCIETY
104-3009 23 AVE SW
CALGARY, AB, T3E 0J3
PHONE: (403) 249-0696
FAX: (403) 286-1321

SASKATCHEWAN MOTION PICTURE ASSOC. (SMPIA)
2347 BROAD ST
REGINA, SK, S4P 1Y9
PHONE: (306) 525-9899
FAX: (306) 569-1818

SASKATCHEWAN WRITERS GUILD
BOX 3986
REGINA, SK, S4P 3R9
PHONE: (306) 757-6310

STUNTS CANADA
BC: PHONE: (604) 872-2031
AB: PHONE: (403) 243-5132

TEAMSTERS LOCAL 155
490 EAST BROADWAY
VANCOUVER, V5T 1X3
PHONE: (604) 872-0151

UNION OF B.C. PERFORMERS
#904-525 SEYMOUR ST
VANCOUVER, V6B 3H7
PHONE: (604) 689-0727
FAX: (604) 689-1145

VANCOUVER MUSICIANS ASSOC. LOCAL 145
AMERICAN FEDERATION OF MUSICIANS OF THE U.S. AND CANADA
100-925 W. 8TH AVE
VANCOUVER, V5Z 1E4
PHONE: (604) 737-1110
FAX: (604) 734-FAXX

VANCOUVER WOMEN IN FILM AND VIDEO
BOX 1238 - STATION A
VANCOUVER, V6C 2T1
PHONE: (604) 685-1152

CANADA - MANITOBA

SPECIAL EFFECTS

VIDEO POOL INC.
300-100 ARTHUR ST.
WINNIPEG R3B 1H3
PHONE: (204) 949-9134
FAX: (204) 942-1555

FILM COMMISSIONS

FILM MANITOBA
177 LOMBARD AVE,
4TH FLOOR
WINNIPEG R38 2W5
PHONE: (204) 945-8827

UNIONS AND ASSOCIATIONS

ALLIANCE OF CANADIAN CINEMA, TELEVISION & RADIO ARTISTS (ACTRA)
SUITE 110, PHOENIX BLDG., 388 DONALD ST
WINNIPEG R3B 2J4
PHONE: (204) 943-1307/2365
FAX: (204) 947=5664

MANITOBA FILM PRODUCERS ASSN.
67 SHERBROOK ST
WINNIPEG R3C 2B2
PHONE: (204) 774-5521

MANITOBA MOTION PICTURE INDUSTRIES ASSOCIATION
145 SHERBROOK ST
WINNIPEG R2C 2B5
PHONE: (204) 783-5228
FAX: (204) 2160

CANADA-NEWFOUNDLAND

FILM COMMISSIONS

DEPARTMENT OF DEVELOP-MENT AND TOURISM
P.O. BOX 8700
ST. JOHN'S A1B 4J6
PHONE: (709) 576-4079
TELEX: 016-4949
FAX: (709) 576-5936

UNIONS AND ASSOCIATIONS

ACTRA - NEWFOUNDLAND
210 WATER ST., P.O. BOX 575
ST. JOHN'S A1C 5K8
PHONE: (709) 722-0430

ALLIANCE OF CANADIAN CINEMY, TV & RADIO ARTISTS (ACTRA)
P.O. BOX 575, 210 WATER ST
ST. JOHN'S A1C 5K8
PHONE: (709) 722-0430

CANADA - NOVA SCOTIA

SPECIAL EFFECTS

ATLANTIC ILLUMINATION
23 SHERIDAN ST
DARTMOUTH, B3A 2C9
PHONE: (902) 463-7418
FAX: (902) 469-3255

CONNORS DIVING SERVICES LIMITED
BOX 314, LAKESIDE INDUSTRIAL PARK
HALIFAX, B0J 1Z0
PHONE: (902) 876-7078
FAX: (902) 876-7079

CUSTOM SOUND AND LIGHT
TANTALLON MALL, HWY 103 & HAMMOND PLAINS RD, RR 2
86 SOUTHWOOD RD
HALIFAX, B4A 3Y1
PHONE: (902) 826-7758/
835-9246

DALHOUSIE THEATRE DEPARTMENT
6101 UNIVERSITY AVE
HALIFAX, B3H 3J5
PHONE: (902) 494-7067

FIREWORKS UNLIMITED INC.
P.O. BOX 295
KENTVILLE, B4N 3X1
PHONE: (902) 542-2292
FAX: (902) 678-2428

PRODUCTION SERVICES ATLANTIC
BOX 1086
ARMDALE, B3L 4L5
PHONE: (902) 450-5009
FAX: (902) 450-5140

SIGMA ELECTRONICS LIMITED
135 ILSLEY AVE
DARTMOUTH, B3B 1T1
PHONE: (902) 458-2727

PROPS

CANADIAN BROADCASTING CORPORATION (CBC)
P.O. BOX 3000
HALIFAX, B3J 3E9
PHONE: (902) 420-8311

HOUSTON NORTH GALLERY
1919 UPPER WATER ST
HALIFAX
PHONE: (902) 423-5424
FAX: (902) 634-3572

STUDIOS & SOUND STAGES

ABS PRODUCTIONS LIMITED
196 JOSEPH ZATZMAN DR
DARTMOUTH, B3B 1N4
PHONE: (902) 468-4336

ATLANTIC TELEVISION SYSTEM (ATV)
P.O. BOX 1653
HALIFAX, B3J 2Z4
PHONE: (902) 453-4000

CANADIA BROADCASTING CORPORATION (CBC)
P.O. BOX 3000
HALIFAX, B3J 3E9
PHONE: (902) 420-8311

CUNARD STREET THEATRE
ADMINISTRATIVE OFFICE, 5516 SPRING GARDEN RD., SUITE 304
HALIFAX
PHONE: (902) 421-1902

HALIFAX CABLEVISION LIMITED
5841 BILBY ST
HALIFAX, B3K 1V7
PHONE: (902) 453-2800

LEASECOM INCORPORATED
SUITE 102, 1819 GRANVILLE ST
HALIFAX, B3J 3R1
PHONE: (902) 422-1422
FAX: (902) 429-9866

MITV (INDEPENDENT TELEVISION FOR THE MARITIMES)
14 AKERLEY BLVD.
DARTMOUTH, B3B 1J3
PHONE: (902) 494-5200
FAX: (902) 468-2154

NEEDHAM GATE PRODUCTIONS LIMITED
2176 WINDSOR ST
HALIFAX, 3K 5B6
PHONE: (902) 421-1022

ROBINSON/ CAMPBELL & ASSOCIATES LTD.
10 RAGGED LAKE BLVD., SUITE 7
HALIFAX, B3S 1C2
PHONE: (902) 450-1050
FAX: (902) 450-1052

SERIES ONE PHOTOGRAPHY
61 RADDALL AVE., SUITE N,
BURNSIDE INDUSTRIAL PARK
DARTMOUTH, B3B 1T4
PHONE: (902) 468-1250

UNIONS AND ASSOCIATIONS

ACADEMY OF CANADIAN CINEMA AND TELEVISION
5211 BLOWERS ST., SUITE 32
HALIFAX, B3J 1T6
PHONE: (902) 425-0124

ACADEMY OF CANADIAN CINEMA & TELEVISION
1574 ARGYLE ST., P.O. BOX 1647, STN. M.
HALIFAX, B3J 2Z1
PHONE: (902) 422-5929

ALLIANCE OF CANADIAN CINEMA, TV & RADIO ARTISTS (ACTRA)
5510 SPRING GARDEN DR
HALIFAX, B3J 1G5
PHONE: (902) 420-1404

ALLIANCE OF CANADIAN REGIONAL MOTION PICTURE INDUSTRY ASSOCS. (ACRMPIA)
5211 BLOWERS ST., SUITE 32
HALIFAX, B3J 1T6
PHONE: (902) 425-0124

ATLANTIC FEDERATION OF MUSICIANS (LOCAL 571)
6307 CHEBUCTO RD
HALIFAX, B3L 1K9
PHONE: (902) 422-6492

ATLANTIC INDEPENDENT FILM & VIDEO ASSOCIATION (AIFVA)
5211 BLOWERS ST., SUITE 32
HALIFAX, B3J 1T6
PHONE: (902) 425-0124

CANADIAN ACTORS EQUITY ASSOCIATION
260 RICHMOND STREET E.
TORONTO, ON, M5A 1P4
PHONE: (416) 867-9165
FAX: (416) 867-9246

DGC
DIRECTORS GUILD OF CANADA
NOVA SCOTIA COUNCIL
2085 MAITLAND ST
HALIFAX, B3K 2Z8
PHONE: (902) 492-3424
FAX: (902) 425-7339
IATSE LOCAL 849
ROBIE RPO, BOX 31122
HALIFAX, B3K 5T9
PHONE: (902) 453-1453
FAX: (902) 454-7671
INTERNATIONAL ALLIANCE OF THEATRICAL STAGE EMPLOY-EES (IATSE)
P.O. BOX 711
HALIFAX, B3J 2G3
PHONE: (902) 463-4539
NATIONAL ASSOCIATION OF BROADCAST EMPLOYEES & TECHNICIANS
(NABET)
6080 YOUNG ST., SUITE 313
HALIFAX, B3K 5L2
PHONE: (902) 453-5415
NOVA SCOTIA FILM & VIDEO PRODUCERS ASSOCIATION
PHONE: (902) 421-1326
PERFORMING RIGHTS ORGANIZATION OF CANADA
6080 YOUNG ST., SUITE 300
HALIFAX, B3J 5L2
SOCAN
SOCIETY OF COMPOSERS, AUTHORS & MUSIC PUBLISHERS OF CANADA
45 ALDERNEY DR., SUITE 602, QUEEN SQUARE
DARTMOUTH, B2Y 2N6
PHONE: (902) 464-7000
FAX: (902) 454-9696

CANADA-ONTARIO

SPECIAL EFFECTS

CARERE ENTERPRISES
P.O. BOX 5818, STN. A
TORONTO, M5W 1P2
PHONE: (416) 463-6656
KEILLOR FILM INDUSTRIES
68 BROADVIEW AVE
TORONTO, M4M 2E6
PHONE: (416) 465-9767
MAE STUDIOS/MFG STUDIOS
511 KING STREET WEST, SUITE 300
TORONTO, M5V 1K4
PHONE: (416) 977-5308
FAX: (416) 971-7069
PIERSIG & ASSOC.
290 CARLAW AVE
TORONTO, M4M 3L1
PHONE: (416) 462-9757
SMITH UNLIMITED
68 CLAREMONT ST
TORONTO, M6J 2M5
PHONE: (416) 363-1934
UPLIS LTD
237 CLINTON ST
TORONTO, M6G 2Y7
PHONE: (416) 535-8600

MAKE-UP SUPPLIERS

MALABAR LTD (KRYOLAN AGENTS)
14 MCCAUL ST
TORONTO, M5T 1V6
PHONE: (416) 593-2581

MAVIS THEATRICAL SUPPLIES INC
697 GLASGOW ST
KITCHENER, N2M 2N7
PHONE: (519) 745-3331

PROPS

EDGE & BRATTON SCENERY & DISPLAY
258 WALLACE AVE
TORONTO, M6P 3M9
PHONE: (416) 532-4451
MILSPEC SUPPLIES CO.
BOX 456, STATION K
TORONTO, M4P 2G9
PHONE: (416) 484-4065
PIERSIG & ASSOC.
290 CARLAW AVE
TORONTO, M4M 3L1
PHONE: (416) 462-9757
PROP HOUSE, THE
910 QUEEN ST. WEST
TORONTO, M6J 1G6
PHONE: (416) 588-7725

STUDIOS/SOUND STAGES

ANNEX STUDIOS LTD
176 BEDFORD RD
TORONTO, M5R 2K9
PHONE: (416) 922-8270
CINESPACE STUDIOS
629 EASTERN AVE
TORONTO, M4M 1E4
PHONE: (416) 465-2464
CINEVILLAGE
65 HEWARD AVE
TORONTO, M4M 2T5
PHONE: (416) 461-8750
FAX: (416) 466-9612
GLEN-WARREN PRODUCTIONS LTD
9 CHANNEL NINE COURT
TORONTO, M1S 4B5
PHONE: (416) 291-7571
TELEX: 065-25161
GLOBAL TV NETWORK
81 BARBER GREENE RD
DON MILLS, M3C 2A2
PHONE: (416) 446-5311
LAKESHORE STUDIOS
2264 LAKESHORE BLVD WEST
TORONTO, M8V 1V1
PHONE: (416) 252-5212
MPSL GROUP, THE
9 PRINCE ANDREW PLACE
DON MILLS, M3C 2H2
PHONE: (416) 449-7614
FAX: (416) 449-9239
MCWATERS & ASSOCIATES
491 KING ST. E
TORONTO, M5A 1L9
PHONE: (416) 366-9158
MARC PRODUCTIONS
1163 PARISIEN ST
OTTAWA, K1B 4W4
PHONE: (613) 741-9851
OXFORD STAGE
577 OXFORD ST
ETOBICOKE, M8Y 1E6
PHONE: (416) 252-3726
PANAVISION CANADA LTD, TORONTO
629 EASTERN AVE
TORONTO, M4M 1E4
PHONE: (416) 778-7262
FAX: (416) 778-7270
RAWI SHERMAN & COMPANY
41 PETER ST
TORONTO, M5V 2G2
PHONE: (416) 593-5969
SCANTEC VIDEO
9 ST. JOSEPH ST, SUITE 412
TORONTO, M4Y 1J6
PHONE: (416) 975-1686

SHOWLINE LTD
915 LAKESHORE BLVD. EAST
TORONTO, (416) 778-7379
FAX: (416) 778-7380
STUDIO ARTS CENTRE
15 POLSON ST
TORONTO, M5A 1A4
PHONE: (416) 465-9933
STUDIOASIS MEDIA CORP
793 PHARMACY AVE
SCARBOROUGH, M1L 3K2
PHONE: (416) 977-9740
TORONTO INTERNATIONAL STUDIOS
11031 HIGHWAY 27, BOX 430
KLEINBURG, L0J 1C0
PHONE: (416) 857-3090
TRACE PRODUCTIONS LTD
90 SHEPPARD AVE. EAST
NORTH YORK, M2N 3A1
PHONE: (416) 733-1605
FAX: (416) 733-2008
WALLACE STUDIOS INC
258 WALLACE AVE
TORONTO, M6P 3M9
WHITE, WILLIAM F. LTD
36 PARK LAWN RD
TORONTO, M8Y 3H8
PHONE:(416) 252-7171
FAX: (416) 252-6095

UNIONS AND ASSOCIATIONS

ACADEMY OF CANADIAN CINEMA & TELEVISION
653 YONGE ST, 2ND FLOOR
TORONTO, M4Y 1Z9
PHONE: (416) 967-0315
FAX: (416) 967-3351
ALLIANCE OF CANADIAN CINEMA, TV & RADIO ARTISTS (ACTRA)
130 SLATER ST, SUITE 808
OTTAWA, K1P 6E2
PHONE: (613) 230-0327
FAX: (613) 230-2473
ALLIANCE OF CANADIAN CINEMA, TV & RADIO ARTISTS (ACTRA)
2239 YONGE ST
TORONTO, M4S 2B5
PHONE: (416) 489-1311
FAX: (416) 489-1435
ASSOCIATION OF CANADIAN FILM & TELEVISION PRODUC-ERS (ACFTP)
2040 YONGE ST., SUITE 300
TORONTO, M4S 1Z9
PHONE: (416) 481-5232
FAX: (416) 481-3262
ASSOCIATION OF CANADIAN FILM CRAFTSPEOPLE (ACFC)
65 HEWARD AVE., SUITE 105
TORONTO, M4M 2T5
PHONE: (416) 462-0211
FAX: (416) 462-3248
C.A.M.E.R.A. LOCAL 81
(CANADIAN ASSN. OF MP & ELECTRONIC RECORDING ARTISTS)
181 CARLAW AVE, SUITE 302
TORONTO, M4M 2S1
PHONE: (416) 462-1022
FAX: (416) 461-4869
CANADIAN ACTORS EQUITY ASSOCIATION
260 RICHMOND ST. E
TORONTO, M5A 1P4
PHONE: (416) 967-4252
CANADIAN CABLE TELEVISION ASSOCIATION (CCTA)
360 ALBERT ST., SUITE 1010
OTTAWA, K1R 7X7
PHONE: (613) 232-2631
FAX: (613) 232-2137

CANADIAN FILM & TV ASSOCIATION (CFTA)
663 YONGE ST., SUITE 401
TORONTO, M4Y 2A4
PHONE: (416) 927-3942
FAX: (416) 922-4038
CANADIAN INDEPENDENT FILM CAUCUS
490 ADELAIDE ST. W. SUITE 304
TORONTO, M5V 1T2
PHONE: (416) 362-9822
FAX: (416) 362-3608
CANADIAN MOTION PICTURE DISTRIBUTORS ASSOCIATION (CMPDA)
22 SAINT CLAIR AVE E., SUITE 1703
TORONTO, M4T 2S4
PHONE: (416) 961-1888
FAX: (416) 968-1016
CANADIAN SOCIETY OF CINEMATOGRAPHERS (CSC)
72 FRASER AVE., SUITE 203
TORONTO, M6K 3E1
PHONE: (416) 538-3155
FAX: (416) 538-8821
DIRECTORS GUILD OF CANADA
3 CHURCH ST., SUITE 500
TORONTO, M5E 1M2
PHONE: (416) 364-4185
FAX: (416) 364-1985
DIRECTORS GUILD OF CANADA
ONTARIO DISTRICT COUNCIL
3 CHURCH ST., SUITE 202
TORONTO, M5E 1M2
PHONE: (416) 364-0122
FAX: (416) 364-6353
163 WEST HASTIGS ST., SUITE 339
VANCOUVER B.C. V6C 1A5
PHONE: (604) 688-2976
FEDERATION OF CANADIAN GUILDS & UNIONS IN FILM & TELEVISION
65 HEWARD AVE., SUITE 105
TORONTO, M4M 2T5
PHONE: (416) 462-0211
FAX: (416) 462-3243
GUILD OF CANADIAN FILM COMPSERS
C/O 20 ST. JOSEPH ST
TORONTO, M4Y 1J9
PHONE: (416) 929-9314/781-4191
IATSE LOCAL 667
793 PHARMACY AVE., SUITE 401
SCARBOROUGH, M1L 3K3
PHONE: (416) 759-4108
FAX: (416) 759-7111
IATSE LOCAL 873
2200 LAKE SHORE BLVD. W., SUITE 210
ETOBICOKE, M8V 1A4
PHONE: (416) 252-5025
FAX: (416) 252-3923
ITVA CANADA (INTERNATIONAL TELEVISION ASSOCIATION CANADA)
BOX 1156, ADELAIDE P.O. STN
TORONTO, M5C 2K5
PHONE: (416) 733-3757
FAX: (416) 733-3757
MOTION PICTURE THEATRE ASSOCIATION OF ONTARIO
21 DUDAS SQUARE SUITE 1210
TORONTO, M5B 1B7
PHONE: (416) 368-1139
NABET 700
1179A KINGS ST. WEST, SUITE 102
TORONTO, M6K 3C5
PHONE: (416) 536-4827
FAX: (416) 536-0859

NATIONAL ASSOCIATION OF CANADIAN FILM & VIDEO DISTRIBUTORS
10 OLIVE AVE
TORONTO, M6G 1T8
PHONE: (416) 535-9614
FAX: (416) 368-7519
ONTARIO FILM ASSOCIATION
2 - 1750 THE QUEENSWAY, SUITE 1341
ETOBICOKE, M9C 5H5
PHONE: (416) 761-6056
OTTAWA FEATURE FILM PRODUCERS ASSOCIATION
212 JAMES ST
OTTAWA, K1R 5M7
PHONE: (613) 230-9769
FAX: (613) 230-6004
OTTAWA-HULL FILM & TELEVISION ASSOCIATION
P.O. BOX 142, STN B
OTTAWA, K1P 6C3
PHONE: (613) 233-3836
FAX: (613) 233-0698
PERFORMING RIGHTS ORGANIZATION OF CANADA LTD (PROCAN)
41 VALLEYBROOKE DR
DON MILLS, M3B 2S6
PHONE: (416) 445-8700
FAX: (416) 445-7108
SOCIETY OF MOTION PICTURE & TELEVISION ENGINEERS (SMPTE)
C/O PFA FILMS & VIDEO, 330 ADELAIDE ST. W.
TORONTO, M5V 1R4
PHONE: (416) 593-0556
FAX: (416) 593-7201
UNION DES ARTISTES (UDA)
2 COLLEGE ST., SUITES 206-207
TORONTO, M5G 1K3
PHONE: (416) 967-4408
FAX: (416) 967-6898

CANADA-PRINCE EDWARD ISLAND

UNIONS AND ASSOCIATIONS

UNIONS AND ASSOCIATIONS IATSE LOCAL 90
A THECONFED
RATION CENTRE, P.O. BO
848 CHARLOTTETOWN, C
A 7L9 PHONE: (902)
66-2464 FAX: (902)
566-4648 ISLAND MEDIA CO-OP
P.O. BOX 2726, 92 QUEE
ST. CHARLOTTETOWN, C
A 8C3 PHONE: (902)

92-3131 FAX:

902) 566-1724 CANADA

QUEBEC STUDIOS/SO
UND STAGES 1001 SENES VI
EO 7054, RUE MOLSO
, BUR. 2 MONTREAL, H2
3KI PHONE: (514
376-7001 ANDR
PERRY VIDEO 1501
RUE BARRE MONTREAL, H
C 4HI PHONE: (514)
32-0303 FAX: (514)   93
-308 CHAMPLA
N PRODUCTIONS 405
OGILVY ST MONTREAL, H
N 1M4 PHONE: (514)

CLUB METROPOLIS INC
1413 ST DOMINIQUE
MONTREAL, H2X 3P5
PHONE: (514) 288-2020
CONCORDIA UNIVERSITY
1455 OUEST, BOUL., DE MAISONNEUVE
MONTREAL, H3G 1M8
PHONE: (514) 848-3448
GRAETZ, INC
1501 RUE BARRE
MONTREAL, H3C 4JI
PHONE: (514) 989-9551
FAX: (514) 989-5054
INTER-TEL IMAGE
1310 LARIVIERE (PAR PANET)
MONTREAL, H2L 1M8
J.P.L. PRODUCTIONS INC
1600 DE MAISONNEUVE BLVD. EAST, SUITE 610
MONTREAL, H2L 4P2
PHONE: (514) 526-2881
FAX: (514) 526-3740
KELVIN CLUB
910 EST, DE LA GAUCHEIERE, BUR. 300
MONTREAL, H3L 2N4
PHONE: (514) 935-3548
PANAVISION CANADA LTD. MONTREAL
2170 AVENUE PIERRE DUPUY, CITE DU HAVRE
MONTREAL, H3C 3R4
PHONE: (514) 522-5533
FAX: (514) 522-5971
PRODUCTIONS ASSAM INC. (LES)
2189 LEON HARMEL
MONTREAL, G1N 4N5
PHONE: (418) 681-3537
FAX: (418) 681-3537
PRODUCTIONS SUPERTEL INC
350 MONTPELLIER
MONTREAL, H4N 2G7
PHONE: (514) 748-1442
FAX: (514) 748-1256
SERVICE VIDEO SENECA INC. (LES)
430 OLD CHATEAUGUAY RD
KAHNAWAKE, J0L 1B0
PHONE: (514) 632-8133
FAX: (514) 632-8448
SOCIETE RADIO CANADA
1400 BOUL. RENE-LEVESQUE
MONTREAL, H2L 2M2
PHONE: (514) 597-6323
SONOLAB INC
1500 PAPINEAU ST
MONTREAL, H2K 4L9
PHONE: (514) 527-8671
TELEX: 055-61722
SPECTEL VIDEO
355, STE-CATHERINE OUEST, BUR. 700
MONTREAL, H3B 1A5
PHONE: (514) 288-5363
FAX: (514) 499-0956
STUDIO ARIZONA
3906 RUE CLARK
MONTREAL, H2W 1W6
PHONE: (514) 843-5055
FAX: (514) 844-4263
STUDIO DU HAVRE
2295 RUE ST.-MARC, BUR 305
MONTREAL, H3H 2G9
PHONE: (514) 932-1012
STUDIO IMAGE
1600 BOUL. DE MAISONNEUVE EST. BUR. A646
MONTREAL, H2L 4P2
PHONE: (514) 526-2881
FAX: (514) 526-3740
STUDIO WHITE
715, RUE ST. MAURICE
MONTREAL, H3C 1L4
PHONE: (514) 866-3323
FAX: (514) 866-8856

STUDIOS YVON BLAIS (LES)
2200 GARDENVALE
ST BRUNO DE MONTARVILLE, J3V 1K1
PHONE: (514) 653-8652
SUPERTEL PRODUCTIONS
350 RUE MONTPELLIER
MONTREAL, H4N 2G7
PHONE: (514) 748-1371
FAX: (514) 748-1256
TAQRAMIUT NIPINGAT INC
185 AV. DORVAL, BUR. 501
DORVAL, H9S 5J9
PHONE: (514) 631-1394
TECHNER GROUP
1600 DE MAISONNEUVE BLVD. EAST
MONTREAL, H2L 4P2
PHONE: (514) 598-2950
FAX: (514) 526-1871
TECHNOVIDEO LTEE
267 BOUL. LAURIER
ST. BASILE LE GRAND, J0L 1S0
PHONE: (514) 461-3895
TRANSIMAGE LTD
2327 DU VERSANT BLVD. N., SUITE 150
STE-FOY, G1N 4C2
PHONE: (418) 681-0872
UNIVERSITE DU QUEBEC A MONTREAL
1495 RUE ST DENIS, BUR 2707
MONTREAL, H3C 3P8
PHONE: (514) 282-4450

UNIONS AND ASSOCIATIONS

ASIFA-CANADA/ASSOCIATION INTERNATIONALE DU FILM DANIMATION
10707 RUE GRANDE-ALLEE, BUREAU 3
MONTREAL, H3L 2MB
PHONE: (514) 842-9763
FAX: (514) 842-1816
ACADEMY OF CANADIAN CINEMA & TELEVISION
3603 SAINT-DENIS ST, SUITE 302
MOTREAL, H2X 3L6
PHONE: (514) 849-7448
FAX: (514) 849-5069
ALLIANCE OF CANADIAN CINEMA, TV & RADIO ARTISTS (ACTRA)
1450 CITY COUNCILLORS ST., SUITE 530
MONTREAL, H3A 2E6
PHONE: (514) 844-3318
FAX: (514) 844-2068
J6J 5R4
PHONE: (514) 691-2334
ASSOC. CANADIENNE DE LA RADIO ET DE LA T.V. DE LANGUE
FRANCAISE INC. (ACRTF)
1600 BOUL. DE MAISONNEUVE E., BUREAU A762
MONTREAL, H2L 4P6
PHONE: (514) 522-2783
ASSOC. DES REALISATEURS ET REALISATRICES DE FILMS DU QEUBEC
(ARRFQ)
1600 AVENUE DE LORIMIER, BUREAU 122
MONTREAL, H2K 3W5
PHONE: (514) 527-2197
FAX: (514) 521-7081
ASSOCIATION DES DIRECTEURS DE CASTING DE QUEBEC
1043 BERRI, SUITE 300
MONTREAL, H2L 4C4
PHONE: (514) 659-6243

ASSOCIATION DES PRODUCTEURS DE FILMS ET DE
VIDEO DU QUEBEC (APFVQ)
430 RUE SAINTE-HELENE, BUREAU 201
MONTREAL, H2Y 2K7
PHONE: (514) 284-9444
FAX: (514) 843-8084
COMPOSERS, AUTHORS & PUBLISHERS ASSOCIATION OF CANADA LTD
(CAPAC)
1245 RUE SHERBROOKE O., BUREAU 1470
MONTREAL, H3G 1G2
PHONE: (514) 288-0828
DIRECTORS GUILD OF CANADA
QUEBEC DISTRICT COUNCIL, 2250 GUY STREET, SUITE 506
MONTREAL, H3H 2M3
PHONE: (514) 989-1714
FAX: (514) 989-1732
INDEPENDENT FILM & VIDEO ALLIANCE
397 SAINT-JOSEPH BLVD. W., SUITE 1
MONTREAL, H2V 2P1
PHONE: (514) 277-0328
FAX: (514) 274-2261
INDEPENDENT FILM & VIDEO ALLIANCE
P.O. BOX 545, DESJARDINS STN
MONTREAL, H5B 1B6
NABET
INTERNATIONAL OFFICE, 6845 SAINT-DENIS STREET
MONTREAL, H2S 2S3
PHONE: (514) 276-8591
FAX: (514) 276-6413
PERFORMING RIGHTS ORGANIZATION OF CANADA LTD (PROCAN)
625 AV. DU PRESIDENT-KENNEDY, BUREAU 1200
MONTREAL, H3A 1K2
PHONE: (514) 849-3294
FAX: (514) 849-8446
REGROUPEMENT DE DISTRIBUTEURS TELEVISION DU QUEBEC
(RDTVQ)
A/S CINAR, 1207 RUE SAINT-ANDRE
MONTREAL, H2L 3SB
PHONE: (514) 273-4251
FAX: (514) 276-5130
S.T.C.Q.
SYNDICAT DE STECHNICIENNES ET TECHNICIENS DU CINEMA DU QUEBEC
4115 AVENUE PAPINEAU
MONTREAL, H2K 4K2
PHONE: (514) 525-8428
SOCIETE DES AUTEURS, RECHERCHISTES DOCUMENTALISTES
ET COMPOSITEURS (SARDEC)
1229 RUE PANET
MONTREAL, H2L 2Y6
PHONE: (514) 526-9196
FAX: (514) 526-4124
SOCIETE FILM ET TELEVISION GROUPE QUEBEC INC
872 RUE CHARLES-GUIMOND
BOUCHERVILLE, J4B 3Z5
PHONE: (514) 641-4691
FAX: (514) 449-2651
SOCIETE POUR LADVANCEMENT DES DROITS EN AUDIOVISUEL
LTEE (SADA)
5225 RUE BERRI
MONTREAL, H2J 2S4
PHONE: (514) 273-4231
TELEX: 055-62281
FAX: (514) 276-5130

SYNDICAT DES TECH. ET TECH. DU CINEMA ET DE LA VIDEO DU QUEBEC (STCVQ)
4115 AVENUE PAPINEAU
MONTREAL, H2K 4K2
PHONE: (514) 525-8428
FAX: (514) 522-1140

SYNDICAT GENERAL DU CINEMA ET DE LA TELEVISION - SECTION ONE
1285 RUE HODGE, BUREAU 315
SAINT-LAURENT, H4N 2B6
PHONE: (514) 744-4989

UNION DES ARTISTES (UDA)
580 RUE SAINT-DENIS,
6TH FLOOR
MONTREAL, H2X 3J7
PHONE: (514) 288-6682
FAX: (514) 288-7150

UNION DES ARTISTS (UDA)
580 AVENUE GRANDE-ALLEE E.,
BUREAU 430
QUEBEC, G1R 2K2
PHONE: (418) 523-4241
FAX: (418) 523-0168

CHILE

FILM COMMISIONS

CHILE FILM COMMISSION
PO BOX 2829,734 SUECIA AVE
SANTIAGO
CHILE
PHONE: (2)251 3466 EXT.
FAX: (2)331871

CZECHOSLAVAKIA

STUDIOS/SOUND STAGES

FILMMOVE STUDIO BARRANDOV (BRANDOV FILM STUDIOS)
PRAHA 5
KRIZENCEKEHO NEM 322
PHONE: 541640/9

FILMOVE STUDIO GOTTWALDOV (GOTTWALDOV FILM STUDIO)
GOTTWALDOV-KUDLOV
SVERMOVA 174
PHONE: 2301/2

STUDIO VIDEOFILMU
BRATISLAVA-KOLITE
PHONE: 331807

UNIONS AND ASSOCIATIONS

CESKOSLOVENSKY (FILM EXPORT)
VACLAVSKE NAMESTT 28
1145 PRAHA
PHONE: 246741/5

CESKOSLOVENSKY FILMOVY USTAV
NARODNI TRIDA 40
11000 PRAHA
PHONE: 260087

SLOVENSKA FILMOVA TVORBA
BRATISLAVA CERCELJ ARMADY 32
PHONE: 51789

USTREDIE SLOVENSKEHS FILMU
BRECTANOVA 1
BRATISLAVA-KOLIBA
PHONE: 42193

DENMARK

EQUIPMENT RENTALS

MORTEN JACOBSEN TRADING APS
VIBENSHUS. OLIEMOLLEGADE 12 (4-4)
2100 COPENHAGEN
PHONE: 31-180455
TELEX: 15456 DANCAN DK

PERF-FIX COMPANY
DISTRIBUTORS: MORTON
JACOBSEN TRADING APS
VIBENSHUS, OLIEMOLLEGADE 12 (4-4)
2100 COPENHAGEN O
PHONE: 31-180455

RESEARCH TECHNOLOGY INT.
REVETLOWSGADE 30, DK-1651
COPENHAGEN V
PHONE: 31-310093

MAKE-UP SUPPLIERS

KRYOLAN PROFESSIONAL MAKEUP
ANKER & LOVBO AMAGERTORV
15, 11 DK-1160
KOBENHAVEN

STUDIOS/SOUND STAGES

BELLEVUE STUDIO A/S VIDEO & FILM
71 DORTHEAVEJ, DK-2400
COPENHAGEN NV
PHONE: 31-196333 22226
FAX: 38-332232

DANISH GOVERNMENT FILM STUDIO, THE
BLOMSTTERVAENGET 52
2800 LYNGBY
PHONE: 42-872700

FOX-HILL STUDIO A/S
RAEVEHOJVEJ 15
28000 LYNGBY
PHONE: 42-878600

KAERNE FILMSTUDIER APS
LIVJAEGERGADE 17
2100 COPENHAGEN
PHONE: 31-264200
31-868659

UNIONS AND ASSOCIATIONS

ASIFA DANMARK
INTERNATIONAL ANIMATED,
ASSOCIATION,
RODHASSTRAEDE 6
1466 COPENHAGEN
PHONE: 31-151981

DANISH FILM INSTITUTE, THE
STORE SONDERVOLD STRAEDE,
DK-1016
COPENHAGEN K
PHONE: 45-156500
TELEX: 31465 DFILM DK
FAX: 45-1576700

DANISH FILM MUSEUM
ST. SOENDERVOLDSTRAEDE
1419 COPENHAGEN K
PHONE: 31-576500
TELEX: 31765 DFILM DK

DANMARKS BIOGRAFTEATER FORENING
BULOWSVEJ 50A
1870 FREDERIKSBERG C
PHONE: 31-372407

DANMARKS FILM AKADEMI
VESTAGERVEJ 5
2100 COPENHAGEN
PHONE: 31-182511

DANSK ARTIST FORBUND
VENDERSGADE 24, 1,
1363 COPENHAGEN V
PHONE: 33-326677

DANSK FILMFOTOGRAF FORBUND
SKOVVEJ 7
2820 GENTOFTE
PHONE: 31-6331=016

DANSK FORFATTERFORENING
TORDENSKJOLDS GRD.,
STRANDAGADE 6 ST,
1401 COPENHAGEN K
PHONE: 31-955100

DANSK TEATERFORBUND
SANKT KNUDS VEJ 26
1903 FR. BERG C
PHONE: 31-242200

DANSKE BORNEFILMKLUBBER
NIELS HEMMINGSENS GADE 20
BAGH 3
1153 COPENHAGEN K
PHONE: 33-156760

DANSKE DRAMATIKERES FORBUND
KLOSTERSTRAEDE 24
1157 COPENHAGEN K
PHONE: 33-110901

DANSKE FILMPRODUCENTER I/S
AXELTORV 7
1609 COPENHAGEN V
PHONE: 31-147606

DANSKE VIDEO - OG KORTFILMPRODUCENTER-DUK
THORAVEJ 24
2400 COPENHAGEN NV
31-191144

FILMARBEJDERFORENINGEN (FAF)
LYKKESHOLMS ALLE 29, ST.
1902 FR. BERG C
PHONE: 31-233066

FILMBRANCHENS HJAELPEFOND
C/O DANMARKS
BIOGRAFTEATER, FORENING,
BULOWSVEJ 50A
1870 FR. BERG C
PHONE: 31-372407

FILMBRANCHENS SAMARBEJDS - OG IDEUDVALG
BULOWSVEJ 50A
1870 FR. BERG C
PHONE: 31-372507

FILM-OG FORLYSTELSES-FUNKTIONAIRERNES FORBUND
NITIVEJ 8. A. ST TV
2000 FR. BERG
PHONE: 31-878213

FILMTONEMESTERFORENINGEN
RODOVREVEJ 408
2610 RODOVRE
PHONE: 31-700076

FORENING FOR MEDIEPAEDAGOGIK - DANSK FILMLAERERFORENING
NIELS HEMMINGSENS GADE 20
1153 KBH K
PHONE: 33-134378

FORENINGEN AF AV - PRODUCENTER 1 DANMARK
JOMFRU ANE CADE 14
9000 OLBORG
PHONE: 98-136900

FORENINGEN AF DANSKE FILMPRODUCENTER
BREDGADE 73, DK-1260
COPENHAGEN K
PHONE: 33-132299

FORENINGEN AF DANSKE VIDEOGRAMDISTRIBUTORER
OSLO PLADS 14
2100 COPENHAGEN O
PHONE: 31-422166

FORENINGEN AF FILMUDLEJERE 1 DANMARK
OSLO PLADS 14
2100 COPENHAGEN D
PHONE: 31-422166

FORENINGEN AF KOMMUNALE BIDGREFLEDER 1 DANMARK COLOSEUM
FREDERISHAJN
PHONE: 98-421389

GOVERNMENT FILM CENSOR
VESTERPROT 1537,
MELDAHLSGDE 5, DK-1513
COPENHAGEN V
PHONE: 33-159639

KOMMUNALE BIOGRAFERS SAMUIRKE
BALLERUP BIO, SANKT JACOBS
VEJ 1
2750 BALLERUP
PHONE: 42-654356

NORDISK FILM/TV UNION (SOCIETY)
STATENS FILM CENTRAL,
VESTERGRADE 27, DK-1756
COPENHAGEN K
PHONE: 33-132686

SAMMENSLUTNINGEN AF DANSKE DISTRIBUTORER AF FILM OG VIDEOGRAMMER TIL IKKE-
IKKE-KOMMERCIALLE,
FOREVISNINGER
BALDERSBAEKVEJ 5
2635 ISHOJ
PHONE: 92-998211

SAMMENSLUTNINGEN AF DANSKE FILMKRITIKERE
SYLOWS ALLE 8
2000 FR. BERG
PHONE: 31-873783

SAMMENSLUTNINGEN AF DANSKE FILMINSTRUKTORER
CLYDENLUNDSVEJ 9B
2920 CHARLOTTENLJND
PHONE: 31-637325

SAMMENSLUTNINGEN AF SCENOFRAFER, INSTRUKTORER, FILMARBEJDERE
LYKKESHOLMS ALLEY 29, ST.,
1902 FR. BERG C
PHONE: 31-233066

STATENS FILMCENTRAL DANISH GOVERNMENT FILM OFFICE
VESTERGRADE 27
COPENHAGEN K,1456
PHONE: 33-132686

EGYPT

EQUIPMENT RENTAL

NEILSON-HORDELL LTD
BUILDING 53,EL,EAM CITY
AQQUZA,CAIRO
EGYPT
PHONE: 811125
PHONE: TX 92592

UNIONS AND ASSOCIATIONS

CAIRO CINEMA CLUB
36 SHRIEF STREET,CAIRO
EGYPT
PHONE: 747460

CHAMBER OF CINEMA & INDUSTRY
33 ORABY STREET,CAIRO
EGYPT
PHONE: 748366
PHONE: 741638
FAX: TX 92624

CINEMA PROFESSIONS UNION
20 ADLY STREET,CAIRO
EGYPT
PHONE: 756687
EGYPTIAN ASSEMBLY FOR CINEMA
WRITERS & CRETICIST
9 ORABI STRET,CAIRO
PHONE: 741112
FILM ASSEMBLY
36 SHRIEF STREET,CAIRO
PHONE: 747460
PROFESSIONAL ACTORS UNION
1 JULY 26TH STREET,CAIRO
PHONE: 912147

FINLAND

SPECIAL EFFECTS

YLEISRADIO OY (FINNISH BROADCASTING COMPANY)
KESAKATU 2
SF 00260 HELSINKI
PHONE: 90-441 141
TELEX: 124735

UNIONS AND ASSOCIATIONS

ELOKUVA-ALAN AMMATTIYHDISTYS RY
11 LINJA 3
SF 00530 HELSINKI
PHONE: 90-701 7722
(TRADE ASSOC. OF FILM INDUSTRY EMPLOYERS)
ELOKUVA-ALAN TYONLANTAJAYHDISTYS-ETA
C/O SEOL KAISANIEMENKATU 3 B 29
SF 00100
PHONE: 90-131 191
(TRADE ASSOC. OF FILM INDUSTRY EMPLOYERS)
ELOKUVA-JA TV-OPISKELIJAT RY
PURSIMIEHENKATU 29-31 B
SF 00150 HELSINKI
PHONE: 90-636 995
(FILM AND TV STUDENTS)
ELOKUVAKONEENHOITAJIEN YHDISTYS RY
11 LINJA 3
SF 00530 HELSINKI
PHONE:
(ASSOC. OF FILM PROJECTOR OPERATORS)
FILMIAURA RY
LAIVURINKATU 10 B
SF 00150 HELSINKI
PHONE: 90-175 944
(AWARDS THE ANNUAL FINNISH FILM AWARDS)
LASTEN-JA NUORTENELOKUVAYHDISTYS RY
ALAKIVENTIE 1 E 63
SF 00920 HELSINKI
(THE ASSOC. OF FILMS FOR CHILDREN AND YOUNG PEOPLE)
MAINOSELOKUVATUOTTAJAIN LIITTO RY
UUDENMAANKATU
4-6 F 24
SF 00120 HELSINKI
PHONE: 90-647 267
FAX: 90-601 526
(ADVERTISING FILM PRODUCERS ASSOC IN FINLAND)
NORDIC FILM-TV SOCIETY/ FINNISH SECTION
P.O. BOX 159
SF 00251 HELSINKI
PHONE: 90-134 171

SUOMEN AUDIOVISUAALINEN YHDISTYS RY
ARINATIE 8
SF 00370 HELSINKI
PHONE: 90-558 182
FAX: 90-558 198
(THE FINNISH AUDIO VISUAL ASSOC)
SUOMEN ELOKUVAAJIEN YHDISTYS RY
C/O CINERENT MERITULINKATU 11
SF 00170 HELSINKI
PHONE: 90-602 306
(THE SOCIETY OF FINNISH CINEMATOGRAPHERS)
SUOMEN ELOKUVA-JA VIDEOTYONTEKIJAIN LIITTO RY MANEESIKATU 4 C
SF 00170 HELSINKI
PHONE: 90-174 460
(TRADE ASSOC OF FINNISH FILM & VIDEO EMPLOYEES)
SUOMEN ELOKUVAKONTAKTI YRJONKATU 11 A 5
SF 00120 HELSINKI
PHONE: 90-607 380
(THE FINNISH FILM CONTACT)
SUOMEN ELOKUVAOHJAAJALIITTO SELO MERITULINKATU 11
SF 00170 HELSINKI
PHONE: 90-461 961
(THE ASSOC OF FINNISH FILM DIRECTORS)
SUOMEN ELOKUVASAATIO K13 KANAVAKATU
SF 00160 HELSINKI
PHONE: 90-177 727
TELEX: 125032 SESFI SF
FAX: 90-177 113
(THE FINNISH FILM FOUNDATION)
SUOMEN ELOKUVATOIMISTOJEN LIITTO SEL
TALLBERGIN PUISTOTIE 7 A 5
SF 00200 HELSINKI
PHONE: 90-636 305
FAX: 90-176 689
(THE FINNISH FILM DISTRIBUTORS ASSOC.)
SUOMEN ELOKUVATUOTTAJIEN KESKUSLIITTO KAISANIEMENKATU 3 B 25
SF 00100 HELSINKI
PHONE: 90-636 305
FAX: 90-176 689
(THE CENTRAL ORGANIZATION OF FINNISH FILM PRODUCERS)
SUOMEN ELOKUVAYHDISTYS RY'
P.O. BOX 17
SF 00241 HELSINKI
PHONE: 90-15 001/345
(THE MOTION PICTURE ASSOC. OF FINLAND)
SUOMEN FILIKAMARI
KAISANIEMENKATU 3 B 29
SF 00100 HELSINKI
PHONE: 90-636 305
(FINNISH FILM CHAMBER)
VALTION AUDIOVISUAALINEN KESKUS
HAKANIEMENKATU 2
SF 00530 HELSINKI
PHONE: 90-7061
(STATE AV-CENTER)
VALTION ELOKUVATAIDETOIMIKUNTA
MARIANKATU 5
SF 00170 HELSINKI
PHONE: 90-134 171
(STATE COMMITTEE FOR CINEMA)

FRANCE

SPECIAL EFFECTS

AAZ CHRISTIAN LABES
40 RUE DE FRESNES
94240 L'HAY-LES-ROSES
PHONE: (1) 46-60-90-23
ACL
2 AV. DES MARONNIERS
94389 BONNEUIL-SUR-MARNE CE-DEX
PHONE: (1) 43-77-56-28
FAX: (1) 43-99-29-74
TELEX: 262685
ACME FILMS
24-26 RUE PAUL FORT
75014 PARIS
PHONE: (1) 45-43-66-57
ACME STUDIO
24-26 RUE PAUL FORT
75014 PARIS
PHONE: (1) 45-43-66-57
AF2
60 RUE D'OSTWALD
67200 STRASBOURGE
PHONE: 88-30-13-13
AFTER MOVIES VIDEO
17 RUE DU COLISEE
75008 PARIS
PHONE: (1) 42-56-35-03
ALCOPLAST
9 ALLEE DES PROGRES
92170 VANVES
PHONE: (1) 46-42-79-21
ANCOR
329 RUE LECOURBE
75015 PARIS
PHONE: 45.54.21.05
AND MAX
28 RUE KLEBER
93100 MONTREUIL
PHONE: (1) 48-57-26-58
AOR ATELIERS DES OUVRIERS REUNIS
16 CHEMIN LATERAL BP 121
93505 PANTIN
PHONE: (1) 48-45-13-39
FAX: (1) 48-46-56-46
ARANE
5 PLACE DU GENERAL LECLERC
92300 LEVALLOIS-PERRET
PHONE: (1) 40-89-03-04
FAX: (1) 47-58-89-08
TELEX: 270105
ARCADY EQUIPE NICOLAS BRACHLIANOFF
135 BD ST. MICHEL
75005 PARIS
PHONE: (1) 43-54-56-31
ATELIER 287
287 RUE DU FG ST-ANTOINE
75011 PARIS
PHONE: (1) 43-72-06-29
FAX: (1) 43-72-25-15
ATELIER D
139 RUE DE CHARONEE
75011 PARIS
PHONE: (1) 43-72-57-43
FAX: (1) 43-72-80-00
ATELIER DAYNES
129 FBG DU TEMPLE
75010 PARIS
PHONE: (1) 42-41-17-36
FAX: (1) 42-41-08-05
ATELIER DEXET
45 RUE DES WATTIGNIES
75012 PARIS
PHONE: (1) 43-41-30-68
ATELIER MICHEL SOUBEYRAND
2 RUE DE TLEMCEN
75020 PARIS
PHONE: (1) 47-97-23-00

ATELIER ROBESPIERRE
99 RUE SAINT-DENIS
93130 NOISY-LE-SEC
PHONE: (1) 48-58-34-26
FAX: (1) 48-91-89-61
ATELIERS DAMALIX
2 RUE ADRIEN DAMALIX
94410 SAINT-MAURICE
PHONE: (1) 48-93-43-44
AUDIO MASTER
26 AVENUE DU MAL LECLERC
64000 PAU
PHONE: 59-84-03-03
AUDIO TECHNIC
BP 5528
3400 MONTPELIER
PHONE: 67-42-75-07
FAX: 67-27-85-64
AUDIOENERGIE
45 BIS AV MAX DORMOY
18000 BOURGES
PHONE: 48-65-22-33
FAX: 48-24-24-95
BARAT JEAN
52 RUE DE CRIMEE
75019 PARIS
PHONE: (1) 42-05-44-99
BARGY IMAGES
31 RUE D'ALLERAY
75018 PARIS
PHONE: (1) 45-33-71-68
TELEX: 260739
BEAUPASCHER
119 RUE ST. DENIS
75001 PARIS
PHONE: (1) 40-26-49-74
BOUDET MARC
9 RUE DES TROIS BORNES
75011 PARIS
PHONE: (1) 43-38-02-89
FAX: (1) 43-38-02-89
BRACHLIANOFF NICOLAS EQUIPE ARCADY
135 BD ST. MICHEL
75005 PARIS
PHONE: (1) 43-54-56-31
CABESTAN LASER MEDIA
35 BD DU ROI
78000 VERSAILLES
PHONE: (1) 39-51-61-50
FAX: (1) 39-50-32-62
CAUQUIL PHILLIPPE
7 RUE DU DR ARNOUDET
92190 MEUDON
PHONE: (1) 45-07-14-36
CAUVY MARC
73 ROUTE DE MONTPELLIER
34730 PRADES-LE-LEZ
PHONE: 67-59-76-10
CINEDIA
55 AVENUE JOFFRE
93800 EPENAY-SUR SEINE
PHONE: (1) 48-41-36-25
FAX: (1) 48-41-38-03
CINEFORMES
7 RUE DES VERTUGADINS
92190 MEUDON
PHONE: (1) 46-26-82-35
CINEMATION
11 RUE D'ODESSA
75014 PARIS
PHONE: (1) 45-48-79-28
COLOMBO JACQUES
15 RUE ADAM
94100 ST. MAUR
PHONE: (1) 48-83-35-48
COMANDA
9 RUE DE LA TOUR
75116 PARIS
PHONE: (1) 45-20-20-59
COMEL
36 RUE DANTON B.P. 162
93103 MONTREUIL CEDEX
PHONE: (1) 48-57-32-92
FAX: (1) 48-57-24-02
TELEX: 233538

CONTRE-JOUR
35 RUE MIGUEL HILDALGO
75019 PARIS
PHONE: (1) 42-06-56-35
FAX: (1) 42-40-27-02
TELEX: 215864

CREA S
7 AV JEAN JUARES
93450 L'ILE ST-DENIS
PHONE: (1) 48-09-87-78
FAX: (1) 48-09-10-72

DARU
5 RUE DES COLONNES DU TRONE
75012 PARIS
PHONE: (1) 43-46-96-28

DEUS
100, RUE DU FG SAINT ANTOINE
75012 PARIS
PHONE: 40.19.95.96
FAX: 40.19.94.38

DIMAPHOT
16 RUE DLEMENT MAROT
75008 PARIS
PHONE: (1) 47-23-98-87
FAX: (1) 47-23-40-44
TELEX: 642113

DOUBLE FACE
151 RUE DU FG ST. ANTOINE
75011 PARIS
PHONE: (1) 43-41-56-79

DOUBOU
58 BD DE CHANZY
93100 MONTREUIL
PHONE: (1) 48-58-22-02
FAX: (1) 48-58-85-27

DURAN
74, RUE JOSEPH DE MAISTRE
75018 PARIS
PHONE: 42.29 00.33

ECA - SON ET LUMIERE PROD
31 AVE. FERNAND LE-FEBREVRE
78300 POISSY
PHONE: (1) 39-79-00-15

ELYSEES RELATIONS CINEMATOGR CLARENS BERNARD
11 RUE SOULNIER
75009 PARIS
PHONE: (1) 45-23-05-76

ERCIDAN FILMS
13-16 RUE DUVIVIER
75007 PARIS
PHONE: (1) 45-56-10-44
FAX: (1) 45-55-58-97

ERE FORCE
21 RUE GEORGES BOISSEAU
92110 CLICHY
PHONE: (1) 47-39-05-10
TELEX: 611837

ESPACE DE LHUILLY
7 BOIS DE L'HUILLY
45170 NEUVILLE-AUX-BOIS
PHONE: 38-39-82-23

ESPACE ET STRATEGIE
15, ALLEE GLUCK
68200 MULHOUSE
PHONE: 89.32.12.10

EURO TITRES
48 RUE LACORDAIRE
75015 PARIS
PHONE: (1) 45-58-49-09
FAX: (1) 45-54-78-49
TELEX: 204011

EUROCITEL
1 QUAI GABRIEL PERI
94340 JOINVILLE-LE-PONT
PHONE: (1) 43-97-25-25
FAX: (1) 43-97-19-23
TELEX: 230118

EURODROP
171 RUE VERON
94140 ALFORTVILLE
PHONE: (1) 43-75-08-31
FAX: (1) 43-68-19-76
TELEX: 213905

EX MACHINA
22, RUE HEGESIFPE MOREAU
75018 PARIS
PHONE: 43.87.58.58
TELEX: 270105

EXCALIBUR
1 QUAI GABRIEL PERI
94340 JOINVILLE-LE-PONT
PHONE: (1) 43-97-22-66
FAX: (1) 43-97-19-23

EXPOSURE
1 BIS RUE DES EPINETTE
94410 ST. MAURICE
PHONE: (1) 48-93-45-01
FAX: (1) 48-93-17-17

EXTERIEUR NUIT
48 RUE MONTMARTRE
75002 PARIS
PHONE: (1) 45-C8-19-19
FAX: (1) 40-26-42-52
TELEX: 213640

FANTOME
71, RUE AMPERE
75017 PARIS
PHONE: 40.53.01.23
FAX: 40.53.02.07

FILMS ET FORMES
7 BIS AV. DE ST-MANDE
75012 PARIS
PHONE: (1) 46-28-28-80
FAX: (1) 43-46-16-31

FILMS MICHEL FRANCOIS FRANCOIS JANINE
7 RUE DES DARNES AUGUSTINE
92200 NEUILLY
PHONE: (1) 47-57-68-05

FOG AND SMOKE
20 BIS RUE DU DOCTEUR ROUX
92110 CLICHY
PHONE: (1) 42-70-99-97

FOUILLET-WIEBER
55 BD FELIX FOURE
93300 AUBERVILLIERS
PHONE: (1) 48-33-47-06
FAX: (1) 48-33-17-07

FROISSAC PIERRE
93 RUE DES GRANDS CHAMPS
75020 PARIS
PHONE: (1) 43-79-11-34
FAX: (1) 40-24-07-26

GAELET SPECIFIC
4 CITE GRISET
75011 PARIS
PHONE: (1) 43-57-97-69
FAX: (1) 43-38-61-58

GL PIPA
7, RUE BARBES
92300 LEVALLOIS PERRET
PHONE: 47.58.46.56
FAX: 47.48.05.19

HUBERT PIERRE-ALAIN
83 CHEMIN DE LA VALBERELLE
13101 MARSEILLE
PHONE: 91-45-32-23
FAX: 91-44-30-10

INA
4, AV. DE L'EUROPE
94360 BRY SUR MARNE
PHONE: 33.1.49.83.24.87
FAX: 33.1.49.83.25.82

LABEL 35
12 RUE GEORGES ENESCO
94000 CRETEIL LECHAT
PHONE: 42.07.61.00

LASER MOVEMENT
7 ROUTE DE TIGERY
91250 ST.GERMAIN-LES-CORBEIL
PHONE: (1) 60-75-67-09
FAX: (1) 69-89-08-43
TELEX: 270105

LMP
4 BIS RUE DE NANTERRE
92150 SURSNES
PHONE: (1) 42-04-45-45

LUMEX CINEMA
24 AVENUE EDOUARD-HERRIOT
92350 LE PLESSIS-ROBINSON
PHONE: 46301326
TELEX: 270105

MAC GUFF LIGNE
4, PASSAGE DE LA MAIN D'OR
75004 PARIS
PHONE: 43.38.44.55

MARCHETTI PAUL MARCHETTI MIRELLE
1 RUE DE CHATILLON
92170 VANVES
PHONE: (1) 46-38-29-00
FAX: (1) 46-44-48-23

MARTIN ET BOSCHET FILMS BOSCHET MICHEL
8 RUE DES MOULINS
75001 PARIS
PHONE: (1) 42-96-98-74

MAX
28 RUE KLEBER
93100 MONTREUIL
PHONE: (1) 48-57-26-58

MEGA HERTZ
35 RUE VOLTAIRE
1000 TROYES
PHONE: 45-74-10-50

MIKROS IMAGE
7 RUE BISCORNET
75012 PARIS
PHONE: (1) 43-42-21-22
FAX: (1) 43-47-44-84
TELEX: 214057

MPS INTERNATIONAL
29 RUE GAMBETTA
93100 MONTREUIL
PHONE: (1) 48-70-08-32
FAX: (1) 48-59-77-10

MOLE RICHARDSON (FRANCE) S.A.
28-28 BIS, RUE MARCELIN, BERTHELOT
92120 MONTROUGE
PHONE: 47359798
TELEX: 632677 F
FAX: 46653355

MOSAIQUE
1, QUAI GABR EL PERI
94340 JOINVILLE LE PONT
PHONE: 48.85.41.25

NEOVISION
14 AV DE LA PAIX
92170 VANVES
PHONE: (1) 46-38-84-84

NEW TONE
2 RUE DE LA ROQUETTE
75011 PARIS
PHONE: (1) 43-38-59-59
FAX: (1) 43-38-11-99

OXYGENE
105 AV. PARMENTIER
75011 PARIS
PHONE: (1) 48-06-06-13
FAX: (1) 43-57-89-38

PASSE-MURAILLE ATELIER
23 RUE DES FEDERES
93100 MONTREUIL
PHONE: (1) 48-59-03-04
FAX: (1) 48-59-53-65

PICHERIT PATRICK
6 ALLEE JBAINVILLE
94300 VINCENNES
PHONE: (1) 43-65-51-44

PIXI BOX
26, RUE BERTHOLLET
94110 ARCUEIL
PHONE: 49.85.17.18
FAX: 49.85.16.96

PLASTIC STUDIO
102 RUE ROBESPIERRE
93190 BAGNOLET
PHONE: (1) 43-64-50-57
FAX: (1) 43-64-26-72

PRODUCTION MEDIA 6
16, RUE BOUSAIROLLES
34000 MONTPELLIER
PHONE: 67.58.17.91
FAX: 67.92.52.8

PROJECT IMAGES
CHEMIN DES PRES - ZIRST
38240 MEYLAN
PHONE: 76.41.14.18
FAX: 76.41.13.76

PUBLICIS
133, CHAMPS ELYSEES
75008 PARIS
PHONE: 47 20.78.00

PYROSCENE
4 RUE BELLANGER
92300 LEVALLOIS
PHONE: (1) 42-70-35-39
FAX: (1) 47-56-98-42
TELEX: 612046

RELIEF (SYN X)
12-20, RUE VOLTAIRE
93100 MONTREUIL
PHONE: 48 57.91.59

RIFF PRODUCTION
39, RUE CENSIER
75005 PARIS
PHONE: 43 31.44.11

ROUSSEL ALAIN
34 RUE CLAUDE TERRASSE
75016 PARIS
PHONE: (1) 45-25-31-89
FAX: (1) 40-50-01-82

ROUX DIDIER
19 RUE PIERRE GUIGNOIS
94200 IVRY-SU-SIENNE
PHONE: (1) 45-82-95-25
FAX: (1) 46-70-59-09

RUGGIERI
164 ROUTE DE REVEL
31029 TOLLOUSE CEDEX
PHONE: 61-20-11-24
FAX: 61-34-99-74
TELEX: 520959

RUGGIERI
BD DE VELLERON
84170 MONTEUX
PHONE: 90-366-32-22
TELEX: 431380

S F X 12A
11 RUE DE CAMRAI
75019 PARIS
PHONE: 40383000
FAX: 40381650

SNARC STE NELLE APPLIC. LUCIEN PERINI
1 RUE DAMIENS
92100 BOULOGNE
PHONE: (1) 46-21-17-13

SOCIETE POUR L'EXPLOITATION DES EFFETS SPECIAUX
50 QUAI DU POINT DU JOUR
92100 BILLANCOURT
PHONE: 6099324

SOIREE SERVICE
18 RUE DE LA VEGA
75012 PARIS
PHONE: (1) 43-44-26-26

STUDIO BASE 2
134, ROUTE DE BORDEAUX
16000 ANGOULEME
PHONE: 45.92.84.11

STUDIO FOUCHER
32 RUE JEAN JAURES
92260 FONTENAY-AUX-ROSES
PHONE: (1) 46-83-91-16

TALENTON CHRISTIAN
20 BIS RUE DR EMILE ROUX
92110 CLICHY
PHONE: (1) 42-70-99-97

TEST
31 RUE LEDRU ROLLIN
92150 SURESNES
PHONE: (1) 47-72-80-05

TOMAWAK
15 RUE DU BUISSON ST. LOUIS
75010 PARIS
PHONE: (1) 42-01-56-49

TRIELLI PAUL
28 RUE GAGNEE
94400 VITRY
PHONE: (1) 46-71-58-14

TRUQUE
10, RUE BELLANGER
92300 LEVALLOIS PERRET
PHONE: 47.39.06.37

UPTECH
49 RUE CALMETTE GUERIN
78500 SARTROUVILLE
PHONE: (1) 39-57-64-69
FAX: (1) 39-57-36-29

VICTORINE COTE D'AZUR, LA
16 AVENUE EDOUARD-GRINDA
06200 NICE
PHONE: 93725454
TELEX: 93719173

VIDEO CENTRE INTERNATIONAL
13 RUE BEETHOVEN
75016 PARIS
PHONE: (1) 45-24-43-13
TELEX: 630487

VIDEO RIFF
39 RUE CENSIER
75005 PARIS
PHONE: (1) 43-31-44-11
FAX: (1) 47-07-58-22
TELEX: 206470

VIDEOSYSTEM
107, RUE DU FG SAINT HONORE
75008 PARIS

VIVA SYNTHESE
90 AVE ANDRE MORIZET
92100 BOULOGNE
PHONE: 46.03.20.50

WICHEGROD LAURENCE
75 RUE GEORGES LARDENNOIS
75019 PARIS
PHONE: (1) 42-39-14-29

Z PROFESSIONNEL
8 RUE BORIE
33300 BORDEAUX
PHONE: 56-81-49-99
FAX: (56-79-16-10

ZA PRODUCTION
128, BD RICHARD LENOIR
75011 PARIS
PHONE: 48.06.65.66

SPECIAL EFFECTS: DECORATION

AAZ
40 RUE DE FRESNES
94240 L'HAY-LES-ROSES
PHONE: 46-60-90-23

A.B. DECORATION
32 RUE ANDRE KARMEN
93300 AUBERVILLIERS
PHONE: 43-52-40-75
FAX: 43-52-75-08

A.B.C.D. (ATELIER DE LA BELETERIE, CONCEPT ET DECO)
LA BELETERIE
24300 ST.-FRONT LA RIVIERE
PHONE: (16) 53-56-61-45

AND MAX
28 R. KLEBER
93100 MONTREUIL
PHONE: 48-57-26-58

ARCHIMEDE INTERNATIONAL
7 AV. STEPHANE MALLARME
75017 PARIS
PHONE: 47-66-25-73
FAX: 42-67-87-84

ARDECO
19 R. ROQUEPINE
ART-SCENE
107 R. ANTOLDE DE LA FORGE
59000 LILLE
PHONE: (16) 20-33-21-01

ART TECHNIQUE VISUEL
16 R. DES BLEUETS
94000 CRETIEL
PHONE: 48-98-52-93

ATELIER 287
287 FBG. ST. ANTIONIE
75011 PARIS
PHONE: 43-72-06-29
FAX: 43-72-25-15

ATELIER DYANES
129 FBG DU TEMPLE
75010 PARIS
PHONE: 42-41-17-36

ATELIER FREMENT
3 IMP. DU RUISSUEAU
93170 BAGNOLET
PHONE: 48-57-19-25

ATELIER GILOD-EZZOUBIR
3 BD BIRON
93400 ST.OUEN
PHONE: 40-12-78-80

ATELIERS MICHEL SOUBEYRAND
2 R. DE TLEMCEN
75020 PARIS
PHONE: 47-97-23-00

CAZANEUVE GUY
91 QUAI DE LA GARE
75013 PARIS
PHONE: 45-82-10-70

CHIMENTO YVES
LA MAISON DE L'ABEILLE, RTE DE DIGNE
04500 RIEZ
PHONE: (16) 92-77-84-15

DESIGN AND CO.
175 AV. FRANCIS TONNER
06150 CANNES LA BOCCA
PHONE: (16) 93-48-21-20

EXPLORER FILMS
10 BD DE LA VILLETTE
75019 PARIS
PHONE: 43-85-48-26
FAX: (43-85-48-32

FOG AND SMOKE
20 BIS, R. DU DR. EMILE ROUX
92110 CLICHY
PHONE: 42-70-99-97
FAX: 47-37-62-58

ILHERO
MAISON CAMOU
65120 LUZ ST. SAUVEUR
PHONE: (16) 62-92-71-91

KOPERA ROGER
8 R. DIDOT
75014 PARIS
PHONE: 45-43-96-92

L.M.C.
72 AV. DU MAL FOCHE
78000 VERSAILLES
PHONE: 39-53-88-50

L.M.P.
4 BIS R. DE NANTERRE
92150 SURESNES
PHONE: 42-04-45-45
FAX: 45-06-46-58

LE SINGE QUI BOUGE
73 AV. GALLINE
69100 VILLEURBANNE
PHONE: (16) 78-89-09-60

MAGIC DREAM FACTORY
PLATEAU DU PIQUET
34790 GRABELS
PHONE: (16) 67-63-45-45

MAX R. WHITTLE
28 R. KLEBER
93100 MONTREUIL
PHONE: 48-57-26-58
FAX: 48-57-46-48

PLASTIC STUDIO
102 R. ROBESPIERRE
93170 BAGNOLET
PHONE: 43-64-50-57

PLASTILEX SA
149 R. DE ROSNY
93100 MONTREUIL
PHONE: 48-57-46-07
FAX: 48-57-20-03

PUBLIDECOR
6 R. COLBERT
93100 MONTREUIL
PHONE: 42-87-54-36
FAX: 42-87-01-79

SPECIAL EFFECTS: MAKE-UP (MASQUES, PROTHESIS, ETC)

ATELIER DYANES
129 FBG DU TEMPLE
75010 PARIS
PHONE: 42-41-17-36

ATELIER GILOD-EZZOUBIR
3 BD BIRON
93400 ST. OUEN
PHONE: 40-12-78-80

ATELIER INTERNATIONAL DE MAQUILLAGE
34-36 R. DE LA FOLIE REGNAULT
75011 PARIS
PHONE: 43-48-47-46

ATELIERS MICHEL SOUBEYRAND
2 R. DE TLEMCEN
65020 PARIS
PHONE: 47-97-23-00

CAZANEUVE GUY
91 QUAI DE LA GARE
75013 PARIS
PHONE: 45-82-10-70

KOPERA ROGER
8 R. DIDOT
75014 PARIS
PHONE: 45-43-96-92

PLASTIC STUDIO
102 R. ROBESPIERRE
93170 BAGNOLET
PHONE: 43-64-50-57

SAINT-SEAUVAGE DENIS
74 BD MAGENTA
75010 PARIS
PHONE: 40-34-18-12

SPECIAL EFFECTS: MATERIAL

ADB SOFAIR S.A.
47 RUE DE LA VANNE
92120 MONTROUGE
PHONE: (1) 42-53-14-33
FAX: (1) 42-53-54-76
TELEX: 206428

AMERICAN SUPPLY CORPORATION
19 AVENUE DE BERLINCAN
33160 ST MEDARD EN JALLES
PHONE: 56-05-44-44

ART TECHNIQUE VISUEL
16 T. DES BLEUETS
94000 CRETEIL
PHONE: 48-98-52-93

ATKIS
2 AV. JEAN MOULIN
94120 FONTENAY SOUS BOIS
PHONE: 48-73-83-90
FAX: 43-94-02-61

BICE
ROUTE DE ST BERNARD
06220 VALLAURIS
PHONE: (93-64-29-06
TELEX: 970181

CABESTAN LASE MOUVEMENT
57 Z.A. LE TROU GRILLON
91280 ST. PIERRE DU PERRAY
PHONE: 60-75-67-27
FAX: 69-89-08-43

CINECO
72 AV. DES CHAMPS-ELYSEES
75008 PARIS
PHONE: (1) 53-59-61-59
FAX: (1) 42-25-92-03
TELEX: 640346

DAYNES ATELIER
129 RUE DU FG. DU TEMPLE
75010 PARIS
PHONE: (1) 42-41-17-36

DYNACORD FRANCE
77 BD DE MENILMONTANT
75011 PARIS
PHONE: (1) 43-57-00-30
TELEX: 230798

EURODROP
171 RUE VERON
94140 ALFORTVILLE
PHONE: (1) 43-75-08-31
FAX: (1) 43-68-19-76
TELEX: 213905

FOR-A
C/O TECHNI-CINE PHOT
64 BIS, BD JEAN JUARES BP 90
93402 ST.OUEN CEDEX
PHONE: 40-11-81-81
FAX: 40-10-17-27
TELEX: 234 959 MM L-L

GALATEC
97, R. SENOUQUE ZI CENTRE
78530 BUC
PHONE: 39-56-07-63

LA VICTORINE COTE DAZUR
16 AV. EDOUARD GRINDA
06200 NICE
PHONE: 93-72-54-54
FAX: 93-71-91-73

LUMEX CINEMA
24 AV. EDOUARD HERRIOT
92350 LE PLESSIS ROBINSON
PHONE: 46-30-13-26
FAX: 46-32-85-30

LUMILUX
3 RUE D ALEXANDRIE
75002 PARIS
PHONE: (1) 42-33-15-56
TELEX: 212133

MOLE RICHARDSON INTERNATIONAL
28 BIS, R. M-BERTHELOT
92120 MONTROUGE
PHONE: 47-35-97-98

MULITCOM VIDEO
12 R. DE LORRAINE
92300 LEVALLOIS-TERRET
PHONE: 47-31-21-31
FAX: 47-30-19-15

PLUS TRENTE
37 RUE DES ANNELETS
75019 PARIS
PHONE: (1) 42-02-21-02

PROEQUIPEMENT/SOFRAPAV
20 AV. DE ROSNY
93360 NEUILLY-PLAISANCE
PHONE: 43-09-90-90
FAX: (1) 43-09-91-38

PUBLISON
18 AV. DE LA REPUBLIQUE
93170 BAGNOLET
PHONE: (1) 43-60-84-64
FAX: (1) 43-60-80-31
TELEX: 250303

PULSAR FRANCE
10 AV DU FRESNE
14760 BRETTEVILLE-SUR-ODON
PHONE: 31-74-10-01
FAX: 31-73-47-29
TELEX: 171237

RESONANCES
21 RUE DE QUEBEC
1400 CAEN
PHONE: 31-73-01-95

RUIZ PAUL
53 QUAI BLANQUI
94140 ALFORTVILLE
PHONE: (1) 43-75-79-13

SAC ELEC
73 AV. LAFERRIERE
94000 CRETEIL
PHONE: (1) 42-07-47-15
TELEX: 210311

SOGECOM CIDEL FRANCE
11 RUE FRANQUET
75015 PARIS
PHONE: (1) 45-30-09-60

S.O.N.I.S. LILLE
Z.A. DE LA PLAINE R. DES
ECURIES
59650 VILLENEUVE D'ASCQ
PHONE: (16) 20-64-07-07
FAX: (16) 20-64-03-64

SONOR VIDEO SON
15 PLACE DE REIGNAUX
59800 LILLE
PHONE: 20-06-20-09
TELEX: 120945

STANDBAI
63 RUE MOUZAIA
75019 PARIS
PHONE: (1) 42-49-78-98
FAX: (1) 42-40-97-98
TELEX: 215614

STEFANO NECCHI
13 RUE DUPERRE
75009 PARIS
PHONE: (1) 45-26-35-85

STUDER FRANCE
12-14 RUE DESNOUETTES
75015 PARIS
PHONE: (1) 45-33-58-58
FAX: (1) 45-33-46-07
TELEX: 204744

TECHNI VIDEO
64 BIS BD JEAN JAURES
93402 ST. OUEN CEDEX
PHONE: (1) 40-11-80-80
FAX: (1) 40-10-17-27
TELEX: 650959

3M FRANCE
BD DE L'OISE
95006 CERGY-PONTOISE CEDEX
PHONE: (1) 30-31-64-40
FAX: (1) 30-31 64-67
TELEX: 695185

TRAMETAL
21 QUAI DE BONNEUIL
94100 ST. MAUR
PHONE: 48-83-38-82
FAX: 48-83-09-59

WALSH ELECTRONICS
67 RUE HENRI BARBUSSE
92110 CLICHY
PHONE: (1) 42-70-56-47
FAX: (1) 42-70-92-55
TELEX: 612785

SPECIAL EFFECTS: VIDEO

ACET JEAN-PIERRE CORSIA
10 CITE D'ANGOUISEME
75011 PARIS
PHONE: (1)48-07-27-87
FAX: (1) 40-21-09-40

**ANTIGONE PRODUCTION B.
MOUNIER**
38 BD DE COURCELIES
75017 PARIS
PHONE: (1) 47-66-10-32

ANV PRODUCTOINS
20 RUE DE L'HOTEL DE VILLE
92200 NEUILLY
PHONE: (1) 47-47-77-75
FAX: (1) 47-45-07-76
TELEX: 612491

**ASTV ALBAL SAINVAL JEAN-
JACQUES BRISSIAUD**
14 RUE STE-FAMILLE
78000 VERSAILLES
PHONE: (1) 30-21-36-05

**CENTREVILLE PRODUCTIONS
DANIEL DENIS**
12 RUE LOCUEE
75012 PARIS
PHONE: (1) 43-42-47-44

CHANNEL 80
4 RUE PIERRE EROSSOLETTE
92250 LA GARENNE-COLOMBES
PHONE: (1) 47-80-72-44
TELEX: 614579

CITYMAGE
109 BIS RUE DU MONT CENIS
75018 PARIS
PHONE: (1) 42-23-63-63
FAX: (1) 42-58-62-09

EAG VIDEO
27 RUE DU MANS
92400 COURBEVOIE
PHONE: (1) 43-34-31-10
FAX: (1) 43-34-25-58
TELEX: 613600

**FAHRENHEIT 601 PIERRE
DREYFUS**
10 RUE BOYER BARRET
75014 PARIS
PHONE: (1) 40-44-88-76
FAX: (1) 40-44-91-27
TELEX: 206251

FRANCE CONNECT
851 RUE DE BERNAU
94500 CHAMPIGNY-SUR-
MARNE
PHONE: (1) 47-06-09-09
FAX: (1) 48-82-47-57

GL PIPA
7-9 RUE BARBES
92300 LEVALLOIS
PHONE: (1) 47-58-46-56
FAX: (1) 47-58-05-19
TELEX: 615214

IDENEK
44 RUE DE SILLY
92100 BOULOGNE
PHONE: (1) 46-05-66-66
TELEX: 202753

IMAGE IN
46 RUE DE LA FONTAINE AU
ROI
75011 PARIS
PHONE: (1) 48-05-40-05

IMAGE INTECRALE
27 RUE DU MANS
92400 COURBEVOIE
PHONE: (1) 43-34-31-10

IMAGE RESOURCE
27-29 RUE BLOMET
75015 PARIS
PHONE: (1) 43-06-18-67
TELEX: 206438

IMEXPO
851 RUE DE BERNAU
94500 CHAMPIGNY-SUR-
MARNE
PHONE: (1) 47-06-09-09
FAX: (1) 48-82-47-57

NEOVISION
14 AV. DE LA PAIX
92170 VANVES
PHONE: (1) 46-38-84-84

PIPA VIDEO
14 BIS RUE BARBES
92120 MONTROUGE
PHONE: (1) 46-56-70-17

PRESTIGE VIDEO FILM
14 AV. DE TOURVILLE
75017 PARIS
PHONE: (1) 47-05-35-35

PUMA JEAN MARC RIPERT
133, 135 RUE D'AGUESSEAU
92100 BOULOGNE
PHONE: (1) 46-05-31-29

SOCAM VIDEO
171 AV. CHARLES DE GAULLE
92200 NEUILLY
PHONE: (1) 47-47-05-40
TELEX: 614835

TELETOTA
9 BIS RUE CAT PILOT
92200 NEUILLY
PHONE: (1) 46-24-13-10

TRANS COLOR VIDEO
15 RUE SIMON DERUERE
75018 PARIS
PHONE: (1) 42-55-24-24

TRANSATLANTIC VIDEO
6 RUE DES DEUX PONTS.
75004 PARIS
PHONE: (1) 43-26-14-58
TELEX: 203675

TRUQUE
10 RUE BELLANGER
92300 LEVALLOIS-PERRET
PHONE: (1) 47-39-06-37

VIDEO PLUS
21 RUE DE CLICHY
93584 ST. OUEN CEDEX
PHONE: (1) 40-11-39-39
FAX: (1) 40-11-30-34
TELEX: 234912

**VIDEO SCOPIE PRODUCTION
ELISABETH BUNIO**
10 ALLE VERTE
75011 PARIS
PHONE: (1) 47-00-70-36

EXPENDABLES

CINECAM
147 RUE MICHEL CARRE
95100 ARGENTEUIL
PHONE: 1-30763536
FAX: 1-30760609

**PER-FIX COMPANY, THE
DISTRIBUTORS: ZENON**
4 RUE PHILIDOR
75020 PARIS
PHONE: 43726392

FILM COMMISSIONS

**ACADEMIE DES ARTS ET
TECHNIQUE DU CINEMA**
19 AVENUE DU PRESIDENT,
WILSON
75016 PARIS
PHONE: 47237233

**AGENCE POUR LE
DEVELOPPMENT REGIONAL
DU CINEMA**
29 RUE DU COLISEE
75008 PARIS
PHONE: 45611786

ARCHIVES DU FILM
7 BIS RUE ALEXANDRE
TURPAULT
78390 BOIS D'ARCY
PHONE: 34602050

BIBLIOTHEQUE DE L'IDHEC
9 AVENUE ALBERT DE MUN
75016 PARIS
PHONE: 47270632

C.R.E.T.E. INTERNATIONAL
116 AVENUE DU PRESIDENT
KENNEDY
75790 PARIS CEDEX 16
PHONE: 42302905
45240373
RE NATIONAL D'ART ET DE LA
CULTURE
75191 PARIS CEDEX 04
PHONE: 42771233

**CENTRE NATIONAL DE LA
CINEMATOGRAPHIE (C.N.C.)**
12 RUE DE LUBECK
75784 PARIS CEDEX 16
PHONE: 45051440
TELEX: 650306 C.N.-CINE

CINEMATHEQUE DE TOULOUSE
12 RUE DU FAUBOURG
BONNEFAY
31000 TOULOUSE
PHONE: 61489075

CINEMATHEQUE FRANCAISE, LA
29 RUE DU COLISEE
75008 PARIS
PHONE: 45532186

**COMMISSION DES AVANCES
SUR RECETTES**
11 RUE GALILEE
75016 PARIS
PHONE: 45051440
TELEX: 650306 C.M. -CINE

**FOUNDATION
INTERNATIONALE DU CINEMA
ET DE**
LA COMMUNICATION
AUDIOVISUELLE
68 AV DU PETIT JUAS
06400 CANNES
PHONE: 93683335

**INSTITUT NATIONAL DE LA
COMMUNICATION**
AUDIOVISUELLE
193-197 RUE DE BERCY
75012 PARIS
PHONE: 40046400
48758585

INTERMEDIA
19 RUE DE PASSY
75016 PAR S
PHONE: 42246823

**MINISTERE DE LA CULTURE ET
DE LA COMMUNICATION**
3 RUE DE VALOIS
75001 PARIS
PHONE: 40158000

**U.P.F. - UNION DES
PRODUCTEURS DE FILMS**
1 PLACE DES ECUS
75001 PARIS
PHONE: 40280133

**U.P.F. UNION DES
PRODUCTEURS**
1 PLACE DES ECUS
75001 PARIS
PHONE: 40280138

**UNIFRANCE FILM
INTERNATIONAL**
114 CHAMPS-ELYSEES
75008 PARIS
PHONE: 43590334
TELEX: 290358

MAKE-UP SUPPLIERS

ALMATRONIC (STE)
Z.A. DES PETITS CARREAUX, 4
AVE DES COQUELICOTS
94386 BONNEUIL SUR MARNE
PHONE: 43772506

APPARENCES AGENCY
4 RUE DU FGB ST. HONORE
75008 PARIS
PHONE: 4265001

ATELIER INT. DE MAQUILLAGE
34-36 R. DE LA FOLIE-
REGNAULT
75011 PARIS
PHONE: 43434746

BOGARD S.A.
131 RUE DE L'UNIVERSITE
75007 PARIS
PHONE: 1-45561191
FAX: 1-4551884

DANIEL BLANC PERRIQUIER
11 RUE DES PETITES ECURIES
75010 PARIS
PHONE: 45230283

DENIS ST. SAUVAGE
11 RUE DE MOGADOR
75009 PARIS
PHONE: 48749642

**KRYOLAN PROFESSIONAL
MAKE UP**
34/36 RUE DE LA FOLIE-
REGNAULT
75011 PARIS
PHONE: 43484746

**KRYOLAN PROFFESIONAL
MAKE UP LITHEA**
101 GRAND RUE
F-67000 STRASBOURG
**LA SOCIETE FRANCAIS DE
MAQUILLAGE SA**
61 AVE F. D. ROOSEVELT
75008 PARIS
PHONE: 42563428
LEICHNER (FARDS)
11 BIS RUE DU COLISEE
75008 PARIS
PHONE: 42250541
**MERLE (PARFUMERIE
LUCIENNE)**
9 AVENUE MATIGNON
75008 PARIS
PHONE: 42567070
PARFUMERIE DES VEDETTE
85 RUE DU FAUBOURG
SAINT-DENIS
75010 PARIS
PHONE: 48244372
PARIS BERLIN
30 RUE CHAPTAL
75009 PARIS
PHONE: 45263929
**PRODUITS "MICHEL
DERUELLE"**
STE FRANCAISE DE
MAQUILLAGE
61 AVENUE FRANKLIN
ROOSEVELT
75008 PARIS
PHONE: 42563428
ROSIER (PASCAL)
10 RUE DU RENDEZ-VOUZ
75012 PARIS
PHONE: 43443804
S.V.L. PARIS
22 RUE PROUDHON
93210 LA PLAINE SAINT -DENIS
PHONE: 1-48092020
TELEX: 231302 F
FAX: 1-48098392
**SOCIETE FRANCAISE DE
MAQUILLACE**
61 AVENUE FRANKLIN
ROOSEVELT
75008 PARIS
PHONE: 42560552
      42563428
TELEX: SMF 643701 F
VISIORA
131 RUE DE L'UNIVERSITE
75007 PARIS
PHONE: 45561191
TELEX: 250723 BOGARD PARIS
WIG STUDIOS S.A.
45 RUE DE LILLE
75007 PARIS
PHONE: 42616960

PROPS

ALCOPLAST
9 ALLEE PROGRES
92170 VANVES
PHONE: 6427921
ARTISANS RECUPERATEURS
8 LMP. SAINT-SEBASTIEN
75011 PARIS
PHONE: 47005771
      43556650
ATELIER 287
287 FG SAINT-ANTOINE
75011 PARIS
PHONE: 43720629
ATELIER DENIS POULIN
10 CITE TREVISE
75009 PARIS
PHONE: 47701712
      42469861

ATELEIR GILOD-EZZOUBIR
3 BOULEVARD BIRON
93400 ST-OUEN
PHONE: 40127880
AURAN AUTOMOBILE
13550 NOUES
PHONE: 90940014
AUTO-CARNO
14 RUE J.P. BENARD
91200 ATHIS-MONS
PHONE: 60485045
BOULANGER (GIL)
31 AVENUE FAIDHERBE
93310 LE PRE-SAINT GERVAIS
PHONE: 8447031
CATILLON LOCATION
63 BOULEVARD DE LA LIBERTE
72322 CHATILLON S/BAGNEUX
PHONE: 4253500
CHAMOURAT, PATRICK
116 RUE DE RIVOLI
75001 PARIS
PHONE: 42364021
**CINE LIMOUSINES
AUTOMOBILES**
26 RUE ARMAND SILVESTRE
92400 COURBEWIE
PHONE: 43336030
CINEMATHEQUE FRNCAISE, LA
29 RUE DU COLISEE
75008 PARIS
PHONE: 45532186
CRESPI, JEAN CHARLES
104 AVE. DE LA REPUBLIQUE
94800 VILLEJUIF
PHONE: 1-6022460
FAX: 1-60224164
DEFRISE ET CIE
23 RUE BASFROI
75011 PARIS
PHONE: 43797829
DUBUIS (CHRISTIAN)
9 RUE DE L'EGLISE
77450 LE PLESSI FEU AUSSOUS
PHONE: 64041602
ECURIES HARDY
FREME DE SAUVAGE - EMANCE
78120 RAMBOUILLET
PHONE: 1-34859204
FAX: 1-34859416
ECURIES J.R. COUTURE
77220 LIVERDY EN BRI
PHONE: 1-64255107
      1-48833207
FILMS OSIRIS
42 BIS RUE MOLIERE
93100 MONTREUIL
PHONE: 48578237
GASTINNE-RENETE
39 AVENUE FRANKLIN-
ROOSEVELT
75008 PARIS
PHONE: 43597774
**GERMAIN LOCATION AUTOS-
CINEMA S.A.R.L.**
61 ROUTE D'ARGENTEUIL
95240 CORMEILLES EN PARISIS
PHONE: 1-34509494
FAX: 1-34507301
GUIFFRANCE
88 RUE DES PETITS BOIS
78370 PLAISIR
PHONE: 34811056
JEANNOT
9 RUE LOUISE MICHEL
92300 LEVALLOIS
PHONE: 47575320
KOPERA (ROGER)
8 RUE DIDEROT
75014 PARIS
PHONE: 42054499
LANZANI (GAETAN)
19-21 RUE BASFROI
75011 PARIS
PHONE: 43790074
TELEX: 670360 F LAFILPA

MARY-STAFF
6 RUE DE LA TIRELIRE
5100 REIMS
PHONE: 26473550
MASSONI (JEAN-PAUL)
7 RUE DEFLY
06000 NICE
PHONE: 93-770365
**MUSEE DE L'AUTOMOBILE DE
LA SARTHE**
"CIRCUIT DES 24 HEURES", LES
RAINERIES 19X
72040 LE MANS CEDEX
PHONE: 43725066
PELEGRY (ANDRE)
20 RUE CHASSAGNOLLE
93260 LE LILAS
PHONE: 43616700
PUBLIDECOR
6 RUE COLBERT
93100 MONTREUIL
PHONE: 42875436
FAX: 42870179
RYB (ATELIERS)
36 AVENUE BRIGOLLE
93700 DRANCY
PHONE: 48300115
REPERAGE FRANCAIS, LE
18 RUE VIGNON
75009 PARIS
PHONE: 42651550
ROSIER (PASCAL)
10 RUE DU RENDEZ-VOUS
75012 PARIS
PHONE: 43443804
RUGGIERI (ETS)
164 ROUTE DE REVEL
31029 TOULOUSE , CEDEX
PHONE: 61-201124
S.A.D.
45 RUE FESSART
75019 PARIS
PHONE: 42094053
SAFARI CINE
59 RUE EDITH CAVELL
9440 VITRY/SEINE
PHONE: 46820624
FAX: 46808538
**SCULPTURE ET DECORATION
(STE FSE DE)**
54 BIS RUE DE L'AMIRAL
ROUSSIN
75015 PARIS
PHONE: 48282671
      49227101
**SOCIETE FRANCAISE
D'ATTELAGES HIPPOMOBILES**
(M. DEMARET JACQUES), 30
RUE GABRIEL REBY
95870 BEZONS
PHONE: 39829029
**SOCIETE PARISIENNE
D'EQUIPEMENT ET DE
DIFFUSION**
35 RUE CREVECOEUR
93120 LA COURNEUVE
PHONE: 48330667
STE MARAJIER, JEAN CHARLES
15 RUE DU PLATEAU
75019 PARIS
PHONE: 42404242
FAX: 42403990
STE PROMOBILE
91 AV. JEAN -BAPTISTE
CLEMENT
92100 BOULOGNE
PHONE: 48258833
FAX: 48250604
TRIELLI (PAUL)
28 RUE GAGNEE
94400 VITRY
PHONE: 46715814
VIDEO FILM DECOR
11 VILLA NIEUPORT
75018 PARIS
PHONE: 45843983

PYROTECHNICS

BERASTEGUI-ROLLET
55 RUE SAINT-ANTOINE
1600 ANGOULEME
BOLLENGIER, ROGER
12 RESIDENCE DES GEMEAUX,
RUE AUGUSTE DAIX
94260 FRESNES
PHONE: 42374996
BOULANGER (GIL)
31 AVENUE FAIDHERBE
93310 LE PRE-SAINT GERVAIS
PHONE: 8447031
DEFRISE ET CIE
23 RUE BASFROI
75011 PARIS
PHONE: 43797829
JEANNOT
9 RUE LOUISE MICHEL
92300 LEVALLOIS
PHONE: 47575320
MARATIER, JEAN CHARLES
15 RUE DU PLATEAU
75019 PARIS
PHONE: 42404242
FAX: 42403990
REGIFILM SARL
60 RUE AMELOT
31029 PARIS
PHONE: 43555255
FAX: 43555146
RUGGIERI (ETS)
164 ROUTE DE REVEL
31029 TOULOUSE, CEDEX
PHONE: 61-201124
TRIELLI (PAUL)
28 RUE GAGNEE
94400 VITRY
PHONE: 46715814

STUDIOS & SOUND
STAGES

ACME FILMS
24-26 RUE PAUL FORT
75014 PARIS
PHONE: 45436657
ASTRE
103 RUE ST. DOMINIQUE
75007 PARIS
PHONE: 45519991
ATELIER BRETAGNE FILM
KERLAMEN
29120 PLONEOUR-LANVERN
PHONE: 98871379
AUDIO VIDEO FRANCE
1, BIS RUE THEODULE-RIBOT
75007 PARIS
PHONE: 47.66.26.66
AUDITORIUM JEAN MERMOZ
7, RUE JEAN-MERMOZ
75008 PARIS
PHONE: 42.25.01.80
AUX SIECLES PASSES
57 RUE DE LA ROQUETTE
75011 PARIS
PHONE: 43558007
AVIA FILMS
31, QUAI D'ANJOU
75004 PARIS
PHONE: 43.29.92.20
**BUREAUX A PARIS: L.T.M. 102-
104 BOULEVARD ST. DENIS**
92400 COURBEVOIE
PHONE: 1-47884450
TELEX: 630277 F
CAIMAN
30 BOULEVARD DE LA BASTILLE
75012 PARIS
PHONE: 4344112
CHRISMAX FILMS
50, QUAI DU POINT-DU-JOUR
92100 BOULOUGNE
PHONE: 46.09.93.24

CIE DE TRAVAUX MECANIQUES
84 RUE PAUL-VALLIANT-
COUTURIER
92300 LEVALLOIS
PHONE: 47374824

COSMOS
7 RUE SENTON
92510 SURESNES
PHONE: 45061880

DAYLIGHT STUDIOS
7 RUE MORET
75011 PARIS
PHONE: 43382416

DOVIDIS
42, BIS RUE LOURMEL
75015 PARIS
PHONE: 45.79.41.89

DURAN DUPON DUSON
74, RUE JOSEPH DE MAISTRE
75018 PARIS
PHONE: 42.29.00.33

DYNASCOPE
12 RUE CLAVEL
75019 PARIS
PHONE: 46078081

ECLAIR (ESTABLISSMENT
CINEMATOGRAPHIQUES)
2 RUE DU BAC
92158 SURESNES CEDEX
PHONE: 40995040
43596072
FAX: 42047201

ECLAIR STUDIOS
2 RUE DU BAC
92158 SURESNES CEDEX
PHONE: 40995040

EURO-STUDIO
15, RUE FOREST
75018 PARIS
PHONE: 42.93.16.16

EXPOSURE
1 BIS RES EPINETTES
94410 SAINT MAURICE
PHONE: 48934501

IMAGES DE FRANCE
29, RUE VERNET
75008 PARIS
PHONE: 47.20.23.23.

JACANA WILDLIFE STUDIOS
30 RUE SAINT-MARC
75002 PARIS FRANCE
PHONE: 42969914

L.T.C.
48, QUAI CARNOT
92210 SAINT-CLOUD
PHONE: 46.02.70.25
FAX: 49.11.11.67
TELEX: 202 352 F

LE STUDIO
36 RUE DU FER A MOULIN
75005 PARIS
PHONE: 43378096

LES STUDIOS DE BOULOGNE
2, RUE DE SILLY
92100 BOULOGNE-
BILLANCOURT
PHONE: (1) 46.05.65.69
FAX: (1) 48.25.23.47
TELEX: 270061F

MAGIC FILMS PRODUCTION
23 BIS RUE H. SAVIGNAC
92190 MEUDON-BELLEVUE
PHONE: 45341311

MAISON DE LA CHIMIE
(CENTRE MARCELIN
BERTHELOT)
28 ET 28 BIS RUE, SAINT-
DOMINIQUE
75007 PARIS
PHONE: 47051073
TELEX: 200351 CHIMIE PARIS

MANUDECORS
90/92 RUE MOLIERE
94200 PARIS
PHONE: 46584424
FAX: 46708766

ODESSA FILMS
60 RUE LAUGIER
75017 PARIS
PHONE: 40540606
TELEX: 649659

PJV S.A. "LES STUDIOS
33 RUE DU REPOS
69007 LYON
PHONE: 78612547

P.J.V. "STUDIOS CANUBIS"
175 TER BD J. JAURES
92100 BOULOGNE
PHONE: 46033055

PARIS-CITE-PRODUCTIONS
13, RUE DE PARIS
92100 BOULOGNE-
BILLANCOURT
PHONE: 46.03.34.34

PARIS STUDIOS BILLANCOURT
S.A.
50 QUAI DU POINT DU JOUR
92100 BOULOGNE
PHONE: 1-46099324
FAX: 1-46202471

PATHE CYRANO & S.O.R.
6, RUE DE FRANCOEUR
75018 PARIS
PHONE: 42.57.12.10

PIN UP
23 AVENUE JEAN MOULIN
75014 PARIS
PHONE: 45422136

P.M. PRODUCTIONS
27, RUE DESPORTES
93400 SAINT-OUEN
PHONE: 40.12.64.64

PRISE DE VUES
59-65 RUE DES RIGONDES
93170 BAGNOLET
PHONE: 43626741
FAX: 43620008

PUBLI ECRAN
65 RUE G. BOIDIN
59130 LAMBERSART
PHONE: 927007

SONODI
1, CHEMIN DES ANCIENS PRES
93800 EPINAY-SUR-SEINE
PHONE: 48.26.86.86

S.F.P. (STE FSE DE
PRODUCTION ET DE CREATION
AUDIOVISUELLES)
STUDIOS TELEVISION, 36 RUES
DE ALOUETTES
75019 PARIS
PHONE: 42039904
TELEX: 240888

S.T.A.R.T.
47, RUE DE PARADIS
75010 PARIS
PHOEN: 48.24.18.28
FAX: 42.46.64.60
TELEX: 281 838 F BARCLAY

SETS
19-25 CHEMIN-DES-FOURCHES
93240 STAINS
PHONE: 48219284
FAX: 48211251

SOCIETE FRANCE REGIONS (FR3)
(CENTRE DE PRODUCTION
TELEVISION)
11 RUE FRANCOIS LER
75008 PARIS
PHONE: 43595750

SOCEITE NOUVELLE PATHE
CINEMA
6 RUE FRANCOURT
75018 PARIS
PHONE: 42571210

STUDIO 91, STUDIOS QUARTE
VINGT ONZE "LES COCHETS"
91290 ST GERMIAN LES
ARPAJON
PHONE: 1-60849740
TELEX: 681107
FAX: 1-60844417

STUDIO D'ARCUEIL
37 AVENUE DE LA REPUBLIQUE
94110 ARCUEIL
PHONE: 5470330

STUDIOS BATAILLE
109 RUE BATAILLE
69008 LYON
PHONE: 78000821

STUDIOS DE BOULOGNE
137 AVENUE JEAN BAPTISTE,
CLEMENT
92100 BOULOGNE
PHONE: 01-46056569
TELEX: 270061

STUDIOS DE FRANCE
50 AVENUE DU PRESIDENT
WILSON
93214 LA PLAINE ST. DENIS
CEDEX
PHONE: 42431111
TELEX: 232164 SROSDF

STUDIOS DE L'EST
17 RUE D'ESTIENNE D'ORVES
93310 LE PRE ST. GERVAIS
PHONE: 48466605

STUDIOS DU TOURNE-VENT
RUE DE LA ROSE DES VENTS
95610 ERAGNY
PHONE: 7701256

STUDIOS FILMS, 2 AVENUE DE
L'EUROPE
94460 BRY-SUE MARNE
PHONE: 48758585

STUDIOS FRANCOEUR
6 RUE FRANCOEUR
75018 PARIS
PHONE: 42571210

STUDIOS MARCADET
34, RUE MARCADET
75018 PARIS
PHONE: 42.55.43.29

STUDIOS NIEUPORT
11 VILLA NIEUPORT
75013 PARIS
PHONE: 45843983

STUDIOS PACIFIC
17, PASSAGE RIOU
92000 NANTERRE
PHONE: 42.04.06.20

S.P.S. SYNCHRO 7
8, RUE LEREDDE
75013 PARIS
PHONE: 45.83.37.77

TADIE CINEMA
61 BIS, RUE DES PEUPLIERS
92100 BOULOGNE-
BILLANCOURT
PHONE: 46.20.35.16

TELFRANCE
20 ROUTE DE HOUDAN
78610 LE PERRAY-EN-YVELINES
PHONE: 34839280

T.T.V.C.
48, QUAI CARNOT
92210 ST. CLOUD, PARIS
PHONE: 49.11.13.07
TELEX: 632 006

UNIVERSITE DE VALENCIENNES
DEPARTEMENT AUDIOVISUEL,
LE MONT HOUY CHEMIN VERT
59326 VALENCIENNES CEDEX
PHONE: 27-466608

V.T.F. (VIDEO TELE FRANCE)
13 RUE BEETHOVEN
75016 PARIS
PHONE: 45244313
TELEX: 612 167 F VTF

VICTORINE COTE D'AZUR, LA
16 AVENUE EDOUARD-GRINDA
06200 NICE
PHONE 93725454
TELEX: 93719173

WEST SIDE STUDIO
5 RUE DES SUISSES
92000 NANTERRE
PHONE: 47210767

GERMANY

SPECIAL EFFECTS

ANY MOTION SYSTEM
LAISTRASSE 9
6257 HUNFELDEN 2
PHONE: 06438-2314

BAUMGARTNER, KARL
(PYROTECHNICS)
OTTILIENSTRASSE 25
8000 MUNICH 82
PHONE: 089-4309482
089-6499317

BORYSENKO, SASCHA
BRUCKNERSTRASSE 24
8070 INGOLSTADT
PHONE: 0841-85928/34150

CINE EFFECTS
RITSCHI RICHTSFELD & WAGGI,
KLEE, LAUFZORNERSTRT 28
8022 GRUNWALD
PHONE: 089-6412146
089-6411152

CITY SOUND
AUERFELDSTRASSE 22
D-8000 MUNICH 90
PHONE: 089-4480833
089-4480132

DIRK HILLER DESIGN
GUT HOFFNUNGSTHAL
5231 OBERLAHR/ WW
PHONE: 02685-8185
FAX: 02685-8237

GUNTHER SCHAIDT
SAFEX-CHEMIE
BLANKENESER CHALSSE 26/32
2000 SCHENEFELD
PHONE: 040-8300025
FAX: 040-8301452

HAEDLER & HAEDLER
FRIEDR. - ENGELS - EOGEN 14
8000 MUNICH 83
PHONE: 089-6708595

HEINZ, LUDWIG
AM FISCHERWINKEL 2
8022 GRUNWALD
PHONE: 089-6411153

HUPPMANN, LUDWIG F.
RAINTALERSTRASSE 12A
8000 MUNICH 90
PHONE: 089-6925123

LANGE, PETER (PYROTECHNICS)
DR. ROSENMEUER-WEC 4
8022 GRUNWALD/MUNICH
PHONE: 089-641720

ROTTER, PTT (PYROTECHNICS)
SENFLVTR 1A
8000 MUNICH 90
PHONE: 089-7236598
FAX: 089-7239587

SPECIAL EFFECT TEAM
DEUTZ MULHEIMERSTR 146C
5000 KOLN 80
PHONE: 0221-818274

TROPP ROLAND
WALDENSERSTR 9
1000 BERLIN 21
PHONE: 030-3956240

EQUIPMENT RENTALS

ARNOLD & RICHTER CINE
TECHNK GMBH & CO
TURKENSTRASSE 89
8000 MUNICH 40
PHONE: 089-38091
TELEX: 524317 SRRI D
FAX: 089-3909244

DEDO WEIGERT FILM GMBH
ROTTMANNSTRASSE 5
8000 MUNCHEN 2
PHONE: 089-525064
TELEX: 529865
FAX: 089-529173

KRASSER & CO OHG
MUNCHENER STRASSE 95
8043 UNTERFOHRING, NR.
MUNICH
PHONE: 089-950193
TELEX: 522509 KAUZ
PER-FIX CO. DISTRIBUTORS: -
DEDO WIEGERT FILM GMBH
THE
ROTTMANNSTRASSE 5
D-8000 MUNICH 2
PHONE: 089-525064
TELEX: 529865

MAKE-UP SUPPLIERS

KRYOLAN GMBH
PAPIERSTRASSE 10
D-1000 BERLIN 51
PHONE: 030-4911249
TELEX: 182995 KRY D
FAX: 4930-4914994
KRYOLAN PROFESSIONAL
MAKE UP
MANUELA KULLMANN,
STARGANDERRSTR. 46A
2000 HAMBURG 73
PHONE: 040-6470892
KRYOLAN PROFESSIONAL
MAKE UP
SCHONER-KOSMETIK,
KAUFINGER STR 33
8000 MUNICH 2
LEICHNER, L
RHEMBABENALEE 9
BERLIN-DAHLEM
PHONE: 030-891882

PYROTECHNICS

BORYSENKO, SASCHA
BRUCKNERSTRASSE 24
8070 INGOLSTADT
PHONE: 0841-85928/34150
HEINZ, LUDWIG
AM FISCHERWINKEL 2
8022 GRUNWALD
PHONE: 089-6411155
HUPPMANN, LUDWIG F.
RAINTALERSTRASSE 12A
8000 MUNICH 90
PHONE: 089-6925 183
LANGE, PETER (PYROTECHNICS)
DR. ROSENMEUER-WEG 4
8022 GRUNWALD/MUNICH
PHONE: 089-641720
RICHTSFEILD, RICHARD -
CONTRUCTIONS- FIR
LAUFZORNERSTRASSE 28
8022 GRUNWALD (MUNCHEN)
PHONE: 089-6412146
ROTTER, PITT (PYROTECHNICS)
SENFLVTR 1A
8000 MUNICH 90
PHONE: 089-7236598
FAX: 089-7239587

STUDIOS AND SOUND
STAGES

AIRPORTSTUDIO GMBH
WEIBENFELDERSTRABE
8011 HEIMSTETTEN/MUNCHEN
PHONE: 089-9032051
TELEX: 17 898096
FAX: 9032054
ARNOLD & RICHTER CINE
TECHNIK GMBH & CO
TURKENSTRASSE 89
8000 MUNICH 40
PHONE: 089-38091
TELEX: 524317 ARRI D
FAX: 089-3909244

BAVARIA FILM GMBH
BAVARIAFILMPLATZ 7
D-8022 GEISELGASTAEIG
PHONE: 089-6499-0
TELEX: 5218886 DESI G
FAX: 089-6492507
BERLINER UNION-FILM GMBH
& CO STUDIO KG
OBERLANDSTR 26-35
1000 BERLIN 42
PHONE: 030-75941
TELEX: 184233
FAX: 030-7594398
BRUNNER & EISENREICH GMBH
OBERE LANGERSTRASSE 19
8039 PUCHEIM BHF
PHONE: 089-806025-26
FAX: 089-800127
BUHNER, EVA METSTUDIO
WINTERHUDERWEG 112
2000 HAMBURG 76
PHONE: 040-225160
CCC CENTRAL-CINEMA-COMP
GMBH
VERL DAUMSTR 16
BERLIN-SPANDAU
PHONE: 030-380201
CINE GROUP STUDIO &
SERVICE
PRIMELWEG 10-12
D-8192 GERESTRIED
PHONE: 08171-63102
FAX: 08171-6539
DEWE STUDIOS GMBH
SCHELMENACKERSTRASSE 28
7041 HILDRIZHAUSEN
PHONE: 07034-122-0
TELEX: 7265684
FILM UND SHOW-LICHT
GUNTER PUTZ
RICHARD-BYRD-STR 10
5000 KOLN 30 (OSSENDORF)
PHONE: 0221-594481-83
FAX: 0221-591542
GMD
FURTWEG 34/38
8044 UNTERSCHLEISSHEIM BE,
MUNCHEN
PHONE: 089-3101049 X 79
TELEX: 5216957 GMD D
INFO STUDIOS GMBH
ROBERT BASCH STRASSE 3
4019 MONHEIM BAUMBERG
PHONE: 02173-61071/73
TELEX: 8-515640
KUBISCH
ELISABETHSTR 84
4000 DUSSELDORF 1
PHONE: 0211-345041
MIETSTUDIO 54
KUNIGUNDENSTRASSE 54
8000 MUNICH 40
PHONE: 089-366545
PRO STUDIOS FILM VIDEO
KONIGSBERGER STR 7
4000 DUSSELDORF 1
PHONE: 0211-73834-0
TELEX: 8582634
STUDIO DIEHL
KILLERSTRASSE 7
8032 GRAFELFING/MUNICH
PHONE: 089-852048
STUDIO HAMBURG, ATELIER
GMBH
JEFELDER ALLEE 80, POSTFACH
701560
D-2000 HAMBURG 70
PHONE: 040-66880
TELEX: 214218 STH D
FAX: 040-665601
TAUNUS FILM GMBH
UNTER DEN EICHEN
6200 WIESBADEN
PHONE: 06121-53-1
TELEX: 4186807 TFGW D
FAX: 06121-522712/531226

TELEFILM SAAR GMBH
SCHBERGERWEG 65
6000 SAARBRUKEN
PHONE: 0681-810200
TELEX: 04428620 TFS D
VIDEOTHEK ELECTRONIK TV-
PRODUCTION
HAVELCHAUSSEE 161
1000 BERLIN 19
PHONE: 030-3045431
TELEX: 182855
JACOB-SCHICK STR. 17
6503 MAINZ-KASTEL
PHONE: 06134-4031
TELEX: 4182023

UNIONS AND ASSOCIATIONS

ARBEITGEMEINSCHAFT ZUR
NACHWUCHSFORDERUNG FUR
FILM UND FERNSEHEN
ROTHENBAUMCHAUSSE 132/134
2000 HAMBURG 13
AUDIO VISUELLES MEDIEN
ZENTRUM - AVMZ DER
UNIVERSITAT ESSEN
POSTFACH 6843,
UNIVERSITATSSTRASSE 12
4300 ESSEN
PHONE: 0201-1831 & 1833436/38
TELEX: 8579091
BVK - BERUFSVERBAND
KAMERA
ADELHEIDSTRASSE 7
8000 MUNICH 40
PHONE: 089-2716370
BERLINER FESTSPIELE GMBH
BUDAPERSTRASSE 50
1000 BERLIN 30
PHONE: 030-26341
TELEX: 185255 FEST
BUNDESVERBAND DER
FERNSEH - UND
FILMREGISSEURE IN
DEUTSCHLAND E.V.
TURKENSTRASSE 93
8000 MUNICH 40
BUNDESVERBAND DER
PHONOGRAPHISCHEN
WIRTSCHAFT E.V.
KATHARINENSTRASSE 11
2000 HAMBURG 11
PHONE: 040-367513
TELEX: 213456
BUNDESVERBAN DEUTSCHER
FERNSEHPRODUZENTEN E.V.
WIDENMAYERSTRASSE 32
8000 MUNICH 22
PHONE: 089-223535
TELEX: 529070
BUNDESVERBAND DEUTSCHER
FILM - UND AV PRODUZENTEN
E.V.
LANGENBECKSTRASSE 9
6200 WIESBADEN 1
PHONE: 06121-306200
TELEX: 4186639
DIZ - DEUTSCHE
INDUSTRIEFILM - ZENTRALE
GUSTAV-HEINMANN -
UFER 84/88
5000 KOLN 51
PHONE: 0221-372017
TELEX: 8882768
DEUTSCHE FILM UND
FERNSEHADADEMIE GMBH
PMMERNALLEE 1
1000 BERLIN 9
PHONE: 030-30631
DEUTSCHES INSTITUT FUR
FILMKUNDE
SCHAUMAINKAI 41
6000 FRANKFURT

DRAMATIKER-UNION E.V.
BISMARCKSTRASSE 17
1000 BERLIN 12
PHONE: 030-3416030
EXPORT-UNION DES
DEUTSCHEN FILM E.V.
TURKENSTRASSE 93
8000 MUNICH 40
PHONE: 089-390095
TELEX: 5215627
FILM FONDS HAMBURG
IN HAMURGER MEDIENHAWS,
FRIEDENSALLE 14-16
2000 HAMBURG 50
PHONE: 040-3905883
FAX: 040-395495
HAMBURGER FILMBURO E.V.
FRIEDENSALLEE 7
D-2000 HAMBURG 50
PHONE: 040-3904040
TELEX: 2165087 FILM D
INSTITUT FUR FILM UND BILD
IN WISSENSCHAFT
UND UNTERRICHT
GEMEINNUTZIGE
GMBH, ROBERT-KOCH STRASSE
19A
8022 GRUNWALD
PHONE: 089-64971
SPITZENORGANISATION DER
FILMWIRTSCHAFT (SPIO)
LANGENBECKSTR 9
WIESBADEN
PHONE: 06121-307084
VERBAND DER
FILMVERLEIHER E.V
DISTRIBUTORS' ASSOC.,
WISBADEN, LANGENBECKSTR 9
PHONE: 306475/300611
DUSSELDORF, 1 HERZOGSTR 18
PHONE: 37 52 13
HAMBURG 39, SIERICHSTR 70
PHONE: 279 22 55
MUNICH 2, HERZOGSPITALSTR 13
PHONE: 26 73 79
FRANKFURT/ M TANUSSTR 52-60
PHONE: 23 53 01
VERBAND DEUTSCHER
FILMPRODUZENTNE E.V.
GERMAN FILM PRODUCERS'
ASSOC., LANGENBECKSTR 9
WIESBADEN
PHONE: 59051
VERBAND DEUTSCHER
SPIELFILMPRODUZENTEN E.V.
FREYSTR 4
8000 MUNICH 40
VERBAND TECHNISCHER
BETRIEBE FUR FILM UND
FERNSEHEN E VTFF
KURFURSTENDAMM 179
1000 BERLIN 15
PHONE: 030-8817209

GREECE

SPECIAL EFFECTS

SAMIOTIS, MICHALIS
OFFICE: 40 PATMOU STREET
ATHENS 11254
STORE: 1 THESPROTEOS STREET
ATHENS 11473
PHONE: 6424017
86545

FILM COMMISSIONS

ETEKT (GREEK UNION OF FILM
& TV-TECHNICIANS)
25 VALTETSIOU STREET
ATHENS 10680
PHONE: 3602379
3615675

GREEK FILM CENTER
10 PANEPISTIMIOU AVENUE
106 71 ATHENS
PHONE: 3631733
3634586
TELEX: 222614 GFC GR
FAX: 3614336
MINISTRY OF INDUSTRY
MICHALAKOPOULOU STREET 80
ATHENS
PHONE: 770 4477

STUDIOS/SOUND STAGES

FINOS FILMS
53 CHIOS STREET
PHONE: 8215-087
8223-566
**KARAYIANNIS, GEORGE & CO.
S.A. (KEY VIDEO)**
PICTURES & TELEVISION,
ENTERPRISES
16-18 SKALIDI STREET
115 25 ATHENS
PHONE: 081-671 7297
081-671 8453
TELEX: 226256 KEY GR
FAX: 01-6475057
TELEVISION ENTERPRISES SA
28 KAPODISTIOU STREET
ATHENS 10682
PHONE: 363-1790
362-6423
VIEW STUDIO LTD
16 KRITONOS STREET
ATHENS 16121
PHONE: 7240553
7223562
FAX: 7249892

HONG KONG

SPECIAL EFFECTS

**ACTION ASSOCIATES NEW
ZEALAND LTD**
P.O. BOX 47395
PONSONBY,AUCKLAND
HONG KONG
PHONE: (09)837 0940 EXT.
PHONE: (09)786 5196 EXT.
FAX: (09)277 7526
AV COMMUNICATIONS LTD
4/F, PENNINGTON COMMERCIAL
BUILDING, 17 PENNINGTON
STREET
CAUSEWAY BAY
HONG KONG
PHONE: (852)8901268 EXT.
PHONE: TX 84836 EXT.
FAX: (852)8903755
DREAM QUEST LTD
FLAT A. FU CHENG CENTRE 3/F
WONG CHUK YEUNG
STREET,FO TAN
SHATIN,NEW TERRITORIES
HONG KONG
PHONE: (0)6992773 EXT.
PHONE: TX 63495 EXT.
FAX: (852)8938549
SALON FILMS (H.K.) LTD
MAYSON GARDEN
BUILDING,2C-E
GROUND FLOOR,WING HING
STREET
CAUSEWAY BAY
HONG KONG
PHONE: 5-781051-4 EXT.
PHONE: TX 75303 EXT.
FAX: 3-7643149

UNLIMITED EFFECT (HK) LTD
UNIT 16-17,7/F PO HING CENTRE
WANG CHIU ROAD,KOWLOON-
BAY
KOWLOON
HONG KONG
PHONE: (3)7999481 EXT.
FAX: (3)7998613

EQUIPMENT RENTALS

J.H. TRACHLSER (H.H.) LTD
GPO BOX 1498,HONG KONG
B1306 WATSON'S
ESTATE,NORTH PTE
HONG KONG
PHONE: (852)5780622 EXT.
PHONE: TX 65184 JHTHK EXT.
FAX: (852)8071224
SALON FILMS (H.K.) LTD
MAYSON GARDEN
BUILDING,2C-E
GROUND FLOOR,WING HING
STREET
CAUSEWAY BAY
HONG KONG
PHONE: 5-781051-4 EXT.
PHONE: TX 75303 EXT.
FAX: 3-7643149

HUNGARY

STUDIOS/SOUND STAGES

**BUDAPEST FILM STUDIO
VALLALAT**
LUMUMBA UTCA 174
H-1145 BUPABEST
PHONE: 631062

ICELAND

FILM COMMISSIONS

ICELANDIC FILM FUN, THE
KVIKMYNDASJODUR ISLANDS
C/O THE MINSTRY OF CULTURE
& EDUCATION
HVERFISGATA 6
101 REYKJAV K
PHONE: 25000

STUDIOS/SOUND STAGES

ADSTADA S.F.
P.O. BOX 4405, VATNAGAROAR 4
104 REYKJAVIK
PHONE: 39835
SAGA FILM LTD
VATNAGARDAR 4, P.O. BOX
4249
104 REYKJAVIK
PHONE: 010354-1 685085
TELEX: 94012576 SAGA G

UNIONS AND ASSOCIATIONS

**FELAG
KVIKMYNDAGERDARMANNA**
(THE ICELANDIC FILM-MAKERS,
ASSOC.)
P.O. BOX 5162
REYKJAVIK
**SAMBAND
KVIKMYNDAFRAMLEIDENDA**
(ASSOC. OF FILM PRODUCERS)
C/O JON HERMANNSSON,
HAFNARSTRAETI 19
101 REYKJAVIK

INDIA

SPECIAL EFFECTS

UNWIN SPECIAL EFFECTS LTD
DSSI,H.O.D.D. UPADHYA MARE
(ROUSE AVENUE)
NEW DELHI
110002
INDIA
PHONE: 278157/9 EXT.
PHONE: TX 3337 EXT.

UNIONS AND ASSOCIATIONS

**ASSOCIATION OF MOTION
PICTURES STUDIOS**
ROOP TARA STUDIOS,
DADASAHEB PHALKE
ROAD,DADAR,
BOMBAY
400 014
INDIA
PHONE: 448230 EXT.
**CINE LABORATORIES
ASSOCIATION "SAHAS"**
414/2 VEER SAVARKAR MARG
ROAD,
PRABHADEVI,BOMBAY
400 025
INDIA
PHONE: 451356 EXT.
**INDIAN MOTION PICTURE
DISTRIBUTORS**
ASSOCIATION, THE
33/1ST FLOOR, VIJAYA
CHAMBERS
TRIBHUVAN ROAD,
BOMBAY
400 004
INDIA
PHONE: 387351 EXT.
**NATIONAL FILM DEVELOPMENT
CORP. LTD**
D-5 SHIV SAGAR ESTATE,DR.
ANNIE
BEJANT ROAD,WORLI
400 018
INDIA
PHONE: 372394 EXT.
PHONE: 473618 EXT.

IRELAND

SPECIAL EFFECTS

**DESIGN & PRODUCTION CO.
LTD, THE**
11 ELMWOOD MEWS
BELFAST, BT4 6BD
PHONE: BELFAST 381905
FILM WORKSHOP LTD
TRASIT HOUSE, SIR JOHN
ROGERSONS QUAY
DUBLIN 7
PHONE: 0001-713913
LIGHTING DIMENSIONS
16 CITY QUAY
DUBLIN 2
PHONE: DUBLIN 710416
SPECIAL EFFECTS (IRL) LTD
SILVERDALE HOUSE, 9
WINDSOR TERRACE
DUN LAOGHAIRE, CO. DUBLIN
PHONE: DUBLIN 801031
FAX: 800902

STUDIOS/SOUND STAGES

ANNER COMMUNICATIONS LTD
UNIT 4, STILLORGAN IND. PARK
BLACKROCK, CO. DUBLIN
PHONE: 0001-952221/3
TELEX: 265871
HETV
CONNSWATER STUDIOS, 79A
BLOOMFIELD AVENUE
BELFAST BT5 5AA
PHONE: BELFAST 450231
FAX: 0232-459459
**M.T.M. ARDMORE STUDIOS
LTD**
BRAY, CO. WICK_OW
IRELAND, DUBLIN
PHONE: 862971
WINDMILL LANE
4 WINDMILL LANE
DUBLIN 2
PHONE: 713444
FAX: 01-718413

UNIONS AND ASSOCIATIONS

**FILM PRODUCERS
ASSOCIATION**
36 MORNINGTON ROAD
RENELAGH, DUBLN 6
IRISH ACTORS' EQUITY GROUP
I.T.G.W.U., LIBERTY HALL
DUBLIN 1
PHONE: 743560
740081
**IRISH FILM & TELEVISION
GUILD**
C/O ROYAL MARINE HOTEL
DUN LAOGHAIRE
CO. DUBLIN
**IRISH TRANSPORT & GENERAL
WORKERS UNION**
DUBLIN NO. 7 BRANCH FILM,
INDUSTRY SECTION, LIEERTY
HALL
DUBLIN 1
PHONE: 744565
740552
**NATIONAL FILM INSTITUTE OF
IRELAND**
65 HARCOURT STREET
DUBLIN 2
PHONE: 753638
**NORTHERN IRELAND FILM
COUNCIL**
7 LOWER CRESCENT
BELFAST, BT7 1NR
PHONE: BELFAST 232444

ISRAEL

FILM COMMISSIONS

ISRAEL FILM CENTRE
MINISTRY OF TRADEAND
INDUSTRY
PO BOX 299,30 AGRON STREET
JERUSALEM
PHONE: (02)210279
PHONE: (02)210433
ISRAEL FILM INSTITUTE
7 ROTHSCHILD BOULEVARD
TEL-AVIV
PHONE: (03)656293
PHONE: (03)65899
ISRAEL FILM SERVICE
PO BOX 1324C,JERUSALEM
PHONE: (02)533223
**ISRAEL TELEVISION
BROADCASTING AUTHORITY**
TELEVISION HOUSE,REMEMA,
JERUSALEM
PHONE: (02)557111
PHONE: TX 25301

PYROTECHNICS

PKD ARMS & PYROTECHNICS
36 SOKOLOV STREET,HOLON
PHONE: (03)806771
STUDIOS/SOUND STAGES
**G.G. ISRAEL STUDIOS
JERUSALEM LTD**
HEAD OFFICE:6 BELINSON
STREET
TEL-AVIV   63567
PHONE: (03)202218
PHONE: (03)290833
FAX: TX L 371699
**UNITED STUDIOS OF ISRAEL
LTD**
HAKESSEM STRET,HERZLIYA
PHONE: (052)550151
PHONE: TX 341878
YORAM POLLAK
56 HAHAGANA
STREET,RA'ANANA
PHONE: (052)35741

UNIONS AND ASSOCIATIONS

**CINEMA OWNERS
ASSOCIATON**
16 PINSKER STREET,TEL-AVIV
PHONE: (03)295138
**FEDERATION OF INDEPENDENT
FILM DIST**
C/O CHAMBER OF COMMERCE,
84 HAHASHMONAIM STREET,
TEL-AVIV
PHONE: (03)288224
**ISRAEL FILM & TV PRODUCERS
ASSOCIATION**
PO BOX 22372,TEL-AVIV
PHONE: (03)225964
PHONE: (03)5464526
**ISRAELI UNION OF PERFORMING
ARTISTS**
DIZENGOFF CENTRE,TEL-AVIV
PHONE: (03)297671
SCREENWRITERS ASSOCIATION
C/O KINERETH PUBLISHING
HOUSE
7 FRANKFURT STREET, TEL-AVIV

ITALY

SPECIAL EFFECTS

BATTISTELLI, RAFFAELE
LARGO APPIO CLAUDIO
385-00174 ROME
PHONE: 742281
**CRS (CENTRO RICERCHE
SCENICHE)**
VIA PALOMBARESE 794
ST. LUCIA, ROMA
PHONE: 06-6878496/0774-
303483
CINERADIO MODEL
VIA PIANSANO 26
00189 ROME
PHONE: 3664343
CONTESINI ARTE
VICOCO AGNAGNINO
ROME
PHONE 6878496/6869553/
7246408
CORRIDORI, G A
VIA DELLE CAPANNELLE 114
00178 ROME
PHONE: 06-7185043
FAX: 06-7184357
FILM STUDIO 83
VIA ASMO 88
00198 ROME
PHONE: 865622

GENERAL LASER, THE
VIA LAURENTINA KM. 23,500,
POMEZIA
00040 ROME
PHONE: 06-9126136
GIEMME COLOR
VIA DEPLI SCIPIONI 256/F
00192 ROME
PHONE: 314387
GRILLI, ARMANDO
VIA S BARZILAI 225
00173 ROME
PHONE: 6130045
MOVIECAM 2000
VIA PASSAGLIA 1/B
00136 ROME
PHONE: 312977
PASSERI, ALVARO
VIA FILACCIANO
3-00189 ROME
PHONE: 6913371 & 8816221
PUBBLICINE
VIA ARNO 88
00198 ROMA
PHONE: 8441632/5865622
TELECINEMA 83 S.R.L.
VIA MEDA
61-00137 ROMA
PHONE: 4514180
FAX: 4510783
V.G.
VIA EZIO 30B
00192 ROMA
PHONE: 381597/314387

MAKE-UP SUPPLIERS

BAULE DI BERTINI-RODITI, IL
S. MARCO 583
30124 VENICE
PHONE: 041-84488
**INDIO-LABORATORIO
COSMETICI CINETELETEATRALL**
VIA PORTUENSE 195
ROME
PHONE: 5561621
**KRYOLAN PROFESSIONAL
MAKE-UP**
KRYOLAN ITALIA, MARTIN
MILLER, FREIHEITSSTR. 96
39012 MERAN (BZ)
PHONE: 0473/36092
**ROCCHETTI - CARBONI
COSMETICI**
VIA C. CITERNI 31
ROME
PHONE: 576433
**STUDIO MASCHER D'APPOLO
DI ALESSANDRO**
IAGOPINI & D SAS (COSMETICI
& TRUCE PROFESSIONALE)
VIA FLAVO STILICONE 92
00175 ROMA
PHONE: 7662393

PYROTECHNICS

CORRIDORI, G A
VIA DELLE CAPANNELLE 114
00178 ROME
PHONE: 06-7185043
FAX: 06-7184357

STUDIOS/SOUND STAGES

A.D. SYSTEM SRL
VIA A. SILVANI
29-00138 ROMA
PHONE: 8125228
FAX: 8106992
CTC - PALAZZO
VIALE LEGIONI ROMANE
3-20147 MILAN
PHONE: 02-403061
TELEX: 326562 CTCMI I
FAX: 02-48700476

CENTRO MEDIA
VIA BORRA 26
57100 LIVORNO
PHONE: 0586-25070
CENTRO PALATINO GAUMONT
P. ZZA SS. GIOV E PAOLO 8
00184 ROME
PHONE: 77081
CINECITTA S.P.A.
VIA TUSCOLANA 1055
00174 ROME
PHONE: 722931
TELEX: 620478 CINCIT I
FAX: 7222155
**CINESTABILIMENTO DONATO
S.P.A.**
VIA MUSSI 24
20154 MILAN
PHONE: 02-3452416
TELEX: 332543 DONATO I
FAX: 02-33105056
CORRIDORI STUDIO
VIALE RODI
91-20126 MILAN
PHONE: 06-7185043
DE ANGELIS STUDIO
VIALE RODI
91-20126 MILAN
PHONE: 02-6424183/6470404
DE PAOLIS - IN C.I.R.
VIA TIBURTINA 521
00159 ROME
PHONE: 4385341
DEAR INTERNATIONAL
VIA NOMENTANA 883
00137 ROME
PHONE: 826801/02
TELEX: 610634 DEARIN I
FILM STUDIO 80 S.N.C.
VIA LEGNONE
34-20158 MILANO
PHONE: 02-6686793
GUICAR TELEVISION
VIA FARUFFINI 25
20149 MILANO
PHONE: 02-4696641
ICET STUDIOS S.R.L.
VIA P. ROSSI
3-20093 COLOGNO MONZESE
(MI)
PHONE: 02-2543184 & 2544073
FAX: 02-27300641
I.P.CREN. STUDIO
VIA CHIABRERA 54/D
00145 ROME
PHONE: 5140841/5 & 5141641/5
TELEX: 630138 IPCREN I
FAX: 5128853
INTER TV
VIA GORIZIA 4
20144 MILAN
PHONE: 02-8376951
TELEX: 340319 ITV-I
FAX: 02-8377890
JAMBO FILM S.R.L. SERVICE
VIA CARROCCIO 6
20123 MILAN
PHONE: 02-8356129 & 8356164
FAX: 02-8358246
**MERCURIO CINEMATOGRAFICA
S.R.L.**
VIA ANDREA VERGA 5
20144 MILAN
PHONE: 02-4980241/4988251
TELEX: 351189 MR FILM I
FAX: 02-48008319
MOVIE PEOPLE
VIALE BERBERA 49
20162 MILAN
PHONE: 02-6473877 & 6429420
FAX: 02-66100938
OLIMPIA TEATRO DI POSA
VIA GIUDICI
8-20093 COLOGNO MONZESE
(MI)
PHONE: 02-2538782 & 2543120

R.P.A. ELIOS
VIA TIBURTINA, KM 13, 600
00131 ROME
PHONE: 4090195/424/200
RAP - ELIOS S.A.S.
RAPPER LEG: DR. FILIBERTO,
BANDINI, VIA TIBURTINA
KM 13600
PHONE: 4090195
SBP S.P.A.
VIA E JENNER, 147
00151 ROME
PHONE: 06-5315041
TELEX: 616164 SBP I
FAX: 06-5312367
**STABILIMENTI
CINEMATOGRAFICI PONTINI**
VIA PONTINA KM 23, 700
CASTEL ROMANO (RM)
PHONE: 6490242
FAX: 6490119
TEATRO 3 SRL
VIA GIULIETTI 14
20132 MILAN
PHONE: 2592751/2/3
FAX: 02-2591580
VIDES INTERNATIONAL
VIA CONCESIO, KM 1,800
001881 PRIMA PORTA, ROME
PHONE: 06-691089
VILALGE SRL
VIA CARROCCIO 6
20123 MILAN
PHONE: 02-8370495/8372583/
8322855
FAX: 02-8358246

UNIONS AND ASSOCIATIONS

**A.C.D. (ASSOCIAZIONE
CINEMA DEMOCRATICO)**
VIALE G. CESARE
71-00192 ROMA
PHONE: 388160
**A.C.E.C. (ASSOCIAZIONE
CATTOLICA ESERCENTI
CINEMA)**
VIA NOMENTANA 251
00161 ROME
PHONE: 4402280 & 4402273
**A.I.A.C.E. (ASS. ITALIANA
AMICI DEL CINEMA D'ESSAI)**
VIA GAETA
23-00186 ROMA
PHONE: 4740905 & 4814959
**A.I.C.C.A. (ASSOCIAZIONE
INTERNAZ CINEMA COMICO
D'ARTE)]**
CASELLA POSTALE 6104
00195 ROMAS
PHONE: 3580266
**A.I.C.E.D.(ASSOCIAZIONE
ITALIANA CINEM. EDUCATIVA E
DIDATTICA)**
VIALE REG. MARGHERITA 286
00198 ROMA
PHONE: 841271
**A.N.E.C. (ASSOCIAZIONE
NAZIONALE ESERCENTI
CINEMA)**
VIA DI VILLA PATRIZI 10
00161 ROMA
PHONE: 06-884731
FAX: 06-8848079
**A.N.I.C.A. (ASSOCIAZIONE
NAZIONAL**
INDUSTRIE
CINEMATOGRAFICHE
AUDIOVISIVI)
VIALE REGINA MARGHERITA 286
00198 ROME
PHONE: 8841271
TELEX: 624659 ANICA
FAX: 8848789

A.R.C.I. - UNIONE CIRCOLI CINEMATOGRAFICI
VIA F. CARRARA 24
00196 ROME
PHONE: 36108800,3216878 &
3610731

A.T.I.C. (ASSOCIAZIONE TECNICA ITALIANA PER LA
CINEMATOGRAFIA E PER LA
TELEVISIONE)
V. LE REGINA MARGHERITA
286-00198, ROMA
PHONE: 8841271

A.I.C. (ASSOCIAZIONE ITALIANA AUTORI DELLA
FOTOGRAFIA
CINEMATOGRAFICA)
VIA TUSCOLANA 1055
00173 ROME
PHONE: 72293289

C.C.R. (COMITATO CINEMATOGRAFIA RAGAZZI)
VIA TRIBUNA TOR DI SPECCHI
18-4-00186 ROMA
PHONE: 6794268

C.I.C.T. (CONSEIL INTERNA-TIONAL DU CINEMA ET DE LA
TELEVISION)
LUNGOTEVERE DEI VALLATI
2-00186 ROMA
PHONE: 06-6875891

C.I.C.C.E. (COMITE DES INDUSTRIES CINEMATOGRAPHIQUES
DE LA COMMUNAUTES
EUROPEENNE)
286 VIALE REGINA MARGHERITA
00198 ROMA
PHONE: 8841271
5 RUE DES CIRQUE
75008 PARIS
PHONE: 000331-42 257063

CONSEIL INTERNATIONAL DU CINEMA ET DE LA TELEVISION (CICT)
LUNGOTEVERE VALLATI 2
00186 ROMA
PHONE: 6875898

UNIONE NAZIONALE PRODUTTORI FILM - U.N.P.F.
VIALE REGINA MARGHERITA 286
00198 ROMA
PHONE: 8841271
FAX: 8848789

JAPAN

UNIONS AND ASSOCIATIONS

JAPAN AUDIO-VISUAL EDUCATION ASSOCIATION
1-17-1 TORANOMON,MINATO-KU
TOKYO
105
PHONE: (03)432-6548
FAX: (03)597 0564

JAPAN CAMERA & OPTICAL INSTRUMENTS
INSPECTION & TESTING
INSTITUTE
25 ICHIBAN-CHO,CHIYADA-KU
TOKYO
102
PHONE: (03)263 7111
PHONE: TX 2324229
FAX: (03)234 4624

MOTION PICTURE & OPTICAL INSTRUMENTS
INSPECTION & TESTING
INSTITUTE
25 INCHIBAN-CHO
CHIYADA-KU,TOKYO
102
PHONE: (03)263 7111
PHONE: TX 3234229
FAX: (03)234 6624

MOTION PICTURE & TELEVISION ENGINEERING
SOCIETY OF JAPAN INC.
SANEI BUILDING,BEKKAN,
1-7-2 OTEMACHI,CHIYODA-KU
TOKYO
100
PHONE: (03)231 6417

KENYA

EQUIPMENT RENTALS

NIMROD AFRICA LTD
PLESSEY HOUSE.UHURU
HIGHWAY
PO BOX 45738
NAIROBI
PHONE: 554888
PHONE: 552453
FAX: TX 2405

MAKE-UP SUPPLIERS

RUNE KIM
PO BOX 373,THIKA
PHONE: (0151)22197

MEXICO

FILM COMMISSIONS

BANCO NACIONAL CINEMA TOGRAFICO SA
AV DIVISION, DEL NORTE 2462,
MEXICO 13 D.F.
TEL: 5396955

ESTUDIOS AMERICA
CALSACA DE TLALPAN 2818
MEXICO 21 D.F.
TEL: 5490761

ESTUDIOS CHURUBUSCO
ATLETAS 2
MEXICO 21 D.F.
TEL: 5490761

PELUCULAS MEXICANAS SA DE C.V.
DIVISION DEL NORTE 2462
MEXICO 13 D.F.
TEL: 5392175

PROCINEMAX SA
AVENUE UNIVERSIDAD 1330
MEXICO 21 D.F.
TEL: 5348280

THE NETHERLANDS

SPECIAL EFFECTS

CREATIX
VREDENHOFWEG 3A
1051 LM AMSTERDAM
PHONE: 020-880139
FAX: 020-863985

GREAT AMERICAN MARKET, THE - AGENTS: CORNE-LIGHT BV
KEIENBERGWEG 69
1101 GE AMSTERDAM,
BULLEWIJK
PHONE: 020-979686
FAX: 020-977986

NLF NEDERLANDS LABORATORIUM VOOR FILM & VIDEOTECHNIEK BV
KERKLAAN 5
3632 AK LOENEN A/D VECHT
PHONE: 02943 1855
FAX: 02943-4187

ROSEPORT STUDIOS
T.T. VASUMWEG 102
1033 SH AMSTERDAM
PHONE: 020-334747
FAX: 020-312591

SOLUTIONS
ONDERLANGS 36M
1097 ZK AMSTERDAM
PHONE: 020-949434

SPECIAL EFFECTS
CHROOMSTRAAT 5 C
3067 GN ROTTERDAM
PHONE: 010-4558524
FAX: 010-4213264

SPECIAL EFFECTS (H.W.)
STOCKHOLMPAD 23
3067 DK ROTTERDAM
PHONE: 010-4205037 & 010-4558524

VAN DEN AKKER HENNIE
SOLUTIONS, ONDERLANGS 36M
1097 ZM AMSTERDAM
PHONE: 020-199438

WEVO ARM RENTAL B.V.
HAVEN STRAAT 5 (D), POSTBUS
16537
1001 RA AMSTERDAM
PHONE: 020-735381
FAX: 020-263774

EQUIPMENT RENTALS

CINE-60 - AGENT: CORNE-LIGHT BV
KEIENBERGWEG 69
1101 GE AMSTERDAM,
BULLEWIJK
PHONE: 020-97968698
FAX: 020-977986

HOLLAND EQUIPMENT BV
H.J.E. WENCKEBACHWEG 49
1096 AK AMSTERDAM
PHONE: 020-943575
FAX: 020-6685381

MAKE-UP SUPPLIERS

KRYOLAN PROFESSIONAL MAKE UP
LEON VAN VOLEN EN ZONEN,
V. WOUSTRAAT 8
AMSTERDAM

KRYOLAN PROFESSIONAL MAKE UP
COELHO, NIEUW
PRINSENGRAACHT 7
AMSTERDAM

KRYOLAN PROFESSIONAL MAKE UP
PARTYHAUS, ROSENGRACHT 68
AMSTERDAM C
PHONE: 020-221267

KRYOLAN PROFESSIONAL MAKE UP
P.E. VAN CLADDER,
UTRECHTSESTR 47
NL 1855 AMSTERDAM

STUDIOS/SOUND STAGES

AALSMEER TV STUDIO
VAN CLEFFKADE 15
1431 BA AALSMEER
PHONE: 02977-51711/51722

AMSTERDAM STUDIOS
DUIVENDRECHTSEKADE 83-85
1096 AJ AMSTERDAM
PHONE: 020-930960
FAX: 020-938938

ANGEL STUDIOS
MINERVAHAVENWEG 3-4
1013 AR AMSTERDAM
PHONE: 020-825224/822128
FAX: 020-843737

CINEVIDEO GROUP HOLLAND BV
P.O. BOX 125, AMBACHTSMARK 3
1355 EA ALMERE-HAVEN, NR.
AMSTERDAM
PHONE: 0-3240 12524
FAX: 03240-48212

FIRST FLOOR FEATURES
JARMUIDE 13
1046 AC AMSTERDAM
PHONE: 020-6647471/137231

FIRST FLOOR FEATURES BV
POSTBUS 53221
1007 RE AMSTERDAM
PHONE: 020-6647471

HOLLAND EQUIPMENT BV
H.J.E. WENCKEBACHWEG 49
1096 AK AMSTERDAM
PHONE: 020-943575
FAX: 020-6685381

JONGENELEN VIDEO ROOSENDAAL
RECPHENSEBAAN 52, POSTBUS
1038
4700 BA ROOSENDAAL
PHONE: 01650-81000
TELEX: 78202 JOLEN
FAX: 01650-81347

KOMMER, BOB STUDIO'S BV
VAN DE SPIEGHELSTRAAT 9-11
2518 ES'S-GRAVENHAGE
PHONE: 070-3469572
FAX: 070-4631132

KOMMER STUDIOS BV
VAN DER SPIEGELSTRAAT
2518 ES DEN HAAG
PHONE: 070-3469572
FAX: 070-4631132

LING, SYDNEY FILM STUDIOS
P.O. BOX 1416
7500 BK ENSCHEDE
PHONE: 05428-3007

NEDERLANDS OMROEPPRODUKTIE BEDRIJF NV
POSTBUS 10
1200 JB HILVERSUM
PHONE: 035-775299
FAX: 035-775444

PIET'S POST PRODUTIONS B.V.
PLANTAGE MIDDENLAAN 6
1018 DD AMSTERDAM
PHONE: 020-275711
FAX: 020-208661

Q POINT STUDIO'S NV
HENGEVELDSTRAAT 29,
POSTBUS 15127
3501 BC UTRECHT
PHONE: 030-736536
FAX: 030-713890

ROSEPORT STUDIOS
T.T. VASUMWEG 102
1033 SH AMSTERDAM
PHONE: 020-334747
FAX: 020-312591

STUDIO 49 (HOLLAND EQUIPMENT BV)
H.J.E. WENKEBACHWEG 49
1096 AK AMSTERDAM
PHONE: 020-6650784
FAX: 020-6685381

STUDIO 5 B.V.
DUIVENDRECHTSEKADE 86
1096 AJ AMSTERDAM
PHONE: 020-952359/942232
FAX: 020-6683627
STUDIO HONINGSTRAAT
HINGSTRAAT 14 6
1211 AW HILVERSUM
PHONE: 035-215755
FAX: 035-232527
STUDIO PEULENSCHIL
JARMUIDEN 15
1046 AC AMSTERDAM
PHONE: 020-149200/149177
TV-STUDIO AALSMEER
VAN CLEEFFKADE 15
1431 BA AALSMEER
PHONE: 02977-51711/51722
FAX: 02977-51700
V.P.C.
HONINGSTRAAT 14B
1211 AW HILVERSUM
PHONE: 035-15755
FAX: 035-232527
VIDEO HILVERSUM B.V
CATHARINA VAN RENNESLAAN 20
1217 CX HILVESUM
PHONE: 035-217951
FA X: 035-216181

UNIONS AND ASSOCIATIONS

AVOC AUDIO VISUEEL
ONDERWIJS COMMITEE
P/A A.J. THIECKE,
HYACINEENLAAN 7
6866 DV HEELSUM
PHONE: 08373 1449
ARBEIDSBEDMIDDELING
FILMERS
NOORDHOEK-HEGT SINGEL
202-204
1016 AA AMSTERDAM
PHONE: 020-5200408
ASSOCIATIE VAN
NEDERLANDSE FILMTHEATER
PRINSENGRACHT 770-111
1017 LE AMSTERDAM
PHONE: 020-267602
AUDIOVISUELE
BEROEPSVERENIGING NBF
(PROFESSIONAL
ORGANISATIONS FILM & TV)
KEIZERSGRACHT 58
1018 DT AMSTERDAM
PHONE: 020-246633
CAWO CONTACTGROEP
AUDIOVISUELE CENTRA
WETENSCHAPPELIJK
ONDERWIJS
P/A CENTRALE AUDIOVISUELE,
DIENST, UNIVERSITEIT VAN,
AMSTERDAM, MEIBERGDREEF 15
1005 AZ AMSTERDAM
PHONE: 020-5664704
CENTRAAL OVERLEG VOOR
AUDIOVISUELE
POSTBUS 30, MEDIA, COVAM
3879 CA HOEVELAKEN
PHONE: 03495-34214
CINE Q
P/A RVD ANNA
PALOWNSTRAAT 76
2518 BH 'S-GRAVENHAGE
PHONE: 070-614181
CINEMIEN
AMSTEL 256 A
1017 AL AMSTERDAM
PHONE: 020-279501
DE REGIE
ROZENGRACHT 207 D
1016 LZ AMSTERDAM
PHONE: 020-226548
FAX: 020-224967
(AGENCIES FOR FILM & TV
DIRECTORS)

FID FILM INFORMATIE EN
DOCUMENTATIEDIENST
P/A LOK V GANZENMARKT 6,
POSTBUS 805
3500 AV UTRECHT
PHONE: 030-332328
FILM INTERNATIONAL
EENDRACHTSWEG 21
3012 LB ROTTERDAM
PHONE: 010-133399
TELEX: 21378 FINTR
FILMFESTIVAL ROTTERDAM
EENDRACHTSWEG21
3012 LB ROTTERDAM
PHONE: 010-4130595/4133399
FILMZIEN HET
JEUGDFILMCIRCUIT
P/A LOK V GANZEMARKT 6,
POSTBUS 805
3500 AV UTRECHT
PHONE: 030-332328
GNS GENOOTSCHAP VAN
NEDERLANDSE
SPEELFILMMAKERS
POSTBUS 581
1000 AN AMSTERDAM
PHONE: 020-274339
HOLLAND FILM PROMOTION
JAN LUYKENSTRAAT 2
1071 CM AMSTERDAM
PHONE: 020-799261
TELEX: 12151 NBBL
FAX: 020-750398
ITVA INTERNATIONAL
TELEVISION ASSOCIATION
AFD. NEDERLAND
HYACINTENLAAN 7
6866 DV HEELSUM
PHONE: 08373-14449
NBB BOND VAN BIOSCOOP EN
FILMONDERNEMINGEN
JAN LUYKENSTRAAT 2
1071 CM AMSTERDAM
PHONE: 020-799261
NCGV FILM-EN VIDEOTHEEK
DA COSTAKADE 45
3521 VS UTRECHT
PHONE: 030-941899
FAX: 030-961020
NBF BEROEPSVERENIGING VAN
FILM-EN TELEVISIEMAKERS
DONKER CURTIUSTRAAT 7
1051 JL AMSTERDAM
PHONE: 020-881670
NEDERLANDS FILMMUSEUM
VONDELPARK6
1071 AA AMSTERDAM
PHONE: 020-831646
FAX: 020-833401
NEDERLANDSE FILMKEURING
P/A MINISTERIE VAN WVC,
POSTBUS 525
2280 AM RIJSWIJK
PHONE: 070-406274
SVDN SAMENWERKENDE
VIDEODEALERS NEDERLAN
NASSAU ZUIOENSTEINSTRAAT 15
2596 CA'S-GRAVENHAGE
PHONE: 070-262011
SAM/STICHTING AUDIOVISUELE
MANIFESTATIES
COEHOORNSTRAAT 5
1222 RR HILVERSUM
PHONE: 035-833742
STICHTING NEDERLANDSE
FILMDAGEN
HOOGT 4
3512 GW UTRECHT
PHONE: 030-322684
TECHNISCH FILMCENTRUM
ARNHEMSESTRAATWEG 17
6881 NB VELP
PHONE: 085-629188

VAMG VERENIGING VAN
AUDIOVISUELE
MEDEWERKERS
IN DE GEZONDHEIDSZORG
POSTBUS 64711
2506 CC'S GRAVENHAGE
VEVAM VERENIGING TER
EXPLOITATIE VAN
VERTONINGSRECHTEN OP
AUDIOVISUEEL MATERIAL
POSTBUS 581
1000 AN AMSTERDAM
VNVI VERENIGING
NEDERLANDSE VIDEOTEX
INFORMATIELEVERANCIERS
REITSEPLEIN 1, POSTBUS 581
LG TILBURG
PHONE: 013-654111
VIDEO EDUCATION CENTER
BERGSTRAAT 51-53
7411 ES DEVENTER
PHONE: 05700-13113

NEW ZEALAND

SPECIAL EFFECTS

ACTION ASSOCIATES NEW
ZEALAND LTD
P.O. BOX 47395
PONSONBY, AUCKLAND
PHONE: (09)837 0940
PHONE: (09)786 5196
FAX: (09)277 7526
AVALAON TELEVISION
PRODUCTION
CENTRE (DIV. OF TVNZ LTD)
P.O. BOX 30945, LOWER HUTT
PHONE: (09)666 969
PHONE: TX NZ3867
FAX: (09)678 959

EQUIPMENT RENTALS

ACTION ASSOCIATES NEW
ZEALAND LTD
P.O. BOX 47395
PONSONBY, AUCKLAND
NEW ZEALAND
PHONE: (09)837 0940
PHONE: (09)786 5196
FAX: (09)277 7526
FILM FACILITIES LIMITED
51 MACKELVIE
STREET, PONSONBY
PO BOX 47091
AUCKLAND
PHONE: (09)789 493
FAX: (09)765 207
FILM FACILITIES LIMITED
THE PRODUCTION VILLAGE,
26 WRIGHT STREET,
PO BOX 6698
WELLINTON
PHONE: (04)844 192
FAX: (04)843 774
FIZZGIG
AUCKLAND
PHONE: (09)766 623
GNOME PRODUCTIONS LTD
P.O. BOX 825, WELLINGTON
PHONE: (09)857 552
FAX: (09)857 564

MAKE-UP SUPPLIERS

KRYOLAN PROFESSIONAL
MAKE-UP
SELECON LTD,40 DRAKE STREET
FREEMANS BAY, AUCKLAND
PHONE: (09)792 583
PHONE: TX NZ 2204
FAX: (09)770 116

UNIONS AND ASSOCIATIONS

INDEPENDENT PRODUCERS &
DIRECTORS GUILD
PO BOX 3969,WELLINGTON
PHONE: (04)842350

NORWAY

STUDIOS/SOUND STAGES

CAPRINO FILMSENTER A/S
MARIO CAPRINOSV. 3
1335 SNAROYA
PHONE: 02-533195
INFO FILM PRODUCTION A/S
POSTBOKS 469
1301 SANDVIKA
PHONE: 02-49320
NORSK FILMSTUDIO A/S
WEDEL JABLSBERGSVEI 36
1342 JAR
PHONE: 02-121070
FAX: 02-591370
TEAM-FILM A/S
KEYSERSGT 1
0165 OSLO 1
PHONE: 02-207072

UNIONS AND ASSOCIATIONS

KOMMUNALE
KINEMATOGRAFERS
LANDSFORB
STORTINGSGT. 16
0161 OSLO 1
PHONE: 02-411617
NORDISK FILM OG TV-UNION
C/O OPPLYSNINGSFILM,
PARKVN. 62C
0254 OSLO 2
PHONE: 02-550028
NORSK FILMBYROERS
FORENING
POSTBOKS 1814, VIKA
0123 OSLO 1
PHONE 02-566467
NORSK FILMFORBUND
STORENGVN. 8 B
1342 JAR
PHONE: 02-591000
NORSK FILMINSTITITT
GREV WEDELSPL.
0151 OSLO 1
PHONE: 02-428740
NORSK FILM-OG
VIDEOGRAMPROD
DRAMMENSVN. 30/V
0255 OSLO 2
PHONE: 02-446926
NORSK FILMROD
POSTBOKS 482, SENTRUM
0105 OSLO 1
PHONE: 02-428740
NORSK KORTFILMFOND
POSTBOKS 6824, ST. OLAVSPL.
0130 OSLO 1
PHONE: 02-202742
STATENS FILMSENTRAL
PB. 2655, ST. HANSHAUGEN
0131 OSLO 1
PHONE: 02-602090

PORTUGAL

FILM COMMISSIONS

CENTRO DE AUDIO-VISUAIS
DA FORCA AEREA
ALFRAGIDE
2700 AMADORA
PHONE: 972383

**CENTRO DE AUDIO-VISUAIS
DO EXERCITO**
RUA CONDE NOVA GOA
1000 LISBOA
PHONE: 656011
**INSTITUTO DE TECNOLOGIA
EDUCATIVA**
AV PADRE MANUEL DA
NOBREGA, 8-B/2
1000 LISBOA
PHONE: 803832 & 885468

STUDIOS/SOUND STAGES

**CINEMATE MATERIAL
CINEMATOGRAFICO IDS**
AV RAINHA D. AMELIA 10-A,
12-A
1600 LISBON
PHONE: 7592357 & 7587049 &
7595471
TELEX: 13471 RIMPEF-P
FAX: 7592303
EDIPIM STUDIO
ABRUNHEIRA
2710 SINTRA
PHONE: 9259321/2
INTERFILME
CALCADA DOS 7, MOINHOS
LOTE
1000 LISBOA
PHONE: 680862 & 680870 &
680885
LIBOA FILMES
PRACA BERNARDINO
MACHADO
1700 LISBON
PHONE: 7595396/7595467 &
7595242
FAX: 7590127
MEDIMAGE LDA
RUA COELHO DA ROCHA, 69-
PORTA 8
1300 LISBON
PHONE: 602582 & 661847
**PANORAMICA 35-FILM
PRODUCTION**
RUA AQUILES MONTEVARDE,
10/1
1000 LISBOA
PHONE: 545154 & 555113
TELEX: 15446 P
**TELECINE MORO-SOC
PRODUCTORA DE FILMS
S.A.R.L.**
RUA DE D PEDRO V, 56
1200 LISBOA
PHONE: 361732
TOBIS PORTUGUESA S.A.
ALAMEDA DA LINHAS DE
TORRES, 156-B
1700 LISBON
PHONE: 7595425
TELEX: 64009 TOB S P
FAX: 351 1 7589622

UNIONS AND ASSOCIATIONS

**ASSOCIACOA DE PRODUTORES
DE FILMES**
RUA DUQUE DE LOULE, 86, 2
DTO
1000 LISBOA
PHONE: 01-557501
**ASSOCIACOA PORTUGUESA DE
REALIZAORES DE FILMES**
RUE DA PALMEIRA 7 R/C
1200 LISBOA
PHONE: 01-366631
**INSTITUTO PORTUGUES DE
CINEMA**
RUE S. PEDRO DE ALCANTRA,
45/1
1200 LISBOA
PHONE: 368485 & 366634

**SINDICATO DA ACTIVIDADE
CINEMATROFAFICA**
RUA D. PEDRO V, 60/1 ESQ
1200 LISBOA
PHONE: 322660 & 326943

SINGAPORE

EQUIPMENT RENTALS

**YEOW KONG ELECTRICFAL CO
(PTE) LTD**
5 TUAS DRIVE,2263
SINGAPORE
SINGAPORE
PHONE: 0862-1501 EXT.
FAX: 0862-0182

SOUTH AFRICA

EQUIPMENT RENTALS

**BRIGADIERS MOTION PICTURE
CO.(PTY)LTD.**
PO BOX 65,HONEYDEW
PHONE: (011)7953781

MAKE-UP SUPPLIERS

KRYOLAN OF GERMANY
PO BOX 87239, HOUGHTON
2041
PHONE: (646)4842
**MIKE & LIZ HAIRPIECES AND
PAKO**
TAPESTRIES
PO BOX 44382,LINDEN
JOHANNESBURG
2104
PHONE: 8723907

UNIONS AND ASSOCIATIONS

**SA ASSOCIATION OF
PROFESSIONAL RECORDING
STUDIOS**
P.O. BOX 9190
JOHANNESBURG 2000
PHONE: (011) 299736
**SA FILM & TELEVISION
TECHNICIANS ASSOCIATION**
P.O. BOX 41357
CRAIGHALL 2040
PHONE: (011) 8021730
SA FILM & THEATRE UNION
P.O. BOX 31313
BRAAMFONTEIN 2017
PHONE: (011) 7262122
SA FILM & VIDEO INSTITUTE
P.O. BOX 39639
BRAMLEY 2018
PHONE: (011) 3181349
**SA SCRIPT WRITERS
ASSOCATION**
P.O. BOX 41334
CRAIGHALL 2024
PHONE: (011) 6482642
**SA SOCIETY OF
CINEMATOGRAPHERS**
P.O. BOX 17465
SUNWARD PARK 1470
PHONE: (011) 8931342

SPAIN

SPECIAL EFFECTS

**ARCHIVO CINEMATOGRAFICO
GAN**
ESPOZ Y MINA 5
MADRID 28012
PHONE: 91-2313915
BASILLO CORTIJO
DOCTOR MENDIGUCHI A 19
LEGANES 28913 (MADRID)
PHONE: 91-6930614
CARLOS DE MARCHIS
CUESTA DEL CERRO 1
LA MORALEJA ALCOBENDAS
MADRID
PHONE: 91-6501403
CINEFEC
CALLES NUEVA S/N/
AJALVIR
MADRID
PHONE: 91-8843208
**EFECTOS ESPECIALES Y
ARMAS**
C/ BARACALDO 33
ESQUINA ALVAREZ
PHONE: 315 06 32
**LORENZO CANAS EFFECTOS Y
TECHNICAS AUDIOVISUALS**
CASTILLO DE PINEIRO 3
MADRID 28039
MAQUETA DE TREN DE VAPOR
MOLE RICHARDSON (ESPANA)
LTD.
GUSTAVO FERNANDEZ
BALBUENA 11
MADRID 28002
PHONE: 91-4157254
MOLINA EFECTOS ESPECIALES
CERAMICA 10 Y 12
28038 MADRID
PHONE: 2669175 & 5010169
MOLINARE S.A.
BRESCIA 19
MADRID 28028
PHONE: 010-341 255 9207
MONSTRUOS S.A.
GOMEZ ORTEGA 22
MADRID 28002
PHONE: 91-4155586
PEREZ EZCURRA, FERNANDO
POBLADURA DEL VALLE 5
MADRID 28037
PHONE: 91-2131503
PEREZ MUNOZ, PABLO
NAVALAFUENTE 43
MADRID 28002
PHONE: 91-4155586
REYES ABADES TEJEDOR
CAMINO DE LA VEGA, NAVE 18,
NO, 13
TORREJON DE ARDOZ 28850
MADRID
PHONE: 91-6563059 & 6761470
ROASA S.L.
VIRIATO 71 2 F
MADRID 10
PHONE: 91-4466719
SANCHEZ CALERO, JUAN
PLAZA DE BLANES 3, 3, 4A
PRAT DE LLOBREGAT
BARCELONA
PHONE: 3794298
SAYANS MANUEL
SANCHO DAVILA 27
MADRID 28
PHONE: 2455240 & 2450459
SETLA, S.A.
NTR. SRA. NIEVES 60
BARCELONA 08031
PHONE: 4292519

**TELSON TELEVISION Y SONIDO
S.A.**
PRADILLO 46-64
28002 MADRID
PHONE: 4134463
TELEX: 44122 TELSO E
FAX: 4157572

EQUIPMENT RENTALS

**TELSON TELEVISION Y SONIDO
S.A.**
PRADILLO 46-64
28002 MADRID
PHONE: 4134463
TELEX: 44122 TELSO E
FAX: 4157572

MAKE-UP SUPPLIERS

**KRYOLAN MAQUILA E
PROFESSIONALES**
PERFUMERIA LOVEL, ANTONIO
LOPEZ VEGA, LIBERTAD 5
MADRID 4
PHONE: 2325498
LAURENDOR
BARCELONA
PHONE: 93-2256194
**PERFUMERIA ARENAS
DROGUERIA**
PRINCIPE 16
MADRID
PHONE: 2215164 & 2299297

PYROTECHNICS

**ARCHIVO CINEMATOGRAFICO
GAN**
ESPOZ Y MINA 5
MADRID 28012
PHONE: 91-2313915
BASILIO CORTIJO
DOCTOR MENDIGUCHI A 19
LEGANES 28913 (MADRID)
PHONE: 91-6930614
CARLOS DE MARCHIS
CUESTA DEL CERRO 1
LA MORALEJA ALCOBENDAS
MADRID
PHONE: 91-6501403
CINEFEC
CALLE NUEVA S/M
A.ALVIR
MADRID
PHONE: 91-8843208
**LORENZO CANAS EFFECTOS Y
TECNICAS AUDIOVISUALES**
CASTILLO DE PINEIRO 3
MADRID 28039
PHONE: 91-2537027
MAQUETA DE TREN DE VAPOR
MOLE RICHARDSON (ESPANA)
LTD, GUSTAVO FERNANDEZ
BALBUENA 11
MADRID 28002
PHONE: 961-4157254
**MOLINA EFFECTOS
ESPECIALES**
CERAMICA 10 Y 1
2
28038 MADRID
PHONE: 2669175 & 5010169
MONSTRUOS SA
GOMEZ ORTEGA 22
MADRID 28002
PHONE: 91-4155586
PEREZ EZCURRA, FERNANDO
POBLADURA DEL VALLE 5
MADRID 28037
PHONE: 91-2131503
PEREZ MUNOZ, PABLO
NAVALAFUENTE 43
MADRID 28002
PHONE: 91-4155586

REYES ABADES TEJEDOR
  CAMINO DE LA VEGA, NAVE 18,
  NO. 13
  TORREJON DE ARDOZ 28850
  MADRID
  PHONE: 91-6563059 & 6761470
ROASA S.L.
  VIRIATO 71, 2 F
  MADRID 10
  PHONE: 91-4466719
SANCHEZ CALERO, JUAN
  PLAZA DE BLANES 3, 3, 4A
  PRAT DE LLOBREGAT
  BARCELONA
  PHONE: 3794298
SAYANS MANUEL
  SANCHO DAVILA 27
  MADRID 28
  PHONE: 2455240 & 2450459
SETLA, S.A.
  NRT. SRA. NIEVES 60
  BARCELONA 08031
  PHONE: 4292519
TELSON TELEVISION Y SONIDO S.A.
  PRADILLO 46-64
  28002 MADRID
  PHONE: 4134463
  TELEX: 44122 TELSO E
  FAX: 4157572

STUDIOS/SOUND STAGES

ARGIPLATO
  APARTADO 398
  SAN SEBASTIAN 20080
  PHONE: 943-454061
BARAJAS
  CAMPEZO 3
  POLIGONO LAS MERCECES,
  MADRID 28022
  PHONE: 91-7475087
CAR'S STUDIO S.A.
  ETRURIA 30
  MARDID 28022
  PHONE: 91-7415459
CINEARTE S.A.
  PLAZA CONDE DE BARAJAS 5
  MADRID 28005
  PHONE: 91-2665407 & 2664507
CTV S.A.
  PLAZO DE VIGO 2
  15703 SANTIAGO DE
  COMPOSTELA
  PHONE: 594405
CYCLOP
  P EXTREMADURA 153
  MADRID 28011
  PHONE: 91-4641092
DELTA S.A.
  ESTUDIOS Y PLATOS, AVAIDA
  DE SALAMANCA, S/N
  SAN SEBASTIAN DE LOS REYES
  (MADRID)
  PHONE: 91-6541746, 6541748 &
  6542813
EFISA
  AVERIDA DE LA FABREGADA 18
  L'HOSPITALET, BARCELONA
  PHONE: 93-3378466
ESTUDIS IDEAL
  WAD-RAS 196-198
  BARCELONA 08005
  PHONE: 93-3093127
GALAXIA TELEVISION S.A.
  POLIGONO FUENTE DEL JARRO,
  (2 FASE), C/EIBAR 4
  46988 PATERNA VALENCIA
  PHONE: 1322413
  FAX: 1323960
LUIS BUNUEL
  AVENIDA DE BURGOS 5
  MADRID 28036

MOLINO DEL MANTO
  ALCALA 174
  MADRID 28028
  PHONE: 91-2566510
PLATO 601
  CARRER VALENCIA 601 INTERIO
  BARCELONA 08026
  PHONE: 93-2463600
PLATO S.L.
  CALLE ESCOMBRERAS 10
  PARACUELLOS DE JARAMA,
  MADRID
  PHONE: 91-6580864 & 6581040
PRODUCCIONES LUIS.A.
  GRANADA 25 Y JUAN DE
  URBIETA 13-15
  MADRID 28007
ROMA S.A.
  CARRETERA DE IRUN KM 11,
  700
  MADRID 28049
  PHONE: 91-7342050, 7342054,
  7342058 & 7342062
TELSON TELEVISION Y SONIDA S.A.
  PRADILLO 46-64
  28002 MADRID
  PHONE: 4134463
  TELEX: 44122 TELSO E
  FAX: 4157572
**VIDEOBAI PRODUCCION
VIDEO TV. S.A.**
  CONDE DE VISTAHERMOSA 9
  28019 MADRID
  PHONE: 4697263 & 4697298
  FAX: 4690173
VIDEOSTUDIO 20 X 30 SA
  MISTERIOS 77
  MADRID 28027
  PHONE: 91-2675700
ZOOM TELEVISION S.A.
  PORT BOU 6
  08028 BARCELONA
  PHONE: 34 8 3394150
  FAX: 34 3 4110200

UNIONS AND ASSOCIATIONS

**A.V.E. (ASOCIACION
VIDEOFRAFICA ESPANOLA)**
  GOYA 115
  MADRID 28009
  PHONE: 4012388
ASOCIACION DE ACTORES
  TEATRO DE LAVAPIES, C/
  TRIBULETE 16
  MADRID 12
  PHONE: 91-4678554
**ASOCIACION DE EMPRESAS DE
PRODUCCIONES Y
SERVICIOS**
  AUDIOVISUALES (APSOVAM)
  CONSEJO DE CIENTO 345
  BARCELONA 08007
  PHONE: 2160024
**ASOCIACION INDEPENDIENTE
DE PRODUCTORES**
  CINEMATOGRAFICOS
  ESPANOLES (ASOCINE)
  GRAN VIA 50, 3 DCHA
  MADRID 13
  PHONE: 2415903 & 2419607
**ASOCIACON DE LABORATORIOS
CINEMATOGRAFICOS
ESPANOLES**
  TRAVESERA DE DALT 117-119
  BARCELONA 24
  PHONE: 2131700
**ASOCIACON GALLEGA
PROFESIONALES DEL CINE Y
LA TELEVISION**
  REPUBLICA ARGENTINA 42
  SANTIAGO DE COMPOSTELA
  BARCELONA
  PHONE: 981-592593

**ASOCIATION DE
PROFESSIOANLES DE LA
PRODUCTION**
  CINEMATOGRAFICA DE
  CATALNA (ASPROCE)
  DUPUTACION 155
  BARCELONA
  PHONE: 4232023
CINITALIA FILMS
  RELACIONES
  CINEMATOGRAFICAS, CALLE
  CAVANILLES 9
  MADRID 7
  PHONE: 2517925
**DELEGACION DEL CINE
ALEMAN**
  ALMAGRO 26
  MADRID 4
  PHONE: 4196882
**FEDERACION DE
AGRUPACIONES DE
PRODUCTORES
CINEMATOGRAFICOS DEL
ESTADO ESPANNOL**
  GRAN VIA 50, 3 DCHA
  MADRID 13
  PHONE: 2415903 & 2419607
**FEDERACION DE
DISTRIBUIDORES DE CINE**
  VALEZQUEZ 10
  MADRID 1
  PHONE: 2762774 & 2769913
**FEDERACION ESPANOLA DE
PRODUCTORAS DE CINE**
  PUBLICITARIO Y
  CORTOMETRAJE
  SANCHEZ PACHECO 64
  MADRID 2
  PHONE: 4132454
FILMOTECA NACIONAL
  CARRETERA DE LA DEHESA DE
  LA VILLA
  MADRID S/N 35
  PHONE: 2434790, 2434763
**GABINETE DE
CINEMATOGRAFIA DEL
MINISTERIO DE
ECONOMIA**
  Y COMERCIO
  PASEO DE LA CASTELLANA 162, 3
  MADRID 16
**INSTITUTO DE LA
CINEMATOGRAFIA Y DE LAS
ARTES**
  AUDIOVISUALES
  PLAZA DEL REY 1
  MADRID 28004
  PHONE: 4292444
**INSTITUTO OFICIAL DE
RADIODIFUSION Y TELEVISION**
  CARRETERA DE LA DEHASA DE
  LA VILLA
  S/N MADRID 35
  PHONE: 4492250
**PRODUCTORES
CINEMATOGRAFICOS
ASOCIADOS**
  PINTO JUAN GRIS 5
  MADRID 20
  PHONE: 4554477, 4556382 &
  4567297
**SERVICIO DE
CINEMATOGRAFICA DEL
MINISTERIO DE
AGRICULTURA**
  DON QUIJOTE 3- PLANTA BAJA
  MADRID
**TECNICOS ASOCIADOS
CINEMATOGRAFICOS
ESPANOLES - TACE**
  GRAN VIA 62, 7 DSPCHO
  MADRID 13
  PHONE: 2479555

SWEDEN

PYROTECHNICS

PANORAMA FILM & VIDEO AB
  BOX 92
  S-13106 NACKA
  PHONE: 08-6437476
  FAX: 08-6402880
PYRO
  BOX 15
  360 42 BRAOS
  PHONE: 0474-310 13

STUDIOS/SOUND STAGES

GSP FILM AB
  ODENGATAN 104
  113 22 STOCKHOLM
  PHONE: 08-736 03 00
  FAX: 08-32 77 22
MEXFILM AB
  MAGASIN 3, FRIHAMNEN
  S-115 56 STOCKHOLM
  PHONE: 08-663 77 20
  TELEX: 14962 MEX S
  FAX: 08-665 39 11
PANORAMA FILM & VIDEO AB
  BOX 92
  S-13106 NACKA
  PHONE: 08-6437476
  FAX: 08-6402880
SONET STUDIOS AB
  BOX 20105
  161 20 BOMMA
  PHONE: 08-764 77 00
  TELEX: 17656 SONSTU
  FAX: 08-29 90 91
STUDIO 24
  SIBYLLEGATAN 24
  114 54 STOCKHOLM
  PHONE: 08-662 57 00
SVENCK FILMINDUSTRI AB (SF)
  S MALARSTRAND 27
  117 88 STOCKHOLMD
  PHONE: 08-58 75 00
  TELEX: 17533 ESSEFF
  FAX: 08-68 50 70
SVENSKA FILMISTITUTET
  BOX 27126
  102 52 STOCKHOLM
  PHONE: 08-665 11 00
  TELEX: 13326 FILMINS
  FAX: 08-61 18 20
TV INTER AB
  VANSAPSVAGEN 33
  112 65 STOCKHOLM
  PHONE: 08-13 03 40
  FAX: 08-56 31 56
TELETV AB
  FARSA TORG 26
  123 86 FARSTA
  PHONE: 08-93 10 35
  FAX: 08-713 90 08
TELETV AB
  SCHEELEGATAN 32
  212 28 MALMO
  PHONE: 040-22 25 50
  FAX: 040-94 98 39
UR MEDIA
  105 10 STOCKHOLM
  PHONE: 08-784 64 10
  FAX: 08-85 62 26
UTBILDNINGSRADION
  105 10 STOCKHOLM
  PHONE: 08-85 04 20

UNIONS AND ASSOCIATIONS

**FSF (CAMERAMEN'S
ASSOCIATION)**
  FINNBERGSVAGEN 22
  131 31 NACKA
  PHONE: 08-42 32 97

IFPI
BOX 1008
171 21 SOLNA
PHONE: 08-730 57 80
OFF (INDEPENDENT FILMMAKERS)
VALHALLAVAGEN 158
115 24 STOCKHOLM
PHONE: 0120-114 75
SVIP (PRODUCER'S ORGANISATION)
MALMVAGEN 15
115 41 STOCKHOLM
PHONE: 08-661 63 37
SVENSKA TEATERFORBUNDET
ROKUBBSGATAN 6
115 28 STOCKHOLM
PHONE: 08-24 81 20

SWITZERLAND

EQUIPMENT RENTALS

BKD CINESHOP AG
LORRAINESTRASSE 30
CH - 3013 BERN
PHONE: 031-415741 & 429295
TELEX: 912968 BKD
FAX: 031-401794
CINE CONSTRUCTION SA
CH FLEURETTES 547
LAUSANNE
PHONE: 261963
TELEX: 26531 CINC CH
CINE SWISS - AV DV
23 RUE PRE-DU-MARCHE
CH-1004 LAUSANNE
PHONE: 021-381151-8
TELEX: 26 531 CINC CH
PER-FIX - AGENTS: CINE SWISS DIVISION
23 RUE PRE-DU-MARCH
CH-1004 LAUSANNE

MAKE-UP SUPPLIERS

BKD CINESHOP AG
LORRAINESTRASSE 30
CH-3013 BERN
PHONE: 031-415741 & 429295
TELEX: 912968 BKD
FAX: 031-401794
KRYOLAN PROFESSIONAL MAKE-UP
ROGER & RUDOLF AG,
STRASSTADER STR. 27
CH-6370 STANS

STUDIOS/SOUND STAGES

CINERENT FILMEQUIPMENT SERVICE AG
BOLGRISTRASSE 20
CH-8008 ZURICH
PHONE: 01-552755
TELEX: 59039 CFS
ELITE FILM AG
MOLKENSTRASSE 21, POSTFACH
8026 ZURICH
PHONE: 01-2428822
TELEX: 55542 UFILM CH
HOFFMAN-LA ROCHE & CO. AG
STUDIO ROCHE,
GRENZACHERSTRASSE 124
4002 BASEL
PHONE: 061-273840 & 271122
88939799
LIMELIGHT STUDIOS AG
WAGISTRASSE 2
CH-8952 SCHLIEREN/ZURICH
PHONE: 01-7303577
FAX: 01-7307497

TRINCOVISION AG
TOBELMOFSTRASSE 344
8044 ZURICH-COCKHAUSEN
PHONE: 01-8215959
SIDERAL FILM SA, J.C. CADOUX AAV
6 RUE DU VIEUX-BILLARD
CH-1205 GENEVE
PHONE: 022-211911
TELEX: 421729
STUDIO BELLERIVE AG
KREUZSTRASSE 2
CH-8034 ZURICH
PHONE: 01-2513080
TELEX: 58208 STUBE CH
STUDIO MAUR MUEHLEMANN
BADANSTALTSTRASSE
CH-8124 MAUR/ZH
PHONE: 01-9800677

UNIONS AND ASSOCIATIONS

AG FUER DAS WERBEFERNSEHEN
GIACOMETTISTRASSE 15
CH-3000 BERN 31
PHONE: 031-432221
ARMEEFILMDIENST SERVICE DU FILM MILITAIRE
PAPIERMUHLESTRASSE 14
CH-3003 BERN
PHONE: 031-672322 & 031-6723674
ASSOCIATION DENEMATOGRAPHIQE SUISSE ROMANDE LICHTSPIELTHEATERVERBAND DER FRANZOESISCHEN
5 PLACE DE LA RIPONNE
CH-1005 LAUSANNE
PHONE: 021-227755
AV-ZENTRALSTELLE AM PESTALOZZIANUM ZUERICH OFFICE MOYENS, AUDIO-VISUELS, PESTALOZZIANUM ZUERICH
BECKENHOFSTRASSE 31
CH-8035 ZURICH
PHONE: 01-3620433/28
CHRISTLICHER FILMDIENST SERVICE DU FILM CHRETIEN
AUSTRASSE 10
CH-8570 WEINFELDEN
PHONE: 072-225363
CINEMATHEQUE SUISSE SCHWEIZERISCHES FILMARCHIV
2512 CASE VILLE
CH-1002 LAUSANNE
PHONE: 021-237406/7
FILM INSTITUT (FI)
SCHULFILMZENTRALE BERN,
CENTRALE DU FILM SCOLAIRE,
BERNE, ERLACHSTRASSE 21
CH-3000 BERN 9
PHONE: 031-230831
FONCTION: CINEMA/ ASSOCIATION GENEVOISE POUR LE INDEPENDANT
CASE POSTALE 127
1211 GENEVE 9
PHONE: 022-280454
GESELLSCHAFT CHRISTLICHER FILM SOCIETE CINEMA CHRETIENNE
BEDERSTRASSE 76, POSTFACH
147
CH-8027 ZURICH
PHONE: 01-2015580
GESELLSCHAFT SCHWEIZERISCHES FILMZENTRUM SOCIETE CENTRE SUISSE DU CINEMA
MUNSTERGASSE 18
CH-8001 ZURICH
PHONE: 01-472860

SCHWEIZ. FILMTECHNIKER-VERBAND SFTV ASSOCIATION SUISSE
DES TECHNICIENS DU FILM ASTF
AUGUSTINERGASSE 6
CH-8001 ZURICH
PHONE: 01-2114525
SCHWEIZ. FILMBERLEIHER-VERBAND ASSOCIATION SUISSE DES DISTRIBUTEURS DE FILMS
SCHWARZTORSTRASSE 7,
POSTFACH 2485
CH-3001 BERN
PHONE: 031-456444
VERBAND FREISCHAFFENDER FILM - UND FOTOSTYLISTEN VFS
SEKRETARIAT: WALCHESTRASSE
17
CH-8006 ZURICH
VERBAND SCHWIEZEISCHER FILMGESTALTER ASSOCIATION SUISSE
DES REALISATEURS DE FILMS
POSTFACH
CH-8027 ZURICH
PHONE: 01-4829807 & 01-9292818

TURKEY

MAKE-UP SUPPLIERS

TOPKAPI ENTERPRISES - TOPKAPI YATIRIM
ALYON SOKAK 18/1
BEYOGLU, 80060 ISTANBUL

UNIONS AND ASSOCIATIONS

FID-MOTION PICTURE IMPORTERS & DISTRIBUTORS ASSOC. OF TURKEY
YESILCAM SOKAK,29,BEYOGLU
ISTANBUL
PHONE: (90)11490986
FIYAP-MOTION PICTURE PRODUCERS ASSOCIATION OF TURKEY
ALYON SOKAK,ERMAN HAN,5/2
BEYOGLU,ISTANBUL
PHONE: (90)11454645
PHONE: (90)1499826

VENEZUELA

EQUIPMENT RENTALS

ARTE VISION U.S.B.
VALLE SARTENEJAS,
EDIF DE
TELECOMUNICACIONES,USB
BARUTA
CARACAS
PHONE: (02)9621117
PHONE: (02)9621101
EFECTOS,JERAN PRODUCTORA CINEMATOGRAFICA,C.A.
3A TRANSV. CON 4A AV.
RES SAN GRABRIEL,P.3.OFIC.7,
LOS PALOS GRANDES
CARACAS
PHONE: (02)2843445
FERNANDEZ,GONZALO "MEGATON"
PHONE: (02)351486

JERAN EFECTOS PROMOTORA CINEMATOGRAFICA,C.A.
3A TRANSV. CON 4A AV.,EDIF.
SAN GABREL,P.3,CFIC 7,
LOS PALOS GRANDES
CARACAS
PHONE: (02)2843445
MORILLO JOSE RANCHO
AV. LOS PALMOS, ED F.
AROMAC
LOS CAOBOS,CARACAS
PHONE: (02)7827122
PHONE: TX 21229
FAX: 5827827588

UNIONS AND ASSOCIATIONS

AFOVEP
1A AV. LOS PALOS GRANDES
QTA. BABLLIK
PHONE: 2845144
PHONE: 2841744
ANAC
ATENEO DE CARACAS PISO 4
PHONE: 5713042
PHONE: 5728068
ANDA
1A AV. STA. EDUVIGIS
RESIDENCIAS PRIMAVERA,PB
PHONE: 2839297
PHONE: 2836553
AVECOFA
APARTADO POSTAL:76000,
CARACAS
1070-A
VENEZUELA
CAMARA DE LA INDUSTRIA CINEMATOGRAFICA
AV. FCO. DE MIRANDA,ED
RORAIMA
PISO 15,PH-A
CAMPO ALEGRE
PHONE: 334256
PHONE: 336505
CAMARA VENEZOLANA DE LA TV
TORRE LA FREVISORA,PISO 7,
SABANA GRANDE
PHONE: 781 4608
PHONE: 781 4886
FEVAP
AV. FCO. DE MIRANDA. ED.,
RORAIMA,FISO 15,FH-A
CAMPO ALEGRE
PHONE: 334256
PHONE: 336505
IAA-ASOCIACION INTERNACIONAL
DE PUBLICIDAD
AV. PPAL,LOS RUCES,ED. ARS
PHONE: 2395053
PHONE: 2396390
SACVEN
AV. ANDRES BELLO,ED VARM,
TORRE OESTE,PISO 9
PHONE: 5732389
PHONE: 5731589

WEST INDIES/ CARIBBEAN/ BAHAMAS

SPECIAL EFFECTS

SPECIAL EFFECTS LTD
10A WEST KING'S HOUSE ROAD
KINGSTON 10, JAMAICA
PHONE: (809) 926-039

FILM COMMISSIONS

FILM UNIT OF JAMAICA NATIONAL INVESTMENT PROMOTION LTD
15 OXFORD ROAD
KINGSTON 5, JAMAICA WEST INDIES
PHONE: (809) 929-7190/5
TELEX: 2222 JANIPRO

U.S. VIRGIN ISLAND OF THE U.S.
81-AB KRONPRINDSENS GADE
ST. THOMAS, VI, 00802
PHONE: (809) 775-1444/
(213) 470-0768
FAX: (809) 774-4390

YUGOSLAVIA

STUDIOS/SOUND STAGES

CFS KOSUTNJAK
11000 BEOGRAD
KNEZA VISESLAVA 88
PHONE: 559 455

CENTRALNA FILMSKA LABORATORIJA
11000 BEOGRAD
KNEZA VISESLAVA 90
PHONE: 557 422

JADRAN FILM
41041 ZAGREB
OPOROUECKA 12
PHONE: 3841-251222
FAX: 3441-251394

VIBA FILM
6100 LJUBLJANA
ZRINJSKEGA 9
PHONE: 325 971

UNITED KINGDOM

SPECIAL EFFECTS

333 SPECIAL EFFECTS
UNIT 1, ASHTON IND. ESTATE,
ASHTON ROADS, LEEDS
U.K. LS8 5BZ
TEL 053-485705 & 482333

6666 LTD.
134 LIVERPOOL RD. LONDON,
U.K. N1 1L A
TEL 071-609 6666
FAX 071-609 4888

AARDMAN ANIMATIONS LTD.
14 WETHERELL PL. CLIFTON,
BRISTOL, U.K. BS8 1AR
TEL 0272-744802
FAX 0272-736281

ACE EFFECTS,LTD.
LEE INTERNATIONAL STUDIOS,
BUILDING 25,
SHEPPERTON, MIDDX. U.K.
TW17 OQD TEL 0932-567778
MOBILE PHONE 0860-441444

ACRICIUS LTD.
86 ALBERT ST. LONDON, U.K.
NW1 7NR TEL 071-387 2183
TELEX 889149 VISAIR G FAX
081-940 4356

AIRSPACE THE INFLATABLE PLAY EQUIPMENT CO. LTD.
UNIT 6 RASSAU IND. ESTATE,
EBBOW VALE, GWENT, U.K.
NP3 5SD
TEL 0495-307677
FAX 0495-309727

ALL EFFECTS
LITTLE ORCHARD, FRAMEWOOD
RD, STOKE POGES,
SLOUGH , BUCKS. U.K. SL3 6PG
TEL 0753-662227

ANIMATED EXTRAS
BUILDING 13, SHEPPERTON
FILM STUDIOS, STUDIOS RD.
SHEPPERTON, MIDDX. U.K.
TW17 0QD
TEL 0932- 562611
EXT. 2347 FAX 0932-568989.

ANY EFFECTS
43 FARLTON RD. LONDON, U.K.
SW18 3BJ
TEL 081-874 0927
FAX 081- 877 1372

ARKADON MOTION CONTROL
8 WESSEX RD. WESSEX IND.
ESTATE, BOUTNE END,
BUCKS. U.K. SL8 5DT
TEL 06285-26995 & 29830
FAX 0628-810062

ART EFFECTS
SHEPPERTON STUDIOS,
STUDIOS RD. SHEPPERTON,
MIDDX.
U.K. TW17 OQD
TEL 0932-562611
TELEX 9249416
FAX 0932-568989

ART ELECTRIC INTERNATIONAL LTD.
2 GANTON ST. LONDON, U.K.
W1V 1LJ
TEL 071-494 3167
& 0666825290
FAX 0932-568989

ART MODELS
42 GRIFFITHS RD. LONDON,
U.K. SW19
TEL 081- 540 2744

ARTEM VISUAL EFFECTS
UNIT 25, 12 WADSWORTH RD.
PERIVALE, MIDDX. U.K.
UB6 7JD
TEL 081-997 7771
FAX 081-0081503

ASYLUM
61-65 ASLETT RD. LONDON,
U.K. SW18 2BE
TEL 081- 871 2988
FAX 081- 874 8186

BICKERS ACTION ENT. LTD.
THE GARAGE, CODDENHAM,
IPSWICH, SUFFOLK U.K.
IP6 9PT TELO44 979-201 & 382
FAX 044 979-614

BOWTELL, ALLISTER & ASSOC. LTD.
59 ROTHERWOOD, RD.
PUTNEY, LONDON, U.K.
SW15 TEL 081-788 0114/5 &
081-788 2314/5
FAX 081-788 4012

BRIAN SMITHIES SPECIAL EFFECTS
THE COW SHED, PINEWOOD
STUDIOS, IVER HEATH, BUCKS.
U.K. SLO ONH
TEL 0753-656577
FAX 0753-656844

THE BRITISH TURNTABLE CO. LTD.
EMBLEM WORKS, EMBLEM ST.
BOLTON, LANCASHIRE, U.K.
BL3 5BW
TEL 0204-25626
TELEX 94011903 BTTC G
FAX 0204-382407

BUSSEKM G & COMPANY LTD
MAXTED CLOSE, HEMEL
HEMPSTEAD, HERTS. HP2 7BS
TEL 0442-69101/2/3
TELEX 826995 NISSEL G
FAX 0442-211804

CEL ELECTRONICS LTD.
CHROMA HOUSE, SHIRE HILL
IND. ESTATE,
SAFFRON WALDEN, ESSEX,
U.K. CB11 3AQ
TEL 0799-23817
TELEX 817807
FAX 0799-28081

BUSSEKM G & COMPANY LTD
MAXTED CLOSE, HEMEL
HEMPSTEAD, HERTS. HP2 7BS
TEL 0442-69101/2/3
TELEX 826995 NISSEL G
FAX
0442-211804

COMPUTERISED CAMERA SERVICES
3-7 KEAN ST. LONDON, U.K.
WC2B 4AT
TEL 071-240 3888

CONCEPT ENGINEERING LTD.
7 WOODLANDS BUSINESS PARK,
WOODLANDS PARK AVE.
MAIDENHEAD, BERKS,U.K. SL6
3UA
TEL 062882-5555
TELEX 849462 TELFAC G
FAX 062882-6261

CONSTRUCTION MATERIALS (SOUTHERN) LTD.
UNIT 3, FITTALLS YARD,
SPURLANDS END RD. GREAT
KINGSHILL, HIGH WYCOMBE,
BUCKS, U.K., HP15 6JA
TEL 0494-715858

CUTTS ARTS
THE HALL, COUSLEY WOOD,
WADHURST, EAST SUSSEX, U.K.
TN5 6EY
TEL 089288-3682

DAWSON, PETER
NO. 5 STUCLEY RD.
HOUNSLOW, MIDDX. U.K.
TW5 0TN TEL 081-572 4959

DISTILLERS COMPANY (CARBON DIOXIDE) LTD.
THE CEDAR HOUSE, 39
LONDON RD. REIGATE, SURREY,
U.K.RH2 9QE
TEL 0737-241133
TELEX 917280
FAX 0737-241842

DISTILLERS MG LTD
CEDAR HOUSE, 39 LONDON RD.
REIGATE, SURREY, U.K.
RH2 9QE
TEL 0737-241133
TELEX 264892
FAX 071-790 6634

DONALD, STEVE ANIMATRONICS
80 GROVE AVE. MUSWELL HILL,
LONDON, U.K.
N10 2AN
TEL 081-444 3567 & 267 8647

DONAHUE, PHIL PRODUCTIONS
UNIT 247, STRATFORD
WORKSHOPS, BURFORD RD.
STRATFORD, LONDON. U.K. E15
2SP
TEL 081-534 8335 &
478 3456
FAX 081-478 2629

DONMAR LTD.
DONMAR HOUSE, 54 CAVELL
ST. WHITECHAPEL, LONDON,
E1 2HP
TEL 071-790 1166
TELEX 264892
FAX 071-790 6634

ELF HOLDINGS LTD.
836 YEOVIL TRADING ESTATE,
SLOUGH, BERKS, U.K.
SL1 4JG
TEL 0753-36123
FAX 0753-693858

E.M.A. MODEL SUPPLIES LTD.
56-60 CENTRE, FELTHAM,
MIDDX. U.K. TW13 4BH
TEL 081-890 5270 & 081-890 8404
TELEX 263439 EMALON G
FAX 081-890 5321

EAGLE MODELS LTD.
8 LEBANON RD. LONDON, U.K.
SW18 1RE
TEL 081-871 0927
FAX 081-871 1271

THE EDGE OF TIME LTD.
GOODYEAR HOUSE, 52-56
OSNABURGH ST. LONDON, U.K.
NW1 3NS
TEL 071-388 3073 & 383 7000
FAX 071- 388 5453

THE (MARY EDWARDS)
10 STANHOPE PLACE, MARBLE
ARCH, LONDON, W2S 2HH
TEL 071-262 0368

EFFECTIVE SERVICES
ST. MARGARET'S, BANGORS RD.
NORTH, IVER HEATH,
BUCKS, U.K. SLO OBN
TEL 0753-651094

EFFECTS ASSOC. LTD.
PINEWOOD STUDIOS,
PINEWOOD RD..IVER HEATH,
BUCKS,
U.K. SLO ONH
TEL 0753-652007
FAX 0753-630127.

ELSEY, DAVE
11 PRIORY CRT. BROOKSBYS
WALK, HOMERTON, LONDON,
E9 6DG
TEL 081-985 0535

EMERGENCY HOUSE
EMERGENCY HOUSE,
MANCHESTER RD. MARSDEN,
HUDDERSFIELD,
WEST YORKS, U.K. HD7 6EY
TEL 0484-846999
FAX 0484-845061.

ENTERPRISES UNLIMITED
PHIL STOKES, THE WORKSHOP,
REAR OF 35 ASHBY RD.
SPILSBY, LINCS PE23 5DW
TEL 0790-52807
FAX 0790-5407

EUGENES FLYING BALLETS AND SPECIAL FLYING WIRE EFFECTS
71 BOLTONS LANE, PYRFORD,
WOKING, SURREY, U.K.
GU22 8TN.
TEL 09323-41616

EVANS, JOHN
SHED 6, PINEWOOD STUDIOS,
PINEWOOD RD. IVER HEATH,
BUCKS, U.K.
TEL 0753-651700

EYEDENTITY PRODUCTS
WALTHAM CROSS, HERTS. EN8
7QT U.K.
TEL 0992-25968
FAX 0992-38825

FX PROJECTS
STUDIO HOUSE, RITA ROAD,
VAUXHALL, LONDON, U.K.
SW8 1JU
TEL 071-582 8750
TELEX 26587 MONREF G

FEGGANS BROWN LTS.
8-10 BOW COMMON LANE, E3
4AU, U.K.
TEL 071-538 8858
FAX 071-515 8797

FITNESS INCORPORATED TRAING LTD.
308A STATION RD. HARROOW,
MIDDX. U.K.
HA1 2DX
TEL 081-427 7183
FAX 081-863 9810

FORMIS
UNIT 40, YOUNGS IND. ESTATE,
PAICES HILL
ALDERMASTON, BERKS, U.K.
RG7 4PQ
TEL 0734-
819639

FONTEYNE DISPLAYS
UNIT 3, 181 VERULAM RD.
ST.ALBANS, HERTS. AL3 4DR
TEL 0727 32146/7
FAX 0727-45064

FREEBORNS
2 COBHAM WAY, EAST
HORSLEY, SURREY KT24 5BH
TEL 04865-2986
FAX 04865-3535

GANT, JOHN
43 MANOR WAY, NORTH
HARROW, MIDDX. HA2 6BZ U.K.
TEL 081-863 2111

**GENERAL SCREEN
ENTERPRISES LTD.**
HIGHBRIDGE ESTATE, OXFORD
RD, UXBRIDGE, MIDDX.
UB8 1LX
TEL 0895-3193
TELEX 934883
FAX 0895-35335

GERRY JUDAH STUDIO LTD.
UNIT A, 1ST. FLOOR, LINTON
HOUSE, 39-51 HIGHGATE RD.
LONDON, NW5 1RT
TEL 071-284 1101
FAX 071-267 0661

GIZZMO
BOUNDARY ROW STUDIOS, 1-7
BOUNDARY ROW, LONDON SE1
TEL 071-928 4744

GRANT & GLASS OPTICIANS
143 THE MARLOWES, HEMEL
HEMPSTEAD, HERTS. HP1 1BB
TEL 0442-40607

**GREENFIELD, STEPHEN
MODELMAKERS**
24 LINTONS LANE, EPSOM,
SURREY KT17 1DD
TEL 0372-721031
FAX 0372-745264

GRIFFIN & GEORGE
GERRARD BIOLOGICAL CENTRE,
WORTHING RD, EAST
PRESTON, WEST SUSSEX BN16
1AS
TEL 0903-772071
TELEX 87323 FSI G
FAX 0903-775616

**GUTTERIDGE, MARTIN
(EFFECTS ASSOC. LTD.)**
PINEWOOD STUDIOS, IVER
HEATH, BUCKS, U.K.SLO ONH
TEL 0753-652007
FAX 0753-630127

GUY LUBBOCK ASSOCIATES
516 WANDSWORTH RD.
LONDON, SW8 3JX U.K.
TEL 071-622 3252
FAX 071-622 7975

HAAS - FALSE EYEMAKER
8 FEATHERSTONE RD, LONDON,
NW7 2BN U.K.
TEL 081-959 4820

HI-FLI LTD.
16 COMPTON CLOSE, KINVER,
STOURBRIDGE, WEST
MIDLANDS, U.K. DY7 6DW
TEL KINVER (0384) 872710
FAX 021-550 7249

HODGKINSON, GUY
THE MALTINGS, 129 IPSWICH
ST. STOWMARKET, SUFFOLK
IP14 1BB
TEL 0449-677554
FAX 0449-676339

**HORTON, JOHN SPECIAL
EFFECTS**
2 SNOWDOWN CLOSE,
ST.LEONARDS HILL, WINDSOR,
BERKS.U.K.
TEL 0753-851327

HUTCHINSON, PETER & CO.
COPSE FARM, MOULSHAM
COPSE LANE, YATELEY, HANTS.
GU17 7RF
TEL 0252-870904

ICE COOLING LTD.
LYSONS AVE. ASH VALE,
ALDERSHOT. HANTS, GU12 5QF
TEL 0252-534028
FAX 0252-524 029

**IDM (DESIGN DEVELOPMENTS
LTD)**
7-8 JEFFREYS PL. JEFFREYS ST.
LONDON NW1 9PP
TEL 071-485 0854
TELEX 27950 REF 1192
FAX 071-482 3970

I-LASER SOUND & VISION LTD.
32 LEXINGTON ST. LONDON,
U.K. W1R 3HR TEL 071-437
FAX 071-494 0386

ILLUMINATI
2 ARTHUR ST. COLCHESTER,
ESSEX , U.K. CO2 7D7
TEL 0206-76776
FAX 0206-578073

IN-VIDEO PRODUCTIONS LTD.
16 YORK PLACE, EDINBURGH,
SCOTLAND. EH1 3EP
FAX 031-557 5465

JACOBSON CHEMICALS LTD.
JACOBSON HOUSE, THE
CROSSWAYS, CHURT,
SURREY, U.K. GU10 2JD
TEL 0428-713637
TELEX 858301 JCHEM G
FAX 0428-712835

K. D. MODELMAKERS
UNIT 19 STAINES CENTRAL
TRADING ESTATE, STAINES,
MIDDX.
TW18 3DB U.K.
TEL 0784-461007

KEIR LUSBY
LEE INTERNATIONAL STUDIOS,
STUDIOS RD. SHEPPERTON
MIDDX. TW17 0QD
TEL 0932-561717 & 561877
FAX 0932-565893

TERRY KEMBLE ASSOCIATES
71 LAMBETH WALK, LONDON,
SE1 6DX
TEL 071-582 430

KIRBYS FLYING BALLETS
REAR OF 83. BRIXTON HILL,
LONDON, SW2 1JE
TEL 081-674 3524

DAVE KNOWLES
1 BEECHFIELD COTTAGES,
CHURCH HILL
NUTFIELD, SURREY,
TEL 073782-2764

LTM (U.K.) LTD. (CINEBUILD)
STUDIO HOUSE, RITA RD.
VAUXHALL, LONDON SW8 1JU
TEL 071-582 8750
TELEX 265871 MONREF G
REF:WXX 044 BT
GOLD 84:WXX 044
FAX 071-439 9520

LAILEY, KEN EFFECTS
COWSLIP FARM HOUSE,
DEVIZES RD. SALISBURY, WILTS.
U.K. SP2 7NB
TEL 0722 331866

LASEFX LTD.
55 MERTHYR TERRACE, BARNES,
LONDON, U.K. SW13 9DL
TEL 081-741 5747
FAX 081-748 9879

LASERGRAFIX LTD.
UNIT 15, ORCHARD RD.
ROYSTON, HERTS.,
GS8 5HD
TEL 0763-248846-24630

LEE FILTERS LTD.
CENTRAL WAY, WALWORTH
IND. ESTATE, ANDOVER, HANTS,
SP10 5AN
TEL 0264-66245
TELEX 477259
FAX 0264-55058

LEE LIFTING SERVICES
12 LONGLEAT WAY, BEDFONT,
FELTHAM, MIDDX. TW14 8JW
TEL 081-844 1646 & 081-890 1280
FAX 081-751 4782

LEGGS ENTERPRISES
4TH FLOOR, CENTRAL
BUILDINGS, 11 PETER ST.
MANCHESTER
M2 5QR
TEL 061-832 8607

LIGHT & SOUND DESIGN LTD
201 COVENTRY RD,
BIRMINGHAM, B10 0RA
TEL 021-766 6400

MACGREGOR INDUSTRIES LTD.
CANAL ESTATE, LANGLEY,
BERKS. SL3 6EQ
TEL 0753-49111 & 42251
TELEX 848028 MACIND G
FAX 0753-46983

MCMAHON, K. & SONS LTD.
UNIT 7, COURT LANE, IVER,
BUCKS,
TEL 0753-652285

MEDIA SERVICES LTD.
32 KINGS RD. FOLKESTONE,
KENT, U.K. CT20 3JZ
TEL 0303-277058

MACGREGOR INDUSTRIES LTD.
CANAL ESTATE, LANGLEY,
BERKS. SL3 6EQ U.K.
TEL 0753-49111 & 42251
TELEX 848028 MACIND G
FAX 0753-46983

MIKKI & DAVID
28 TOLLINGTON RD. LONDON,
N7 6PGG
TEL 071-609 8959

MIMIC MODELS
12/16 LAYSTALL ST. LONDON,
EC1R 4UB
TEL 071-837 8930
FAX 071-250 0134

MIRAGE LTD.
LEE INTERNATIONAL STUDIOS,
SHEPPERTON, MIDDX.
TW17 0QD
TEL 0932-562611

MODEL EFFECTS LTD
UNIT 6, SANDOWN IND.
ESTATE, SANDOWN RD.
WATFORD, HERTS.WC2 4UBD
TEL WATFORD (0923) 226278
FAX 0923-36125

**MOVING MAGIC SPECIAL
EFFECTS**
HARLOW, ESSEX CM19 5JU
TEL 027979-3587
FAX 027979-2698

NEIL PEPPE ASSOCIATES
UNIT 3BB, FARM LANE
TRADING, CENTRE,
101 FARM LANE,
FULHAM, LONDON, SW6 1QJ
TEL 071-381 1175
FAX 071- 381 88931

**NEVE ELECTRONICS
INTERNATIONAL LTD.**
CAMBRIDGE HOUSE,
MELBOURNE, ROYSTAN, HERTS,
SG8 6AV
TEL 0763-260776
FAX 0763-261886

NOTCUTT. W. P. LTD.
25 CHURCH RD. TEDDINGTON,
MIDDX. U.K. TW11 8PB
TEL 081-977 2252/3
FAX 081-977 6423

NEIL PEPPE ASSOCIATES
UNIT 3BB, FARM LANE
TRADING, CENTRE,
101 FARM LANE,
FULHAM, LONDON, SW6 1QJ
TEL 071-381 1175
FAX 071- 381 88931

OXFORD SCIENTIFIC FILMS LTD
LOWER RD, LONG
HANBOROUGH, OXFORD,
OXT 2LD
TEL 0993-881881
FAX 0993 882808

PACKAGE AID LTD
18 MORA ST. LONDON, EC1V 8BT
TEL 071- 253 0343
TELEX 28604 REF 3357
FAX 071-454 1015

PENNICOT PAYNE & LILLIE LTD
10/16 GWYNNE RD. BATTERSEA,
LONDON, SW11 3UW
TEL 071-228 6122
FAX 071-223 3332

POWERHOUSE
482 BROAD LANE BRAMLEY,
LEEDS, LS13 3ER
TEL 0532-551090 & 550918

POWERHOUSE
247 OAKLEIGH ROAD NORTH,
LONDON, U.K. N20 0TX
TEL 081-368 9852 & 361 1144
FAX 081-368 6229

POWERHOUSE
166 SYDENHAM RD, LONDON,
U.K. SE26 5JZ TEL 081-659
9022 & 778 6984

PUREFECT LTD
C/O GOLDCREST ELSTREE
STUDIOS, BOREHAMWOOD,
HERTS WD6 1JG
TEL 081-953 1600 & 081-207 6762
TELEX 922436
EFILMS G FAX 081-207 0860

QUESTECH LTD.
EASTHEATH AVE.
WORKINGHAM, BERKS, RG11 2PP
TEL 0734-787209
TELEX 848976 QUESTEC G
FAX 0734-794766

**R.W. MODELS, PROPS &
SPECIAL EFFECTS**
LEDBURY MEWS NORTH,
LONDON, W11 2AF
TEL 071-229
7089 & 792 1183
TELEX 2622
FAX 071-727 1880

RADAMAC GROUP
BRIDGE WHARF, BRIDGE RD.
CGERTSETY, SURREY,
KT16 8LJ
TEL 09325-61181
TELEX 929945
FAX 0932-568775

RAILFILMS LTD.
26 REGENT RD. ALTRINCHAM,
CHESHIRE, WA14 1RP
TEL 061-926 8401

RICHARDSON, CLIFF LTD
PINEWOOD STUDIOS, IVER
HEATH, BUCKS.SLO ONH
TEL 0753-656537 & 071-625
9172

RISKY VENTURES
45 BEVERSBROOK RD.
LONDON, N19 4QQ
TEL 071- 263 5054

MARK ROBERTS FILM SERVICES LTD.
UNIT 4A, BIRCHES IND. ESTATE, IMBERHORNE LANE, EAST GRINSTEAD, WEST SUSSEX, RH19 1XZ
TEL 0342-313522/3
FAX 0342-327566

ROBERTS, RICHARD (EFFECTIVE EFFECTS)
CAROUSEL, STALL HOUSE LANE, NORTH HEATH, PULBOROUGH, W. SUSSEX RH20 2HR
TEL 07982-3773

ROBINSON, STUART
40 BATTLEDEAN RD. LONDON, N5 1UZ
TEL 071-226 1646

ROSCOLAAB LTD
69-71 UPPER GROUND, LONDON, SE1 9PQ
TEL071-633 9220
TELEX 8953352 ROSLAB
FAX 071-633 9146

SALT EFFECTS
58 KINGSBRIDGE CRESCENT, SOUTHALL, MIDDX. UB1 2DL,
TEL 081-571 0926 (3 LINES)
FAX 081-571 4300

SALTECH INTERNATIONAL LTD.
BOURNE END BUSINESS CENTRE, CORES END RD.
BOURNE END, BUCKS, SL8 5AT
TEL 06285-59131
TELEX 848960 SELTEC G
FAX 062 85-27468

SCULPTURE & CONSTRUCTIONAL DESIGN (DEREK HOWARTH)
COLNEY PARK HOUSE, 100 HARPER LANE, SHENLEY, RADLETT, HERTS. WD7 9HG
TEL 0727-22845

SCULPTURE WORKS
THE WORKS, 29 CONCORD ST. LEEDS, LS2 7QS
TEL 0532-341973
FAX 0532-44959

SHAWCRAFT (MODELS) LTD
THE ROCKINGHAM RD. UXBRIDG, MIDDX. UB8 2UA
TEL 0895-72523 & 72462
FAX 0895-73664

SIGN ABILITY
ADDLESTONE MOOR, WEYBRIDGE, SURREY KT15 2QE
TEL 0932-851896 OR 0932-345709 (24 HRS)
FAX 0932-567189

SMALL WORKS
(DESIGNERS & MAKERS), 2A GREENWOOD ROAD, LONDON, E8 1AB
TEL 071-249 3627
FAX 071-254 0306

SMITHIES, BRIAN
7 SEYMOUR ROAD, EAST MOLESEY, SURREY KT8 OPB
TEL 081-979 5657

SNOWDONIA TAXIDERMY STUDIOS
FRON GANOL, SCHOOK BANK ROAD, LLANRWST, NORTH WALES LL26 OHU
TEL 0492-640664
FAX 0492-641643

SNOWMEC ENGINEERING
BOROUGH FARM, STANE ST. FIVE OAKS, BILLINGHURST, WEST SUSSEX RH14 9AG
TEL 0403-783681
FAX 0403-785396

SPECIAL EFFECTS (WORLDWIDE) LTD
WELDERS HOUSE, WELDERS LANE, CHALFONT ST. PETER, GERRARDS CROSS, BUCKS. SL9 8TT U.K.
TEL 02407-5505 & 4242
FAX 02407-1976

STAGEMATE LTD.
SHEPPERTON STUDIOS, LADBROKE HALL, 85 BARLBY RD. LONDON, U.K. TW17 OQD
TEL 081-969 3601

STEARS, M.J.
WELDERS HOUSE, WELDERS LANE, CHALFONT ST. PETER, GERRARDS CROSS, BUCKS SL9 8TT
TEL 02407-5505 & 4242

T.S.F.X.
UNIT 6, PORTLAND BUSINESS CENTRE, MANOR HOUSE LANE, DATCHET, BERKS.
TEL 0753-584665/7
FAX 0753-582 117

TECHNFIX (UK)
CROMWELL HOUSE, CROMWELL RD. MAIDENHEAD, BERKS, SL6 6BJ
TEL 0628-32306
FAX 0628-773048

TELE SCAFFS SERVICES
UNIT 7B ATLANTIC TRADING ESTATE, BARRY, GLAMORGAN, U.K. CF6 6RF
TEL 0446-749787

TELE SCAFFS SERVICES
UNIT 34-37, SAPCOTE BUSINESS CENTRE, SMALL HEATH BIRMINGHAM, B10 OHR
TEL 021-773 5353

TELE SCAFFS SERVICES
186 ACTON LANE, LONDON, U.K. NW10 7NH
TEL 061-477 48447

TELE-STAGE ASSOCIATES (U.K.) LTD
14 BUNTING RD. MORETON HALL IND. ESTATE, BURY ST. EDMUNDS, SUFFOLK, IP32 7BX
TEL 0284-755512
TELEX 817315
FAX 0284-755516

TETRA ASSOCIATES
UNIT 2, 28A GRAFTON SQUARE, CLAPHAM COMMON, LONDON, SW4 0DB
TEL 071-622 4138
FAX 071-498 0341

THEATRICAL PYROTECHNICS INC.
THE LOOP, MABSTIB AIRPORT, RAMSGATE, KENT, CT12 5DE
TEL 0843-823545
FAX 0843-822655

TRITON PROJECTS
UNIT 3D, 29 BLENHEIM STUDIOS, BLENHEIM GARDENS, LONDON, SW2 5UE
TEL 081-671 3139

UK SYSTEMS SALES
LASERPOINT, 44/45 CLIFTON RD. CAMBRIDGE, CB1 4FD
TEL 0223-212331
TELEX 81728 LASER P-G
FAX 0223-214085

VENDETTA FX
SHEPPERTON STUDIOS, STUDIOS RD. SHEPPERTON, MIDDX. TW17 0QE
TEL 0932-562611
TELEX MOVIES G 929416
FAX 0932-568989

VISION UK LTD
UNIT 4, BENTINCK ST. IND. ESTATE, MANCHESTER, M15 4LN
TEL 061-834 2894

W.B. LIGHTING LTD.
4 TEMTER RD. MOULTON PARK IND. ESTATE NORTHAMPTON, U.K. NN3 1PZ
TEL 0604-499331
TELEX 312320
FAX 499446

WARCOUR MOTION PICTURES LTD
11 WARDOUR MEWS, LONDON, W1V 3FF
TEL 047485-3538 & 071-439 4313
FAX 047485-2436

WEST, KIT
12 GILPIN AVE, EAST SHEEN, LONDON, SW14 8QY
TEL 081-878 6745
FAX 081-392 2948

THE WESTERN ARMS & ENGINEERING CO.
7 COURTSIDE MEWS, REDLAND, BRISTOL, BS6 6PS
TEL 0272-245958

DAVID WILLIAMS
Y FACHWEN GANOL, LLWYDIARTH, LLANGADFAN, WELSHPOOL, POWAYS, SY21 0QG
TEL 093888-595

WINGROVE, IAN
PARK HOUSE, WASH HILL, WOOBURN GREEN, BUCKS. HP10 0JA
TEL 06285-21356

PYROTECHNICS

1ST UNIT FIRE & SAFETY
NO 3-92 NETHER STREET, LONDON; N12 8AD
TEL 081-446 0757

333 SPECIAL EFFECTS
UNIT 1, ASHTON IND. ESTATE, ASTHTON ROAD, LEEDS, 1S8 5BZ
TEL 0532-485705 & 482333

ACE EFFECTS LTD
LEE INTERNATINAL STUDIOS, BUILDING 25, SHEPPERTON, MIDDX. TW17 OQD
TEL 0932-567778
MOBILE PHONE 0860-441444

ALL EFFECTS
LITTLE ORCHARD, FRAMEWOOD ROAD, STOKE PAGES, SLOUGH, BUCKS. SL3 6PG
TEL 0753-662227

ANY EFFECTS
43 FARLTON ROAD, LONDON, SW18 3BJ
TEL 081-874 0927
FAX 081-877 1372

ARTEM VISUAL EFFECTS
UNIT 25, 12 WQADSWROTH ROAD, PERIVALE, MIDDX. UB6 7JD
TEL 081-997 7771